Seismic Risk Analysis and Management of Structure Systems

Seismic Risk Analysis and Management of Structure Systems

Editors

Xiaowei Wang
Chao Li
Jian Zhong
Weiping Wen
Suiwen Wu

MDPI

Basel • Beijing • Wuhan • Barcelona • Belgrade • Novi Sad • Cluj • Manchester

Editors
Xiaowei Wang
Tongji University
Shanghai
China

Chao Li
Dalian University of Technology
Dalian
China

Jian Zhong
Hefei University of Technology
Hefei
China

Weiping Wen
Harbin Institute of Technology
Harbin
China

Suiwen Wu
Hunan University
Changsha
China

Editorial Office
MDPI
St. Alban-Anlage 66
4052 Basel, Switzerland

This is a reprint of articles from the Special Issue published online in the open access journal *Buildings* (ISSN 2075-5309) (available at: https://www.mdpi.com/journal/buildings/special_issues/BG9WT445M8).

For citation purposes, cite each article independently as indicated on the article page online and as indicated below:

Lastname, A.A.; Lastname, B.B. Article Title. *Journal Name* **Year**, *Volume Number*, Page Range.

ISBN 978-3-7258-1385-8 (Hbk)
ISBN 978-3-7258-1386-5 (PDF)
doi.org/10.3390/books978-3-7258-1386-5

Contents

buildings

MDPI

Article

Equivalent Linearization and Parameter Optimization of the Negative Stiffness Bistable Damper

Liming Fan, Chen Huang and Linsheng Huo *

State Key Laboratory of Coastal and Offshore Engineering, Dalian University of Technology, Dalian 116023, China; fanliming421@126.com (L.F.); chenhuang92@126.com (C.H.)
* Correspondence: lshuo@dlut.edu.cn

Abstract: The negative stiffness bistable damper (NSBD) was proposed to suppress structural dynamic responses in our previous study. The vibration mitigation performance of the NSBD is influenced by its design parameters, including negative stiffness, cubic stiffness, and damping coefficients. However, it is extremely challenging to directly acquire the ideal design parameters of the NSBD owing to its inherent nonlinearity. To address this disadvantage, the optimal design approach for the NSBD, based on the equivalent linearization method (ELM) and genetic algorithm (GA), is presented in this paper. The nonlinear NSBD system can be transformed to a linear system utilizing the ELM based on the pseudo-excitation method (PEM). The linearization model that corresponds to the nonlinear NSBD is fairly accurate in its approximation and can be indicated from the numerical results. Then, the main structure's peak response is minimized through the optimization of the design parameters of the NSBD using the $H\infty$ norm and GA. Moreover, the proposed approach's effectiveness is assessed using the optimal parameters to calculate the displacement responses of a tall building equipped with the NSBD during various seismic excitations. As revealed by the numerical results, the displacement of the tall building can be effectively restrained by the optimized NSBD.

Keywords: equivalent linearization; genetic algorithm; Monte Carlo method; negative stiffness bistable damper (NSBD); negative stiffness; optimal design; tall building

Citation: Fan, L.; Huang, C.; Huo, L. Equivalent Linearization and Parameter Optimization of the Negative Stiffness Bistable Damper. *Buildings* **2024**, *14*, 744. https://doi.org/10.3390/buildings14030744

Academic Editor: Harry Far

Received: 2 January 2024
Revised: 5 March 2024
Accepted: 7 March 2024
Published: 10 March 2024

1. Introduction

Structural vibration control is well developed but is still a potentially developing field and is an effective earthquake protection method for building structures in civil engineering [1]. Based on different types of damping strategies, structural control approaches are often categorized as passive control, active control, semi-active control, and hybrid control [2]. Among these approaches, passive control, such as tuned-mass dampers (TMDs) [3–6] and energy dissipation devices [7,8], is the most widely utilized structural control technology owing to its high dependability and effectiveness in real-world applications. Practical applications in civil engineering are demonstrated by examples such as the Sydney Chifley Tower and Shanghai Center Tower [9,10].

Generally, the natural frequencies and models of civil engineering structures play a dominant role in vibration control. The frequency of the conventional tuned-mass damper (TMD) is designed based on the frequency of the main structure, greatly reducing the mitigating effect of the dampers once there is an inconsistency with the frequency of the main structure. Therefore, the detuning of the main frequency is a major disadvantage of the conventional tuned-mass damper (TMD). To address this shortcoming, numerous negative stiffness dampers (NSDs) have been proposed to upgrade the efficacy of structural control systems. Pasala et al. [11] proposed a combined NSD–structure system consisting of an adaptive negative stiffness system (ANSS), a viscous damper, and a combination of NSDs. This novel damper can significantly mitigate the acceleration, displacement, and base shear of the main structure. A further experimental and numerical simulation study demonstrated the effectiveness and superior performance in reducing the vibration responses of the main structure under earthquake excitations [12]. Li et al. [13] and Sun et al. [14] proposed an innovative negative stiffness apparatus and damper system, respectively, to reduce the seismic responses of highway bridges. The experimental results demonstrated the effectiveness of the analytical model of the NSD in retaining the displacements of the base and bridge. Sun et al. [15–17] invented a new passive negative-stiffness-amplifying damper (NSAD) for preserving the significant damping magnification effect and property of the negative stiffness, which can greatly reduce the responses of the structure when exposed to earthquakes. Zhou and Li [18] and Chen et al. [19] introduced negative stiffness devices (NSDs) using a self-contained highly compressed spring to mitigate stay-cable vibrations, as investigated by numerical simulation and experimental tests. Shi and Zhu [20,21] and Shi et al. [20–22] investigated the significant vibration-damping performances of two innovative devices of magnetic negative stiffness dampers (MNSDs) compared with different active control approaches, as successfully verified by laboratory experiments. Furthermore, numerous other innovative negative stiffness dampers [23–28] have been explored for the vibration damping of structures, showing a significant control effect.

Meanwhile, the bistable structure can provide both negative stiffness and cubic stiffness properties, presenting an excellent vibration-damping effect when installed in dampers. The bistable damper installed on the main structure has two stable equilibrium positions, which can maintain a stable motion state without the continuous input of the external energy and can achieve rapid conversion between steady-state configurations under the driving force of the main structure's conduction. These special structural characteristics make the bistable structure extensively utilized in vibration attenuation and vibration isolation. As for the research on vibration reduction, experiments and analytical models have been carried out to investigate the dynamic response control under harmonic and earthquake excitations, and the results indicated that the vibration-reducing effects of a bistable attachment [29] and a bistable tuned-mass damper (BTMD) [30] were effective for a bistable oscillator and a bridge deck, respectively. In addition, bistable vibration isolation (BVI) [31,32] and quasi-zero stiffness (QZS) [33] isolation devices have been developed, and experiments have shown that these devices can significantly improve the vibration isolation effect through the snap-through effect of the bistable structures.

The NSBD has been proposed in our previous study [34] to reduce the structural dynamic responses, inspired by the properties of the negative stiffness and bistable structure, which benefit the passive control. Based on previous experimental research, this paper further studies and proposes a detailed optimization design method to achieve the excellent vibration control of high-rise structures. The design parameters of the NSBD, such as negative stiffness, cubic stiffness, and damping coefficients, can have a significant impact on the vibration control performance. Therefore, to maximize its vibration suppression capability when used to mitigate structural vibrations, it is important to set the optimal design parameters for the NSBD. Since the negative stiffness element installed in numerous dampers to attenuate vibration responses has been proposed by several researchers, various design approaches have been developed to reduce vibrations caused by various types of excitation sources by installing vibration-suppressing devices on various primary structures. Li and Sun [35] invented an optimization objective function for the rail-type negative stiffness control system according to the characteristics of the negative stiffness control system. Dai and Zhao [36] optimized the equal-peak design of a dynamic vibration absorber with negative stiffness and analyzed the effects of the structural parameters on the effective damping frequency's bandwidth, achieving a reduction percentage of over 40% within the effective damping frequency's band range. Li et al. [37] and Ullslam et al. [38] proposed an innovative dynamic vibration absorber (DVA) and a vibration isolator using negative stiffness and focused on examining the optimal frequency ratio and negative stiffness ratio using the fixed-point theory. Zhang et al. [33] invented a tuned-mass damper with a negative stiffness device (TMD_NSD) and verified its excellent vibration-suppressing effect by experimental tests. Charef et al. [39] studied a non-traditional tuned-mass damper (NTTMD) employing negative stiffness to dampen a primary system, achieving better optimal tuning parameters.

Moreover, many other novel and valuable algorithms and methodologies [40–44] can lead to more accurate and automated mechanical systems shortly.

However, the NSBD's inherent nonlinearity makes it impossible to directly compute its transfer function using analytical methods. Although the numerical simulation method is sufficient to compute dynamic responses of the nonlinear system, it is still challenging to optimize the parameters of the NSBD owing to the massive numerical analysis. The optimal design for an NSBD installed on a damped structure cannot be determined through analytical solutions. To address this issue, this paper recommends a strategy using the ELM and GA to optimize the design parameters of the NSBD initially. First, utilizing the ELM and PEM, the NSBD is transformed to a linear system. Second, the optimal design parameters of the NSBD are calculated using the optimal design method to decrease the peak displacement response of the primary structure. The GA is utilized to achieve an optimal solution during the optimization procedure. Furthermore, to evaluate the efficacy of the suggested approach, the NSBD-installed tall building's dynamic responses under different seismic excitations are numerically computed using the optimal parameters.

2. Control Equations of a Single-Degree-of-Freedom (SDOF) Structure Equipped with an NSBD

2.1. Mechanism Characterization of the NSBD

A linear SDOF structure equipped with an NSBD was introduced in our previous study [34]. For the dynamic analysis of beam-like structures, the Galerkin-assumed modal approach [45–48] has been applied in practical applications [48–50] and has led to remarkable results. On this basis, this paper further simplifies the beam-like structure to a cosine beam structure for the mechanical performance analysis. According to a study by Qiu et al. [51], based on the assumption of a small deformation, the bending deformation of the buckling beam is considered, while because the shear deformation has a relatively small impact on the buckling beam's direction of motion, the shear deformation is ignored. Then, the mechanical constitutive equation for the displacement of the buckling beam (as illustrated in Figure 1) under the external force is as follows:

$$F_0 = (\frac{3\pi^4 Q^2}{2})\Delta[\Delta - \frac{3}{2} - \sqrt{\frac{1}{4} - \frac{4}{3Q^2}}] \times [\Delta - \frac{3}{2} + \sqrt{\frac{1}{4} - \frac{4}{3Q^2}}] \quad (1)$$

where $F_0 = \frac{Fl_c^3}{EI\delta}$, $\Delta = \frac{\sigma}{\delta}$, and $Q = \frac{w_x}{\delta}$ are dimensionless normalization parameters. The geometric parameters of the buckling beam include the span (l_c) and the thickness (δ); σ, which is the intermediate deformation of the buckling beam under external force F; and the arching height (w_x).

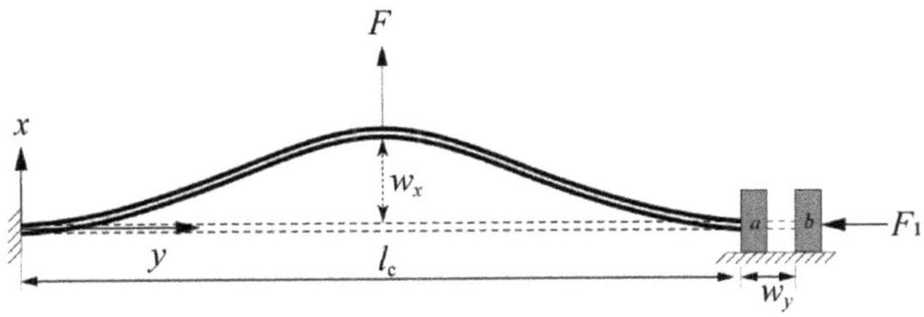

Figure 1. The buckled beam with pre-stress.

As derived in our previous study [34], the mechanical constitutive equation for the bistable buckling beam is

$$F = -(\frac{3\pi^4 w_y{}^2 EI}{2\delta^2 l_c^3} - \frac{2\pi^4 EI}{l_c^3})x + \frac{3\pi^4 EI}{2l_c^3\delta^2}x^3 + \frac{2\pi^4 EIw_y}{l_c^3} \quad (2)$$

As shown in Equation (3), the mechanical constitutive law of the buckling beam consisted of the negative stiffness term and the cubic stiffness term.

$$F_s = k_n x + k_c x^3 \tag{3}$$

where the negative stiffness coefficient is $k_n = -(\frac{3\pi^4 w_y{}^2 EI}{2\delta^2 l_c^3} - \frac{2\pi^4 EI}{l_c^3})$, and the cubic stiffness coefficient is $k_c = \frac{3\pi^4 EI}{2 l_c^3 \delta^2}$.

2.2. Kinematic Equations for an NSBD-Equipped Structure

For brevity, a simplified single-degree-of-freedom (SDOF) structure equipped with an NSBD is illustrated in Figure 2. Parameters m, c, and k represent the mass, damping, and stiffness of the host structure, respectively. Earthquakes are assessed based on the base acceleration ($\ddot{x}_g$). Furthermore, m_s is the NSBD's mass; c_s is the damping of the NSBD; k_n and k_c represent the negative stiffness and the cubic stiffness of the NSBD, respectively; and x and x_s represent the relative displacements of the structure and the NSBD with respect to the ground, respectively.

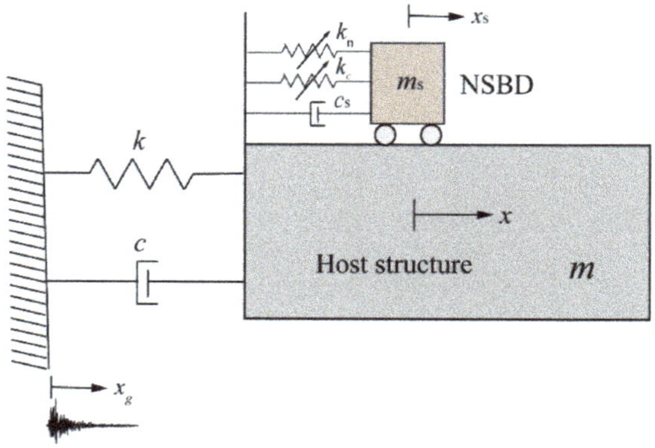

Figure 2. Schematic of the SDOF system with an NSBD.

In structural analysis, the governing differential equations of an SDOF system equipped with an NSBD subjected to base excitation can be directly derived from Newton's second law and expressed as

$$\begin{aligned} m\ddot{x} + c\dot{x} + kx + c_s(\dot{x} - \dot{x}_s) + k_n(x - x_s) + k_c(x - x_s)^3 &= -m\ddot{x}_g \\ m_s\ddot{x}_s + c_s(\dot{x}_s - \dot{x}) + k_n(x_s - x) + k_c(x_s - x)^3 &= 0 \end{aligned} \tag{4}$$

Equations (4) can be expressed in matrix form as follows:

$$\begin{bmatrix} m & 0 \\ 0 & m_s \end{bmatrix} \begin{Bmatrix} \ddot{x} \\ \ddot{x}_s \end{Bmatrix} + \begin{bmatrix} c + c_s & -c_s \\ -c_s & c_s \end{bmatrix} \begin{Bmatrix} \dot{x} \\ \dot{x}_s \end{Bmatrix} + \begin{bmatrix} k + k_n & -k_n \\ -k_n & k_n \end{bmatrix} \begin{Bmatrix} x \\ x_s \end{Bmatrix} + \{Q\} = \begin{Bmatrix} -m\ddot{x}_g \\ 0 \end{Bmatrix} \tag{5}$$

$$\{Q\} = \begin{Bmatrix} k_c(x - x_s)^3 \\ k_c(x_s - x)^3 \end{Bmatrix} \tag{6}$$

where k_n represents the negative stiffness coefficient, and k_c represents the cubic stiffness coefficient.

3. Equivalent Llinearization of the NSBD

3.1. Equivalent Linearization Method (ELM)

The NSBD is a nonlinear system, as outlined in Section 2. The fundamental-frequency-based TMD parameters' design may not be sufficient for NSBDs. To overcome this limitation, in this section, the optimization parameters of the nonlinear NSBD are determined using the ELM to linearize them to a linear system.

In the realm of nonlinear random-vibration-engineering analysis, the ELM has advanced significantly, offering an effective, straightforward approach. This method represents an approximate solution for predicting a nonlinear system's stochastic responses, holding considerable practical application potential. The core principles of the ELM are to substitute a nonlinear NSBD system with a linear one having an exact solution and to minimize their differences statistically. The control equation of the NSBD–structure system is as follows:

$$\mathbf{M}\ddot{X} - \mathbf{C}\dot{X} + \mathbf{K}X + Q = -\mathbf{M}_1 P\ddot{x}_g \tag{7}$$

where Q represents the nonlinear part, and $P = \begin{bmatrix}1 & 0\end{bmatrix}^T$ denotes the position vector of the seismic excitation. Assuming that a stationary response for the system given by Equation (7) is achievable, the NSBD–structure system is approximately replaced with an equivalent linear system, as per the ELM, as expressed in Equation (8).

$$\mathbf{M}\ddot{X} + \mathbf{C}\dot{X} + (\mathbf{K} + \mathbf{K}_e)X = -\mathbf{M}_1 P\ddot{x}_g \tag{8}$$

where $\mathbf{K}_e$ represents the equivalent stiffness matrix.

To render the system given by Equation (7) as being equivalent to the linear system described in Equation (8), one must strive to minimize the mean square value of the difference (e) between these equations.

$$e = Q - \mathbf{K}_e X \tag{9}$$

The necessary condition for this equivalence is

$$\frac{\partial}{\partial[\mathbf{K}_e]_{ij}} E[e^T e] = 0 \tag{10}$$

where $(i, j = 1, 2)$, and E represents the expectation of the variable. Equation (11) can be obtained by solving Equation (10).

$$\mathbf{K}_e = E\begin{bmatrix} 3k_c(E(x^2) - 2E(xx_s) + E(x_s^2)) & -3k_c(E(x^2) - 2E(xx_s) + E(x_s^2)) \\ -3k_c(E(x^2) - 2E(xx_s) + E(x_s^2)) & 3k_c(E(x^2) - 2E(xx_s) + E(x_s^2)) \end{bmatrix} \tag{11}$$

The pseudo-excitation method (PEM) was employed to iteratively compute $\mathbf{K}_e$ in Equation (11). The PEM, as proposed by Lin et al. [52–54], is gaining acceptance as an advanced and reasonable analysis tool that adequately accounts for the statistical probability characteristics of earthquake occurrences. Owing to the PEM's speed and efficient use of storage space, the structure's responses, such as displacement and acceleration, can be efficiently computed on ordinary microcomputers. This advanced structural modeling approach ensures greater accuracy in engineering structural analyses.

In the case of the NSBD–structure system, a pseudo-harmonic excitation can be expressed by Equation (12).

$$\ddot{x}_g = \sqrt{S_{\ddot{x}_g}(\omega)}e^{i\omega t} \tag{12}$$

where $S_{\ddot{x}_g}$ denotes the spectral density function. As given by Equation (13), the Kanai–Tajimi [55] model was employed in this paper.

$$S_{\ddot{x}_g}(\omega) = S_0 \frac{1 + 4\xi_g^2(\omega/\omega_g)^2}{\left[1 - (\omega/\omega_g)^2\right]^2 + 4\xi_g^2(\omega/\omega_g)^2} \tag{13}$$

where S_0 denotes the power spectral density of the bedrock's acceleration, while ξ_g and ω_g represent the damping ratio and rotational angular frequency, respectively, of the site's soil. The parameters of the Kanai–Tajimi spectrum are detailed in Table 1.

Table 1. Parameters of the Kanai–Tajimi spectrum.

Parameter	Type of Site			
	I	II	III	IV
S_0 (m^2/s^3)	0.0072	0.0091	0.0111	0.0166
ζ_g	0.64	0.72	0.80	0.90
ω_g (rad/s)	20.94	15.71	11.42	8.38

I, II, III, and IV represent the first, second, third, and fourth types of sites, respectively.

Therefore, Equation (8) is expressed as follows:

$$-\omega^2 M + i\omega C + (K + K_e))\{X\} = \{P\}\sqrt{S_{\ddot{x}_g}(\omega)}e^{i\omega t} \tag{14}$$

As is known,

$$E(x^2) = \int_{-\infty}^{+\infty} |x|^2 d\omega,\ E(xx_s) = \int_{-\infty}^{+\infty} |x||x_s| d\omega,\ E(x_s^2) = \int_{-\infty}^{+\infty} |x_s|^2 d\omega \tag{15}$$

By iteratively computing Equations (7), (8), (10), (11), (14) and (15), one can derive the equivalent linearization parameters of an NSBD. The calculation details are as follows:

Step 1: Given that the equivalent mass and damping matrices are zero, computing only the equivalent stiffness matrix is required using Equations (7), (8) and (10);

Step 2: Determine the appropriate input power spectrum according to the properties of the external excitation;

Step 3: Initially set the values of the equivalent linearization parameters before employing Equation (14) to compute the pseudo-responses. Subsequently, obtain the variance and covariance by solving Equation (15), leading to the replacement of the equivalent parameter matrices with new ones using Equation (11);

Step 4: Continue with step 3 until the parameters meet the specified criteria (The relative error value of the adjacent variances for each variable is greater than 10×10^6);

Step 5: By solving Equation (11), the equivalent parameter matrices can be determined for the original nonlinear system.

3.2. *Monte Carlo Method*

The Monte Carlo random simulation method [56–58], grounded in the central limit theorem of probability theory, is a widely recognized random method. Theoretically, its accuracy is enhanced with an increment in the number of samples. Applicable to diverse problems, the Monte Carlo method facilitates modeling and solving both mathematical models and complex systems in practical contexts. Particularly for complex problems, the Monte Carlo method often yields relatively accurate estimates. This method is an effective tool widely used in mathematical modeling and practical problem-solving and is noted for its flexibility, accuracy, and interpretability.

For analyzing the dynamic responses of structures with random parameters using the Monte Carlo method, the basic steps include the following:

Step 1: Generate N samples of random parameters reflecting the statistical characteristics prescribed. The stochastic excitation sample can be simulated using Equation (16).

$$\ddot{x}_g = \sum_{k=1}^{N} A_k \cos(\omega_k t + \theta_k) \tag{16}$$

where $\Delta\omega = (\omega_n - \omega_1)/N$, $A_k^2 = 4S_{\ddot{x}_g}(\omega_k) \cdot \Delta\omega$, and $\omega_k = \omega_1 + (k - \frac{1}{2}) \cdot \Delta\omega$ for $k = 1, 2, \cdots, N$. N represents the number of samples for random parameters; ω_1 and ω_n specify the limits within the frequency range; θ_k denotes a random number in [0, 2π];

Step 2: Insert sampled values of structural parameters into the structure's vibration equation;

Step 3: Employ numerical methods to solve the structure's kinematic equation and determine its dynamic responses;

Step 4: Repeat from steps (1) to (3) to generate N samples of structural dynamic responses and subsequently calculate their statistical characteristics.

3.3. Numerical Analysis

To ascertain whether the equivalent linearization of the NSBD–structure coupled system corresponds to the dynamic responses of the original nonlinear system, the dynamic responses are compared between the two systems computed using the Monte Carlo method. As per Equation (16), 1000 seismic excitations were randomly generated, each lasting 200 s. The parameters for the stochastic excitation sample include N = 10,000, $\omega_1 = 0$ Hz, and $\omega_n = 100.0$ Hz. The peak acceleration of the stochastic excitation is set at 1 m/s^2.

Figure 3 depicts the root mean squares (RMSs) of the primary structural displacements, where r_x represents the RMSs of the structural displacements for 1000 stochastic excitations. From Figure 3, it is evident that the RMSs of the dynamic responses of the structure align closely for both calculation methods. The maximum relative error (r_x) between the two systems amounts to 0.7%. The peak structural displacements are presented in Figure 4, where $x_{\max}$ denotes the peak structural displacement at each time. The maximum relative error ($x_{\max}$) between the two systems is as low as 1.7%. Figure 5 displays the variances in the displacements, where σ_x^2 represents the variance in the structure's displacement. The results of the dynamic responses calculated using the equivalent linearization model show remarkable agreement with those of the original nonlinear system, with the maximum relative errors of σ_x^2 between the two systems being just 1.15%. In Figures 3–5, the statistical calculation results of the equivalent linearization model demonstrate high accuracy.

For further verification of the accuracy of the equivalent linearization of nonlinear systems, comparisons were made between the dynamic responses of the two systems subjected to 16 seismic excitations across four site categories. In accordance with the standards in [59], 16 seismic records were selected, with their peak ground acceleration (PGA) uniformly established at 1 m/s^2, as detailed in Table 2.

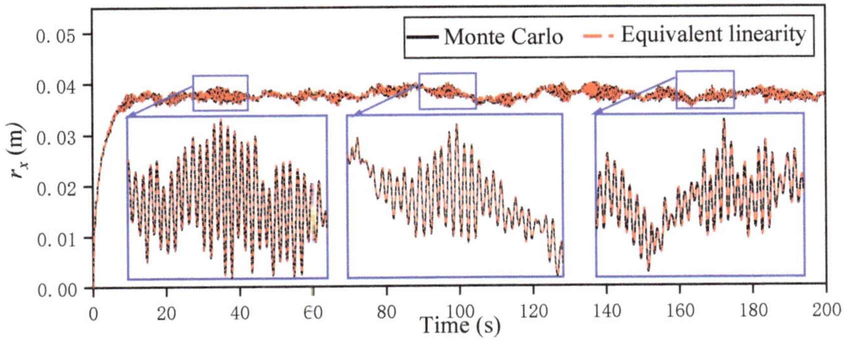

Figure 3. Root mean squares of the structural displacements versus time.

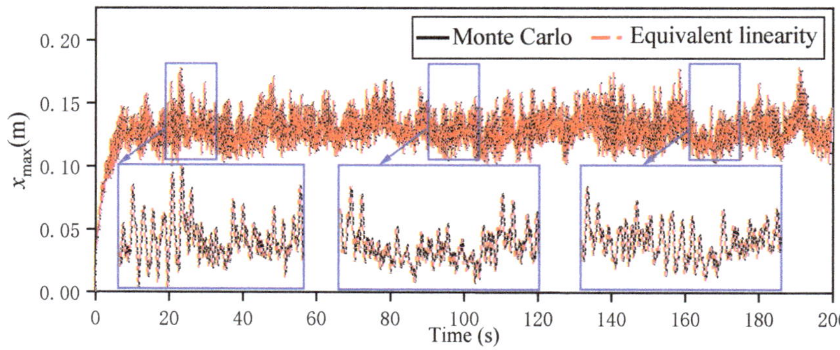

Figure 4. Peak structural displacements versus time.

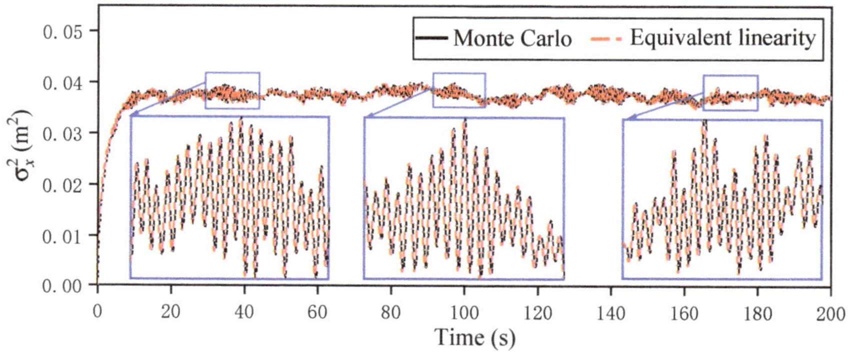

Figure 5. Variances in the structural displacements versus time.

Table 2. Earthquake records.

Site Classification	Earthquake Name	Year	Station Name	Magnitude
I	Northern Calif-07	1975	Cape Mendocino	5.2
	Helena-01	1935	Carroll College	6.0
	San Fernando4	1971	Castaic–Old Ridge Route	6.61
	Parkfield	1966	Temblor Pre-1969	6.19
II	Kern County1	1952	Taft Lincoln School	7.36
	San Fernando2	1971	Gormon–Oso Pumping Plant	6.61
	Borrego Mtn.	1968	El Centro Array #9	6.63
	Northern Calif-03	1954	Ferndale City Hall	6.5
III	Point Mugu	1973	Port Hueneme	5.65
	San Fernando3	1971	Palmdale Fire Station	6.61
	Hollister-02	1961	Hollister City Hall	5.5
	Kern County3	1952	LA–Hollywood Stor FF	7.36
IV	Northern Calif-02	1952	Ferndale City Hall	5.2
	Kern County2	1952	Santa Barbara Courthouse	7.36
	San Fernando1	1971	Lake Hughes #1	6.61
	Kern County4	1952	Pasadena–CIT Athenaeum	7.36

Figure 6 displays the seismic responses of the structure equipped with an NSBD for various earthquakes. Table 3 details the peak structural displacements with an NSBD for 16 varied earthquakes. In all the calculations, a mass ratio of 0.02 is assumed for the NSBD. As demonstrated in these figures, there is a close match between the two systems. Table 3 indicates that the relative error in the seismic responses for various earthquakes remains below 4.5%. The simulation results suggest that the equivalent linearization system equipped with an NSBD exhibits high accuracy. Consequently, the consistency of the calculation results confirms the accuracy of the equivalent linearization system.

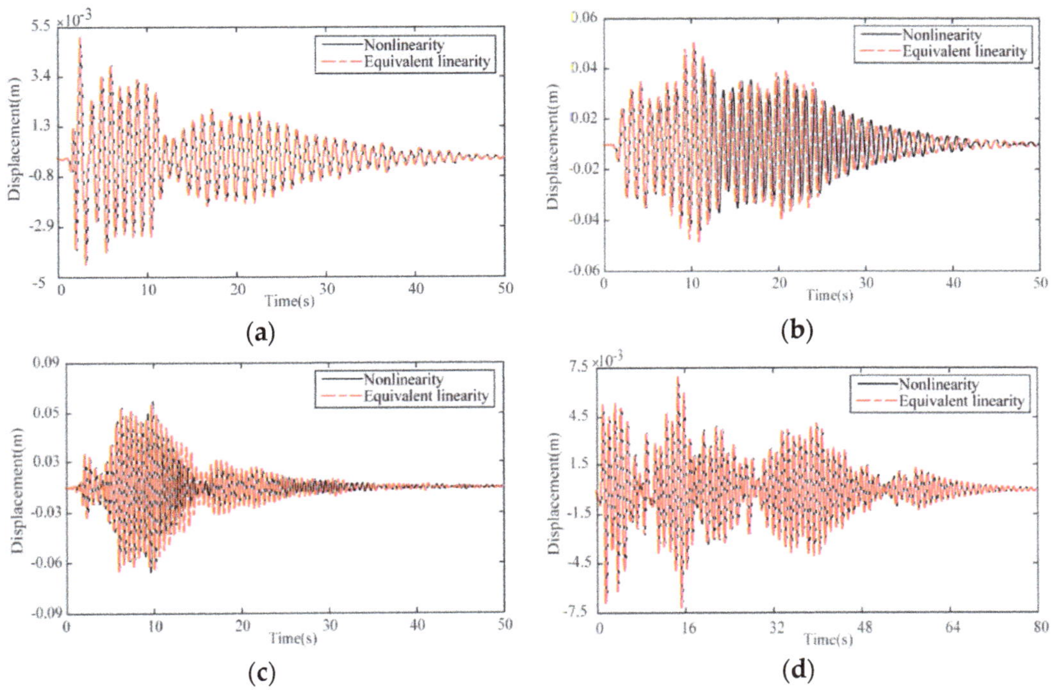

Figure 6. Seismic responses of the structure for (**a**) Helena-1 earthquake, (**b**) Borrego Mtn. earthquake, (**c**) Hollister-2 earthquake, and (**d**) San Fernando earthquake.

Table 3. Seismic responses of the primary structure with an NSBD for various earthquakes.

Earthquake	Peak Displacement (cm)		
	Nonlinearity	**Equivalent Linearity**	**Relative Error (%)**
Northern Calif-07	0.144	0.149	3.0
Helena-01	0.512	0.5099	0.4
San Fernando4	0.6856	0.703	2.5
Parkfield	0.8458	0.8707	3.0
Kern County1	2.158	2.118	1.85
San Fernando2	2.157	2.123	1.6
Borrego Mtn.	4.904	4.783	2.5
Northern Calif-03	4.919	5.018	2
Point Mugu	6.6	6.518	1.24
San Fernando3	6.846	6.631	3.1
Hollister-02	6.761	6.737	0.3
Kern County3	7.13	6.833	4.2
Northern Calif-02	7.704	7.511	2.5
Kern County2	7.999	7.87	1.6
San Fernando1	8.461	8.253	2.5
Kern County4	11.1	11.5	3.5

To explore the influence of the excitation amplitude on the equivalent linearization accuracy of the negative stiffness bistable damper (NSBD), the equivalent linearization results of the structure equipped with a damper for different PGA values are calculated.

As shown in Figure 7, with the increase in the PGA value, the structural displacement responses of the equivalent linearization system and the original nonlinear system are less consistent, which shows that the excitation amplitude becomes an important limiting factor for the negative stiffness bistable structure.

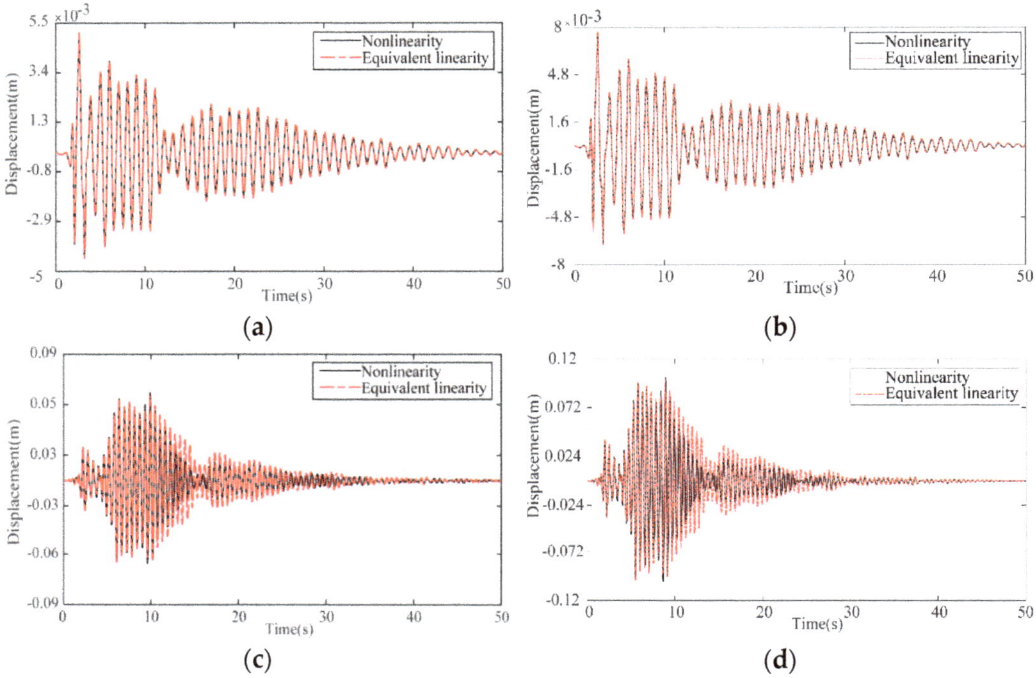

Figure 7. Seismic responses of the structures for (**a**) Helena-1 earthquake, PGA = 0.1 g; (**b**) Helena-1 earthquake, PGA = 0.15 g; (**c**) Hollister-2 earthquake, PGA = 0.1 g; (**d**) Hollister-2 earthquake, PGA = 0.15 g.

4. Optimal Design of the NSBD

The design parameters of the NSBD, such as negative stiffness, cubic stiffness, and damping, have substantial impacts on its vibration suppression. Thus, for the effective enhancement of the NSBD's damping effect, the optimization of the design parameters is necessary. In this study, the ELM and GA are employed to optimally design a damped tall building equipped with an NSBD, using the H∞ norm as the objective function.

4.1. Control Equations of a Tall Building Installed with an NSBD

As depicted in Figure 8, a mathematical model of a tall building controlled by an NSBD is chosen for the vibration control analysis example. The tall building has a total height of 162.15 m and a 49-story framed shear-wall structure. To simplify the numerical analysis, this building is simplified as a benchmark model [60] for the calculations. These 49 degrees of freedom represent the lateral displacement of each layer, and the structural damping is taken as 0.05. The first three natural frequencies of the tall building are 0.32 Hz, 0.84 Hz, and 1.33 Hz. In Figure 8b, m_i, k_i, and c_i are the *i*th floor's mass, stiffness, and damping of the tall building, respectively. As shown in Figure 8b, an NSBD is mounted on the top floor of the primary structure.

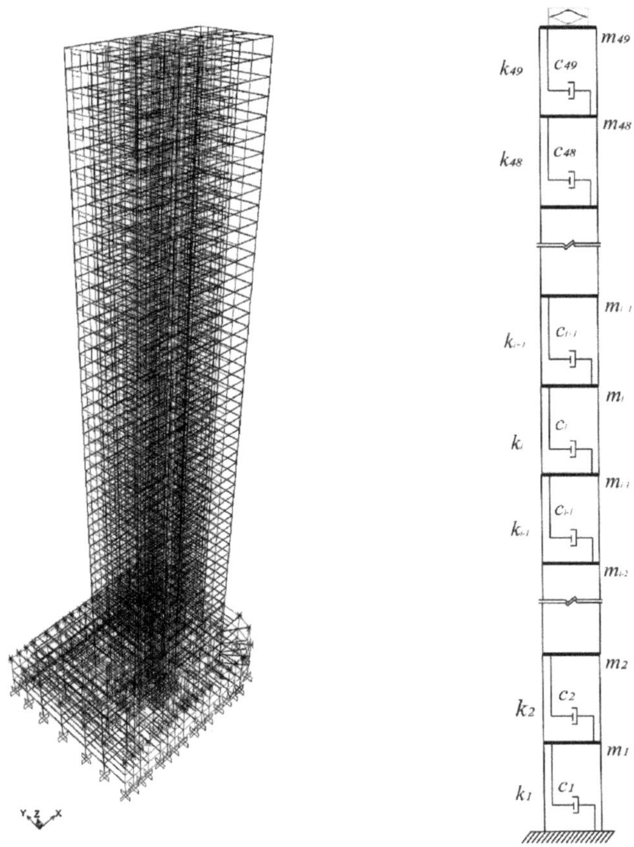

(a) (b)

Figure 8. Analysis models of the tall building: (**a**) structural model; (**b**) simplified calculation model.

The control equations of the primary structure with an NSBD are presented in Equation (17), following the derivation outlined in Section 3.1.

$$\begin{bmatrix} \mathbf{M}_s & \mathbf{0}_{49\times1} \\ \mathbf{0}_{1\times49} & m_s \end{bmatrix} \begin{Bmatrix} \ddot{\boldsymbol{x}} \\ \ddot{x}_s \end{Bmatrix} + \begin{bmatrix} \mathbf{C}_s & \mathbf{0}_{49\times1} \\ \mathbf{0}_{1\times49} & 0 \end{bmatrix} \begin{Bmatrix} \dot{\boldsymbol{x}} \\ \dot{x}_s \end{Bmatrix} + \begin{bmatrix} \mathbf{K}_s & \mathbf{0}_{49\times1} \\ \mathbf{0}_{1\times49} & 0 \end{bmatrix} \begin{Bmatrix} \boldsymbol{x} \\ x_s \end{Bmatrix} + \boldsymbol{Q}_1 = \begin{Bmatrix} -\mathbf{M}_s \boldsymbol{P} \ddot{x}_g \\ 0 \end{Bmatrix} \tag{17}$$

$$\boldsymbol{Q}_1 = \begin{Bmatrix} \mathbf{0}_{48\times1} \\ m_s \ddot{x}_{49} + 3k_c (x_{49} - x_s)^3 \\ 3k_c (x_s - x_{49})^3 \end{Bmatrix} \tag{18}$$

where $\mathbf{M}_s$, $\mathbf{C}_s$, and $\mathbf{K}_s$ represent the mass, damping, and stiffness matrices of the primary structure, respectively. Additionally, $\ddot{\boldsymbol{x}}$, $\dot{\boldsymbol{x}}$, and $\boldsymbol{x}$ denote the acceleration, velocity, and displacement vectors of the primary structure, respectively, and $\boldsymbol{P}$ symbolizes the location vector of the seismic excitation.

4.2. Equivalent Linearization of NSBD–Structure System

Optimizing the NSBD–structure system with the H∞ norm precludes the direct acquisition of the coupled structure's transfer function owing to the damper's unique nonlinearity. Consequently, the initial linearization of the coupled structure via the ELM is essential. The kinematic equations for the NSBD–structure system's equivalent linearization, formulated using the ELM, are presented in Equation (19).

$$\mathbf{M}_t\ddot{X} + \mathbf{C}_t\dot{X} + (\mathbf{K}_t + \mathbf{K}_e)X = -\mathbf{M}_t P\ddot{x}_g \tag{19}$$

where

$$\mathbf{M}_t = \begin{bmatrix} \mathbf{M}_s & \mathbf{0}_{49\times1} \\ \mathbf{0}_{1\times49} & m_2 \end{bmatrix}, \mathbf{C}_t = \begin{bmatrix} \mathbf{C}_s & \mathbf{0}_{49\times1} \\ \mathbf{0}_{1\times49} & 0 \end{bmatrix}, \mathbf{K}_t = \begin{bmatrix} \mathbf{K}_s & \mathbf{0}_{49\times1} \\ \mathbf{0}_{1\times49} & 0 \end{bmatrix}, \text{ and}$$
$$\mathbf{K}_e = \begin{bmatrix} \mathbf{0}_{48\times48} & \mathbf{0}_{48\times2} \\ \mathbf{0}_{2\times48} & \begin{matrix} 3k_c(E(x_{49}^2) - 2E(x_{49}x_s) + E(x_s^2)) & -3k_c(E(x_{49}^2) - 2E(x_{49}x_s) + E(x_s^2)) \\ -3k_c(E(x_{49}^2) - 2E(x_{49}x_s) + E(x_s^2)) & 3k_c(E(x_{49}^2) - 2E(x_{49}x_s) + E(x_s^2)) \end{matrix} \end{bmatrix} \tag{20}$$

where k_c represents the cubic stiffness coefficient.

By building Simulink models in MATLAB(R2020a), the dynamic equations of the main structures with and without the NSBD are analyzed and solved, and the top displacement responses of the main structure under the two working conditions are calculated separately. Figure 9 depicts the top displacement responses of the primary structures, while Figure 10 presents the structures' displacement envelope diagrams. The PGA of the earthquake is adjusted to 1.5 m/s^2, and the NSBD's mass ratio is maintained at 0.02.

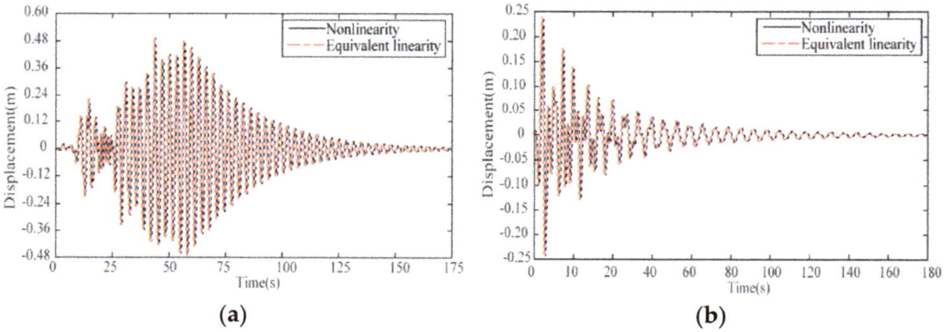

Figure 9. Seismic responses over time: (**a**) Kern County3; (**b**) San Fernando2.

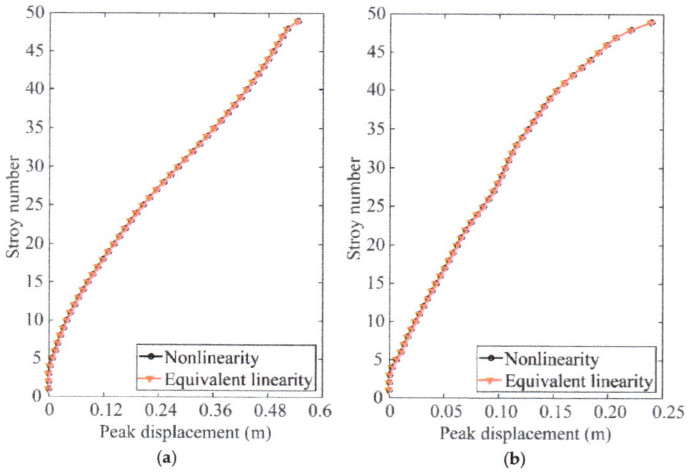

Figure 10. Displacement envelope diagrams of the primary structures for (**a**) Kern County3 and (**b**) San Fernando2 earthquakes.

4.3. H∞ Norm of the NSBD–Tall Building System

Minimizing the peak response of the structure is important in the vibration control of tall buildings. The H∞ norm [61,62], a prevalent indicator, is extensively used in optimizing damper designs. A key advantage for applying the H∞ norm to linear systems is the achievement of the desired results independently of specific excitations. Herein, the H∞ norm is defined as the peak value of the maximum singular value within the frequency domain of the structure. Consequently, a lower H∞ norm suggests the reduced vibration response and output energy of the structure. Thus, the H∞ norm can serve as the main objective function for optimizing parameters that influence the structure's vibration reduction.

The governing equations of the NSBD–tall building system can be replaced by Equation (21).

$$\mathbf{M}\ddot{X} + \mathbf{C}\dot{X} + \mathbf{K}X = -\mathbf{M}_t P \ddot{x}_g \tag{21}$$

where $\mathbf{M} = \mathbf{M}_t$, $\mathbf{C} = \mathbf{C}_t$, and $\mathbf{K} = \mathbf{K}_t + \mathbf{K}_e$.

The state space of Equation (21) is expressed as

$$\begin{aligned} \dot{z} &= \tilde{\mathbf{A}}z + \tilde{\mathbf{B}}w \\ y &= \tilde{\mathbf{C}}z + \tilde{\mathbf{D}}w \end{aligned} \tag{22}$$

where $\dot{z}$ represents the state vector of the system, y is the output vector of the system, and

$$z = \begin{bmatrix} X \\ \dot{X} \end{bmatrix},\ w = -P\ddot{x}_g,\ \tilde{\mathbf{A}} = \begin{bmatrix} \mathbf{0} & I \\ -\mathbf{M}^{-1}\mathbf{K} & -\mathbf{M}^{-1}\mathbf{C} \end{bmatrix},\ \tilde{\mathbf{B}} = \begin{bmatrix} 0 \\ P \end{bmatrix},\ \tilde{\mathbf{C}} = \begin{bmatrix} \mathbf{I} & \mathbf{0} \end{bmatrix},\ \text{and}\ \tilde{\mathbf{D}} = \mathbf{0} \tag{23}$$

The transfer function of Equation (22) is

$$G(s) = \tilde{\mathbf{C}}(sI - \tilde{\mathbf{A}})^{-1}\tilde{\mathbf{B}} + \tilde{\mathbf{D}} \tag{24}$$

The mathematical expression of the H∞ norm is shown as Equation (25) [63].

$$\|G(s)\|_\infty = \sup_{\|w\|_2 \neq 0} \frac{\|y\|_2}{\|w\|_2} = \sup_{\|w\|_2 = 1} \|y\|_2 \tag{25}$$

where y represents the system's output, w represents the seismic input, and $\|y\|_2$ and $\|w\|_2$ are defined as their L_2 norms, symbolizing the system's output energy and the earthquake's energy, respectively. From this derivation, it becomes evident that the NSBD–structure system's output energy diminishes as H∞ norm decreases; thus, achieving the minimum value of H∞ ensures the optimal damping effect of the NSBD. Furthermore, considering the specific spectral characteristics of the seismic excitation, this study employs the Kanai–Tajimi response spectrum to measure the input seismic excitation.

4.4. General Optimization Procedure

The genetic algorithm (GA) [64–66], mimicking natural selection and genetic mechanisms of evolution, serves as a method for finding optimal solutions by simulating these evolutionary processes. This technique operates directly on structural objects, unconstrained by differentiation or functional continuity; it exhibits inherent parallelism and superior global optimization capabilities. Utilizing probabilistic optimization methods, this approach allows for the automatic acquisition and guidance of the optimized search space, facilitating the adaptive adjustment of the search direction. The specific steps to optimize NSBD parameters via the GA include the following:

Step 1: Identify the variables and their optimization ranges; in this case, the key variables include negative stiffness, cubic stiffness, and damping;

Step 2: Choose the optimization model and functions;

Step 3: Establish the chromosome encoding and decoding methods;

Step 4: Set quantitative evaluation criteria for the fitness;

Step 5: Define the parameters of the GA; for this optimization, these include 200 iterations, a population size of 100, and crossover and mutation probabilities of 0.7 and 0.03, respectively;

Step 6: Compute the global optimal solution, which, in this case, involves determining the optimal parameters of the NSBD.

To prevent settling for a local optimal solution, it is crucial to establish the feasible range of optimization parameters based on linear dampers' design principles prior to the computation.

4.5. Numerical Analysis

The primary goal of this optimization endeavor is to minimize the H∞ norm associated with the top floor's displacement. The optimization parameters for the NSBD comprise negative stiffness, cubic stiffness, and damping.

(1) The range of negative stiffness values for the NSBD is set between -9×10^6 and -1×10^6;
(2) The cubic stiffness of the NSBD varies from 2×10^{10} to 3×10^{10};
(3) The damping for the NSBD falls within the range from 0 to 1,716,100.

The convergence trajectory of the H∞ norm is depicted in Figure 11. Initially, the H∞ norm stands at 5.8, but after 100 generations, it converges to 2.4. The optimal parameters obtained through the optimization calculations are as follows: the negative stiffness value is -2.663×10^6 N/s, the cubic stiffness is 6.5045×10^{10} N/m^3, and the damping is 211,485.3 N·s/m.

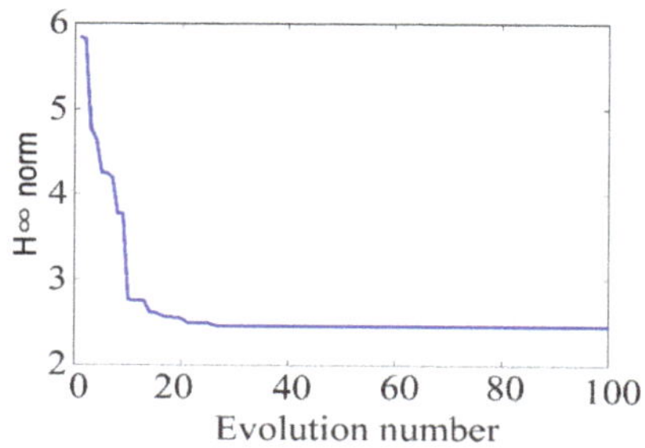

Figure 11. Convergence history of the H∞ norm.

For assessing the accuracy of the optimized design, in this section, we calculate and compare the displacement responses of tall building structures with and without the NSBD, focusing on the top floor's displacements. The top floor's displacement histories for the primary structures with and without dampers are illustrated in Figure 12. It is noted from Figure 13 that the suppression of the peak displacement by the NSBD is apparent at the first-order frequency of the tall building. The peak displacement and vibration reduction ratios for the top floor for various seismic excitations are detailed in Table 4. In these calculations, the peak ground acceleration (PGA) is set at 1.5 m/s^2, and the NSBD's mass ratio remains at 0.02. Having determined all the NSBD's parameters through the proposed optimization methods, it is evident that the top floor of the tall building equipped with the NSBD undergoes significant vibration mitigation for various seismic excitations. The NSBD exhibits its most effective damping for the Hollister-02 earthquake, achieving a displacement mitigation ratio of 52.99%. Furthermore, the vibration mitigation ratios of the NSBD exceed 22% for all the selected earthquakes. Figure 12 illustrates that when the PGA value in the simulation calculation is less than 0.15 g, the displacement response of the top floor of the structure does not exceed 1.0 m, and the same is also reflected in Table 4. The calculation results show that when the PGA value is less than 0.15 g, the negative stiffness bistable structure has a good damping effect.

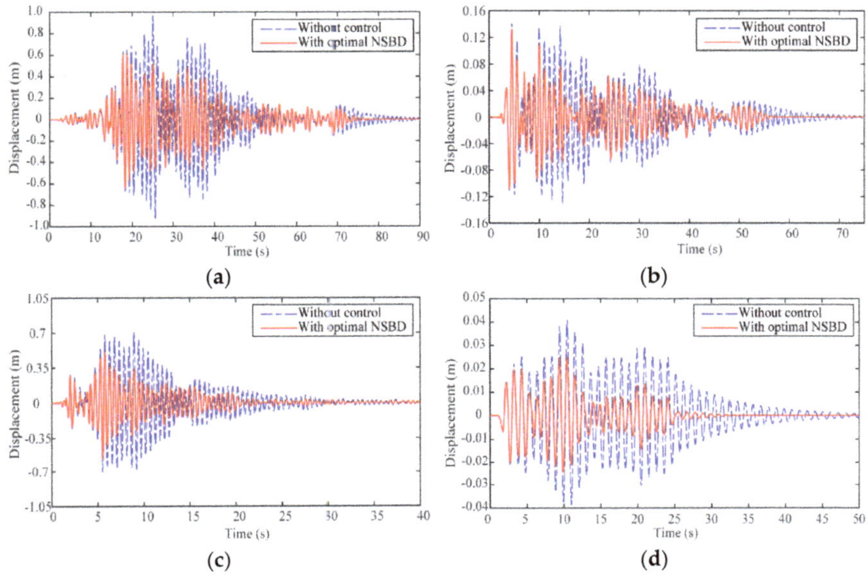

Figure 12. Top floor's displacements in the tall building for (**a**) Kern County4, (**b**) Kern County1, (**c**) Hollister-2, and (**d**) Borrego Mtn. earthquakes.

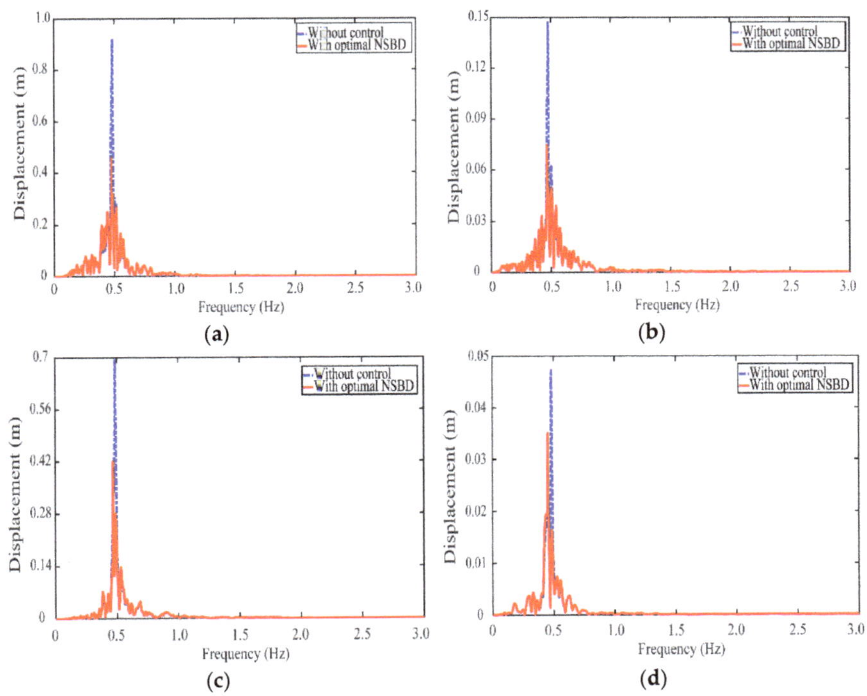

Figure 13. Top floor's displacements in the tall building in the frequency domain for (**a**) Kern County4, (**b**) Kern County1, (**c**) Hollister-02, and (**d**) Borrego Mtn. earthquakes.

Table 4. Peak displacements of the primary structures for various seismic excitations.

Earthquake	Peak Displacement (cm)		Vibration Mitigation Ratio (%)
	Uncontrolled	With Optimal NSBD	
Northern Calif-07	2.151	1.273	40.82
Helena-01	4.073	4.487	38.94
San Fernando4	3.528	2.130	39.63
Parkfield	4.229	3.065	27.52
Kern County1	13.184	6.672	49.39
San Fernando2	20.51	12.64	38.37
Borrego Mtn.	24.75	17.42	29.62
Northern Calif-03	24.48	16.57	32.29
Point Mugu	43.94	26.49	39.71
San Fernando3	54.90	35.42	35.48
Hollister-02	47.33	22.25	52.99
Kern County3	55.64	35.45	36.29
Northern Calif-02	93.18	71.97	22.76
Kern County2	99.00	69.68	29.62
San Fernando1	84.77	57.23	32.49
Kern County4	96.61	51.59	46.60

The numerical simulation results suggest that the NSBD effectively controls the top floor's displacement in tall building structures for diverse seismic excitations, demonstrating that the NSBD parameters, derived from the proposed optimization method, are highly effective in vibration reduction. The vibration mitigation ratio (w) is calculated using Equation (26).

$$w = \frac{m_0}{m_1} \times 100\% \tag{26}$$

where m_0 and m_1 denote the maximum displacements at the top of the structures with and without the NSBD, respectively.

5. Conclusions

This paper introduces a robust method employing the ELM and GA for optimizing NSBD parameters. The efficacy of this approach is evaluated by applying the optimized NSBD to a tall building and assessing the displacement of the tall building for various seismic excitations. The most important findings from the numerical simulations are summarized below:

1. Utilizing the Monte Carlo simulation calculation method, the maximum root-mean-square error for applying the nonlinear NSBD and equivalent linear dampers to the structure is 0.7%, the maximum peak displacement error is 1.7%, and the maximum displacement variance error in the structure is 1.15%. The dynamic responses calculated using the equivalent linearization model show remarkable agreement with those of the original nonlinear system;
2. According to the pseudo-excitation method (PEM), the simulation results suggest that the displacement response's error in the structure will not exceed 4.5% when the building is equipped with the nonlinear NSBD and equivalent linear dampers for different earthquakes. The NSBD can be approximated by a linear system with the help of the ELM, which can be vital for the NSBD's optimal design, as demonstrated by these simulation calculations;
3. As a main objective function, the H∞ norm serves as a very precise method for optimizing parameters that influence the structure's vibration reduction. The genetic algorithm (GA) is perfectly suitable for obtaining the design parameters of the NSBD within an appropriate range. After 100 generations, the H∞ norm converges to 2.4, indicating that the genetic algorithm can simulate and calculate optimization parameters very accurately and quickly;

4. The displacement responses of the tall buildings with and without an NSBD are simulated utilizing the optimized parameters solved through the GA. The best damping for the Hollister-02 earthquake can achieve a displacement mitigation ratio of 52.99%, and the vibration mitigation ratios of the NSBD exceed 22% for all the selected earthquakes. The simulation results suggest that the effective restraint of the structural vibration for different earthquakes can be achieved using the NSBD with the optimal parameters. The proposed method is effective in implementing the optimal design of the NSBD.

There are also some limitations that warrant further investigation and research. This paper focuses solely on optimizing the design analysis for nonlinear NSBDs through numerical simulation calculations. In our future work, detailed experiments will be conducted to verify the accuracy and rationality of the obtained optimized design parameters. Furthermore, the design of the NSBD for high-rise structural vibration reduction control adopts a distributed layout, achieving multi-modal vibration reduction control, which will also be studied in future work.

Author Contributions: Conceptualization, L.F.; methodology, L.F. and C.H.; software, L.F.; validation, L.F. and C.H.; formal analysis, L.F.; investigation, L.F.; resources, L.F.; data curation, L.F.; writing—original draft preparation, L.F.; writing—review and editing, L.F., L.H., and C.H.; visualization, L.F.; supervision, L.H.; project administration, L.H.; funding acquisition, L.H. All authors have read and agreed to the published version of the manuscript.

Funding: This research received no external funding.

Institutional Review Board Statement: Not applicable.

Informed Consent Statement: Not applicable.

Data Availability Statement: The data presented in this study are available in the article.

Conflicts of Interest: The authors declare no conflicts of interest.

References

1. Yao, J. Concept of structural control. *ASCE J. Struct. Div.* **1972**, *98*, 1567–1574. [CrossRef]
2. Soong, T.T.; Spencer, B.F. Active structural control: Theory and practice. *J. Eng. Mech.* **1992**, *118*, 1282–1285. [CrossRef]
3. Hoang, N.; Fujino, Y.; Warnitchai, P. Optimal tuned mass damper for seismic applications and practical design formulas. *Eng. Struct.* **2008**, *30*, 707–715. [CrossRef]
4. Hsiao, F.H.; Chen, C.W.; Liang, Y.W.; Xu, S.D. T-S fuzzy controllers for Nonlinear interconnected systems with multiple time delays. *IEEE Trans. Circuits Syst. I Regul. Pap.* **2005**, *52*, 1883–1893. [CrossRef]
5. Rana, R.; Soong, T.T. Parametric study and simplified design of tuned mass dampers. *Eng. Struct.* **1998**, *20*, 193–204. [CrossRef]
6. Sadek, F.; Mohraz, B.; Taylor, A.W.; Chung, R.M. A method of estimating the parameters of tuned mass dampers for seismic applications. *Earthq. Eng. Struct. Dyn.* **1997**, *26*, 617–635. [CrossRef]
7. Soong, T.T.; Spencer, B.F. Supplemental energy dissipation: State-of-the-art and state-of-the practice. *Eng. Struct.* **2002**, *24*, 243–259. [CrossRef]
8. Symans, M.D.; Charney, F.A.; Whittaker, A.S.; Constantinou, M.C. Energy dissipation systems for seismic applications: Current practice and recent developments. *J. Struct. Eng.* **2008**, *134*, 3–21 [CrossRef]
9. Roffel, A.J.; Narasimhan, S.; Haskett, T. Performance of pendulum tuned mass dampers in reducing the responses of flexible structures. *J. Struct. Eng.* **2013**, *139*, 04013019. [CrossRef]
10. Lu, X.; Chen, J. Mitigation of wind-induced response of Shanghai Center Tower by tuned mass damper. *Struct. Des. Tall Spec. Build.* **2011**, *20*, 435–452. [CrossRef]
11. Pasala, D.T.R.; Sarlis, A.A.; Nagarajaiah, S.; Reinhorn, A.M. Adaptive negative stiffness: New structural modification approach for seismic protection. *J. Struct. Eng.* **2013**, *139*, 1112–1123. [CrossRef]
12. Pasala, D.T.R.; Sarlis, A.A.; Reinhorn, A.M.; Nagarajaiah, S. Simulated bilinear-elastic behavior in a SDOF elastic structure using negative stiffness device: Experimental and analytical study. *J. Struct. Eng.* **2014**, *140*, 04013049. [CrossRef]
13. Li, H.N.; Sun, T.; Lai, Z.L.; Nagarajaiah, S. Effectiveness of negative stiffness system in the benchmark structural-control problem for seismically excited highway bridges. *J. Bridge Eng.* **2018**, *23*, 04018001. [CrossRef]
14. Sun, T.; Lai, Z.; Nagarajaiah, S.; Li, H.N. Negative stiffness device for seismic protection of smart base isolated benchmark building. *Struct. Control Health Monit.* **2017**, *24*, e1968. [CrossRef]
15. Sun, F.F.; Wang, M.; Nagarajaiah, S. Multi-objective optimal design and seismic performance of negative stiffness damped outrigger structures considering damping cost. *Eng. Struct.* **2021**, *229*, 111615. [CrossRef]

16. Wang, M.; Sun, F.F.; Nagarajaiah, S. Simplified optimal design of MDOF structures with negative stiffness amplifying dampers based on effective damping. *Struct. Des. Tall Spec. Build.* **2019**, *28*, e1664. [CrossRef]
17. Wang, M.; Sun, F.F.; Yang, J.Q.; Nagarajaiah, S. Seismic protection of SDOF systems with a negative stiffness amplifying damper. *Eng. Struct.* **2019**, *190*, 128–141. [CrossRef]
18. Zhou, P.; Li, H. Modeling and control performance of a negative stiffness damper for suppressing stay cable vibrations. *Struct. Control Health Monit.* **2016**, *23*, 764–782. [CrossRef]
19. Chen, L.; Sun, L.; Nagarajaiah, S. Cable with discrete negative stiffness device and viscous damper: Passive realization and general characteristics. *Smart Struct. Syst.* **2015**, *15*, 627–643. [CrossRef]
20. Shi, X.; Zhu, S. Magnetic negative stiffness dampers. *Smart Mater. Struct.* **2015**, *24*, 072002. [CrossRef]
21. Shi, X.; Zhu, S. Simulation and optimization of magnetic negative stiffness dampers. *Sens. Actuators A* **2017**, *259*, 14–33. [CrossRef]
22. Shi, X.; Zhu, S.; Spencer, B.F., Jr. Experimental study on passive negative stiffness damper for cable vibration mitigation. *J. Eng. Mech.* **2017**, *143*, 04017070. [CrossRef]
23. Shu, Z.; Zhang, J.; Nagarajaiah, S. Dimensional analysis of inelastic structures with negative stiffness and supplemental damping devices. *J. Struct. Eng.* **2017**, *143*, 04016184. [CrossRef]
24. Hong, N.; Zhao, Z.; Du, Y.; Chen, Q. Energy spectra and performance assessment of isolated structures with a negative stiffness amplification system. *Soil Dyn. Earthq. Eng.* **2023**, *169*, 107857. [CrossRef]
25. Kiran, K.K.; Al-Osta, M.A.; Ahmad, S. Optimum design and performance of a base-isolated structure with tuned mass negative stiffness inerter damper. *Sci. Rep.* **2023**, *13*, 4980. [CrossRef]
26. Su, N.; Bian, J.; Peng, S.; Chen, Z. Analytical optimal design of inerter-based vibration absorbers with negative stiffness balancing static amplification and dynamic reduction effects. *Mech. Syst. Signal Process.* **2023**, *192*, 110235. [CrossRef]
27. Tai, Y.J.; Wang, H.D.; Chen, Z.Q. Vibration isolation performance and optimization design of a tuned inerter negative stiffness damper. *Int. J. Mech. Sci.* **2023**, *241*, 107948. [CrossRef]
28. Ul Islam, N.; Jangid, R.S. Closed form expressions for H2 optimal control of negative stiffness and inerter-based dampers for damped structures. *Structures* **2023**, *50*, 791–809. [CrossRef]
29. Johnson, D.R.; Harne, R.L.; Wang, K.W. A disturbance cancellation perspective on vibration control using a bistable snap-through attachment. *J. Vib. Acoust.* **2014**, *136*, 4026673. [CrossRef]
30. Farhangdoust, S.; Eghbali, P.; Younesian, D. Bistable tuned mass damper for suppressing the vortex induced vibrations in suspension bridges. *Earthq. Struct.* **2020**, *18*, 313–320.
31. Yan, B.; Ma, H.; Zhang, L.; Zheng, W. A bistable vibration isolator with nonlinear electromagnetic shunt damping. *Mech. Syst. Signal Process.* **2020**, *136*, 106504. [CrossRef]
32. Yan, B.; Ling, P.; Zhou, Y.; Wu, C. Shock isolation characteristics of a bistable vibration isolator with tunable magnetic controlled stiffness. *J. Vib. Acoust.* **2022**, *144*, 4051850. [CrossRef]
33. Zhang, Y.; Ye, K.; Nyangi, P. Optimum design of a tuned-mass damper with negative stiffness device subjected to ground excitation. *Struct. Control Health Monit.* **2022**, *29*, e3086. [CrossRef]
34. Fan, L.; Huang, C.; Huo, L. Development of a negative stiffness bistable damper for structural vibration control. *Shock Vib.* **2022**, *2022*, 6397602. [CrossRef]
35. Li, H.; Sun, T. Optimal design for rail-type negative stiffness control system. *Earthq. Eng. Eng. Vib.* **2018**, *38*, 21–27.
36. Dai, H.; Zhao, Y. Equal-peak optimization of dynamic vibration absorber with negative stiffness and delay feedback control. *J. Theor. Appl. Mech.* **2021**, *53*, 1720–1732.
37. Li, J.; Gu, X.; Zhu, S.; Yu, C. Parameter optimization for a novel inerter-based dynamic vibration absorber with negative stiffness. *J. Nonlinear Math. Phys.* **2022**, *29*, 280–295. [CrossRef]
38. Ul Islam, N.; Jangid, R.S. Optimum parameters and performance of negative stiffness and inerter based dampers for base-isolated structures. *Bull. Earthq. Eng.* **2023**, *21*, 1411–1438. [CrossRef]
39. Charef, O.A.; Khalfallah, S. A variant design of tuned mass damper with negative stiffness for vibration control of a damped primary system. *Struct. Control Health Monit.* **2022**, *29*, e3068.
40. Liao, B.L.; Han, L.Y.; Cao, X.W.; Li, S. Double integral-enhanced zeroing neural network with linear noise rejection for time-varying matrix inverse. *CAAI Trans. Intell. Technol.* **2023**, *9*, 197–210. [CrossRef]
41. Alattar, B.; Ghommem, M.; Puzyrev, V. Deep learning for nonlinear characterization of electrostatic vibrating beam MEMS. *Int. J. Bifurc. Chaos* **2023**, *33*, 2330038. [CrossRef]
42. Luo, X.D.; Wen, X.H.; Li, Y.; Li, Q.F. Pruning method for dendritic neuron model based on dendrite layer significance constraints. *CAAI Trans. Intell. Technol.* **2023**, *8*, 308–318. [CrossRef]
43. He, Y.L.; Li, X.; Zhang, M.J.; Fournier-Viger, P. A novel observation points-based positive-unlabeled learning algorithm. *CAAI Trans. Intell. Technol.* **2023**, *8*, 1425–1443. [CrossRef]
44. Waziri, M.Y.; Yusuf, A.; Abubakar, A.B. Improved conjugate gradient method for nonlinear system of equations. *Comput. Appl. Math.* **2020**, *39*, 321. [CrossRef]
45. Xu, B.; Kiani, K. Nonlinear nonlocal-surface energy-based vibrations of a bidirectionally excited nanobeam at its supports. *Phys. Scr.* **2021**, *96*, 025004. [CrossRef]
46. Kiani, K. Nanoparticle delivery via stocky single-walled carbon nanotubes: A nonlinear-nonlocal continuum-based scrutiny. *Compos. Struct.* **2014**, *116*, 254–272. [CrossRef]

47. Kiani, K. Nonlinear vibrations of a single-walled carbon nanotube for delivering of nanoparticles. *Nonlinear Dyn.* **2014**, *76*, 1885–1903. [CrossRef]
48. Kiani, K.; Nikkhoo, A. On the limitations of linear beams for the problems of moving mass-beam interaction using a meshfree method. *Acta Mech. Sin.* **2012**, *28*, 164–179. [CrossRef]
49. Haghpanahi, M.; Nikkhoo, M.; Peirovi, H.; Ghanavi, J.E. Mathematical Modeling of the Intervertebral Disc as an Infrastructure for Studying the Mechanobiology of the Tissue Engineering Procedure. *WSEAS Trans. Appl. Theor. Mech.* **2007**, *2*, 261–273.
50. Nikkhoo, A.; Sharifinejad, M. The impact of a crack existence on the inertial effects of moving forces in thin beams. *Mech. Res. Commun.* **2020**, *107*, 103562. [CrossRef]
51. Qiu, J.; Lang, J.H.; Slocum, A.H. A curved-beam bistable mechanism. *J. Microelectromechanical Syst.* **2004**, *13*, 137–146. [CrossRef]
52. Lin, J.; Song, G.; Sun, Y.; Williams, F.W. Non-stationary random seismic responses of non-uniform beams. *Soil Dyn. Earthq. Eng.* **1995**, *14*, 301–306. [CrossRef]
53. Lin, J.H.; Guo, X.L.; Zhi, H.; Howson, W.P. Computer simulation of structural random loading identification. *Comput. Struct.* **2001**, *79*, 375–387. [CrossRef]
54. Lin, J.H.; Sun, D.K.; Zhong, W.X.; Zhang, W.S. High efficiency computation of the variances of structural evolutionary random responses. *Shock Vib.* **2000**, *98*, 31–35. [CrossRef]
55. Kanai, K. An empirical formula for the spectrum of strong earthquake motions. *Bull. Earthq. Res. Inst.* **1961**, *39*, 85–95.
56. Proppe, C.; Pradlwarter, H.J.; Schuller, G.I. Equivalent linearization and Monte Carlo simulation in stochastic dynamics. *Probabilistic Eng. Mech.* **2003**, *18*, 1–15. [CrossRef]
57. Hammersley, J.M. Simulation and the Monte Carlo Method. *Bull. Lond. Math. Soc.* **2005**, *14*, 278–553. [CrossRef]
58. Takahashi, Y.; Kiureghian, A.D.; Ang, A.H.S. Life-cycle cost analysis based on a renewal model of earthquake occurrences. *Earthq. Eng. Struct. Dyn.* **2010**, *33*, 859–880. [CrossRef]
59. *GB50011-2010*; Code for Seismic Design of Buildings. China Architecture and Building Press: Beijing, China, 2016.
60. Yang, J.N.; Agrawal, A.K.; Samali, B.; Wu, J.C. Benchmark problem for response control of wind-excited tall buildings. *J. Eng. Mech.* **2004**, *130*, 437–446. [CrossRef]
61. Ao, W.K.; Reynolds, P. Analysis and numerical evaluation of H_∞ and H_2 optimal design schemes for an electromagnetic shunt damper. *J. Vib. Acoust.* **2019**, *142*, 021003. [CrossRef]
62. Raze, G.; Kerschen, G. H_∞ optimization of multiple tuned mass dampers for multimodal vibration control. *Comput. Struct.* **2021**, *248*, 106485. [CrossRef]
63. Meinsma, G.; Mirkin, L. H^∞ control of systems with multiple I/O delays via decomposition to Adobe problems. *IEEE Trans. Autom. Control* **2005**, *50*, 199–211. [CrossRef]
64. Bozorgvar, M.; Zahrai, S.M. Semi-active seismic control of buildings using MR damper and adaptive neural-fuzzy intelligent controller optimized with genetic algorithm. *J. Vib. Control* **2019**, *25*, 273–285. [CrossRef]
65. Colherinhas, G.B.; de Morais, M.V.G.; Shzu, M.A.M.; Avila, S. Optimal pendulum tuned mass damper design applied to high towers using genetic algorithms: Two-DOF modeling. *Int. J. Struct. Stab. Dyn.* **2019**, *19*, 1950125. [CrossRef]
66. Li, Z.; Shu, G. Optimal placement of metallic dampers for seismic upgrading of multistory buildings based on a cost-effectiveness criterion using genetic algorithm. *Struct. Des. Tall Spec. Build.* **2019**, *28*, e1595. [CrossRef]

buildings

MDPI

Article

Effects of Embedded Expanded Polystyrene Boards on the Hysteretic Behavior of Innovative Precast Braced Concrete Shear Walls

Yachao Tang [1,*] and Hongnan Li [1,2]

1 State Key Laboratory of Coastal and Offshore Engineering, Faculty of Infrastructure Engineering, Dalian University of Technology, Dalian 116024, China; hnli@dlut.edu.cn
2 School of Civil Engineering, Shenyang Jianzhu University, Shenyang 110168, China
* Correspondence: tangyc.dlut@gmail.com

Abstract: An innovative type of precast braced concrete shear (PBCS) wall has been tested and verified to have comparable shear resistances relative to conventional cast-in-place reinforced concrete (RC) shear walls. The triangular or rectangular embedded expanded polystyrene (EPS) boards in PBCS wall panels can not only considerably reduce concrete use but also reduce the structural weight. To understand the functions of EPS boards in more depth, this paper investigates the effects of the thickness ratio of different shapes of EPS on the hysteretic behaviors of PBCS walls with various shear span ratios (SSRs). The finite element (FE) models of PBCS walls based on the multi-layer shell element are developed and verified to be sufficiently accurate in comparison with the experimental results. The analysis results indicate that the bearing capacity, lateral stiffness and ductility of PBCS walls show a downward trend with the increase in the thickness ratio of EPS boards. The rectangular EPS board has a more pronounced effect on weight reduction as well as concrete use reduction compared to the triangular EPS board under the same thickness ratio. The formulations regarding the bearing capacity are developed and show good agreement with the numerical results. The thickness ratio limit for PBCS walls to satisfy the ductility requirement is addressed. This investigation not only provides insight into the cyclic behavior of PBCS walls with varied thickness ratios but also demonstrates the potential applicability of PBCS walls in precast concrete (PC) structures for both thermal insulation and earthquake resistance purposes.

Keywords: precast braced concrete shear wall; expanded polystyrene board; thickness ratio; shear span ratio; hysteretic behavior

Citation: Tang, Y.; Li, H. Effects of Embedded Expanded Polystyrene Boards on the Hysteretic Behavior of Innovative Precast Braced Concrete Shear Walls. *Buildings* **2024**, *14*, 55. https://doi.org/10.3390/buildings14010055

Academic Editors: Rajai Zuheir Al-Rousan, Dan Bompa and Nerio Tullini

Received: 9 November 2023
Revised: 16 December 2023
Accepted: 21 December 2023
Published: 24 December 2023

1. Introduction

Nowadays, the precast concrete (PC) technology is widely promoted for application in reinforced concrete (RC) structures. It is more compatible with the development of a low-carbon society and the application of sustainable construction relative to the conventional cast-in-place method [1,2]. The utilization of the PC technology in shear walls makes it possible for walls with complex structural layouts and with composite actions to be fabricated beforehand in factory and further assembled on site. Moreover, previous research works showed that lowering the use of concrete can not only reduce carbon emissions but also degrade the structural weight [3,4]. Therefore, the effort related to the weight reduction of PC shear walls, which simultaneously retains sufficient lateral shear resistances and composite actions, is necessary as well as promising.

Considerable attempts have been performed to lighten the shear walls by means of employing various lightweight aggregates or devising novel structural configurations. Mousavi et al. [5] and Lombardi et al. [6] carried out hysteretic cyclic loading tests on lightweight concrete shear walls using expanded glass and expanded polystyrene (EPS) concrete (EPS mortar with low strength) as lightweight aggregates, respectively. According to the findings of their investigations, the overall weight of the panels was decreased by more than 50%, while the hysteretic behavior of the wall, particularly its ductility and energy dissipation capacity, was diminished significantly. Pakizeh et al. [7], Xu et al. [8–10] and Ximei et al. [11] discussed the cyclic behaviors of shear wall casting with high compressive strength lightweight concrete. It was concluded that the high compressive strength of lightweight concrete was finitely beneficial to improving the cyclic behaviors of composite wall systems. In addition to efforts in that regard, many researchers focused on the development of novel wall designs to achieve considerable composite actions of the shear walls, as well as the demands of light weight. Composite timber shear walls [12–16] and RC shear walls with hollow cavities [17–19] are two representatives of lightweight shear wall systems widely applied in engineering practice. However, due to the material's limitations in terms of durability and fireproofing requirements, wood shear walls are mostly used in low-rise and mid-rise constructions.

The sandwich wall panel is another typical composite lightweight wall system, which is commonly composed of two concrete wythes sandwiched by the extruded polystyrene (XPS) or EPS board as the insulation layer [20]. Fernando et al. [21] investigated the cyclic behavior of sandwich wall panels with EPS foam concrete as insulation wythe and found that sandwich wall structures can work as load bearing walls in single-story buildings or as infilled walls in multi-story buildings. To improve the degree of composite action of sandwich wall panels, Yaman [22] and Barbosa et al. [23] developed lightweight sandwich concrete wall panels featuring considerable thermal insulation property, which were proved functionable under push-out tests and axial loading tests, respectively. Lu et al. [24] and He et al. [25] conducted shaking table tests on PC sandwich wall panel structures (with XPS and EPS boards embedded, respectively). Fragility analysis was performed additionally to verify the reliability of the connections of walls in low-rise buildings. However, the composite action depended considerably on the shear transferring mechanism between concrete wythes of sandwich walls, which was generally partly achieved [20]. Thus, the wall integrity as well as stiffness were decreased. Given the increasing building restrictions and the composite actions of the above-mentioned wall systems, the precast braced concrete shear (PBCS) wall system with different shapes of embedded EPS boards and RC bracings was developed [26]. The hysteretic response of PBCS walls with various shear span ratios (SSRs) and RC bracing types was experimentally studied and validated.

Previous research has revealed that the thermal insulation performance of sandwich wall panels with embedded polystyrene boards [27] outperforms that of sandwich wall panels with lightweight thermal insulation concrete as the aggregate [28]. Specifically, XPS boards and EPS boards are generally used as the thermal insulation layer in practical engineering [29,30]. The EPS board outperforms the XPS board in terms of weight reduction, as its density is about 1/3 that of the XPS and 1/150 that of concrete [31]. The XPS board is widely adopted in the foundations of buildings [31] and the cut sections of railways [32] due to its low thermal conductivity and high compressive strength. However, the degree of composite actions of walls with insulation layers is greatly affected by the adhesive bond between the insulation and concrete wythes [33–35]. Naito et al. [36] found that wall panels with EPS layers had higher adhesive forces compared with those of walls with XPS insulation. The PBCS wall studied in this paper also used the EPS board as the insulation material. Moreover, experimental data concluded that the thermal insulating performance [27] of sandwich wall panels increased along with the rise in thickness of the insulation wythe, while the axial bearing capacity [30] as well as lateral shear resistances [37–40] had an obvious decrease. Above all, the thickness of insulation would remarkably affect the structural weight, thermal insulating performance and hysteretic behavior of sandwich shear walls.

It is obvious that increasing the thickness of triangular or rectangular lightweight EPS boards in PBCS wall panels can not only considerably reduce concrete use but also strengthen the thermal insulating performance of PBCS walls, which meets the tendency of developing energy-saving buildings. However, the increase in lightweight EPS board thickness is more likely to lead to weak shear resistance and poor ductility capacity of PBCS walls. The previous study mainly focused on investigating the influences of axial load ratio (ALR) and reinforcement ratio (RR) on the cyclic behavior of PBCS walls [41]. In this paper, the effects of EPS board thickness on the hysteretic behaviors of PBCS walls with various SSRs and bracing types are quantified by numerical simulation methods. Moreover, the formulae predicting the lateral bearing capacity of PBCS walls with different thicknesses of EPS boards are developed. The upper limits of EPS board thickness ratio for different PBCS walls are put forward to satisfy the requirement of displacement ductility. The effects of SSR and bracing types are also taken into consideration. The finite element (FE) models of PBCS walls considering different EPS board thicknesses are established and verified against the test data regarding the initial stiffness. Afterward, FE models with different thickness ratios of EPS boards are determined, established and tested under cyclic loading. The hysteretic behaviors of PBCS walls are comprehensively discussed in terms of stiffness degradation, load bearing capacity, energy dissipation capacity and ductility capacity. Additionally, the thickness ratio limits for PBCS walls are obtained to satisfy the provisions on ductility of shear walls. Finally, recommendations for the practical design of PBCS walls are addressed based on the analytical results.

2. Experimental Work

2.1. PBCS Wall Description

The hysteretic behaviors of PBCS walls with different SSRs and bracing types were experimentally investigated in the previous study performed by Li et al. [26]. The structural layouts of PBCS walls with diagonal braces and cross braces embedded are listed in Figure 1a,b, respectively. There were seven specimens, with SSRs varying from 1.0 to 2.0 (three diagonal PBCS walls, three cross PBCS walls and one monolithic cast in situ RC wall as the counterpart). It should be noted that the number of triangular EPS boards in diagonal PBCS walls remained constant at four, while that in cross PBCS walls ranged from four to eight as the SSR increased, as shown in Figure 1. Diagonal PBCS walls consisted of two diagonal braces, which intersected at the middle of the wall and expanded into the top and bottom beams, respectively. One vertical brace and several horizontal braces were embedded in the cross PBCS wall panels. The number of horizontal braces was determined by the principle of equivalent volumetric ratio of the diagonal braces in the corresponding PBCS walls with the same SSR.

(**a**) Diagonal PBCS wall. (**b**) Cross PBCS wall.

Figure 1. Structural configurations of PBCS walls.

The reinforcement scheme for the two types of PBCS walls is shown in Figure 2. The concealed columns in PBCS walls contained 6C14 steel bars (with yielding strength f_y = 410 MPa and ultimate strength f_u = 610 MPa), which were connected with the base beam to transfer the tensile forces. The concealed RC bracings consisting of 4C8 steel bars were arranged in the panels in order to achieve a more even distribution of damage. There were two concealed beams (6C10 as longitudinal bars) set at the top and bottom of PBCS wall panels, respectively. The Φ6.5 steel bars (with yielding strength f_y = 420 MPa and ultimate strength f_u = 580 MPa) spaced at 150 mm were adopted in the horizontal and vertical web reinforcement. In addition, the compressive strength of concrete in this paper was 28.2 MPa.

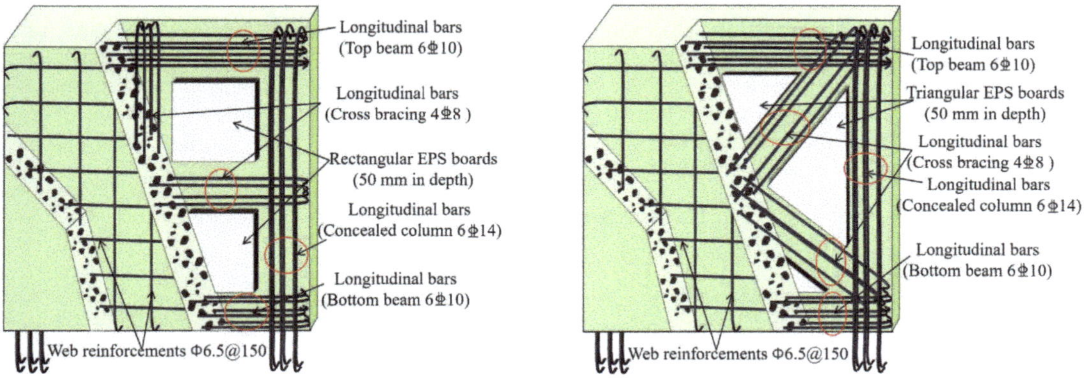

(**a**) Diagonal PBCS walls. (**b**) Cross PBCS walls.

Figure 2. Cutaway view of PBCS walls (SSR = 1.0).

Table 1 presents the specimen information, such as geometric dimensions and reinforcement configurations, in detail. The wall depth and length were 150 mm and 1500 mm, respectively. The wall height varied between 1500 mm and 3000 mm with the increase in SSR. The thickness of EPS boards was 50 mm for both triangular and rectangular boards used in the experiment. The naming convention for each specimen (also applicable for the FE models below) is illustrated in Table 1. The first two letters represent the bracing types of PBCS walls (i.e., CW = PBCS walls with cross braces; DW = PBCS walls with diagonal braces). The first number after the letter denotes the SSR of the specimen, and the last number in the name indicates the thickness of embedded EPS boards in centimeters (cm) for the sake of clarity. The thickness ratio of EPS boards in PBCS walls can be obtained from the ratio of the thickness of EPS board to that of the wall panel. The adoption of thickness ratio can extend the investigation to various wall thicknesses rather than being limited to a fixed wall thickness.

Table 1. Specimen details.

Specimen No.	SSR (*H*/*L*)	Bracing Type	Wall Dimension $H \times L \times T$	Reinforcement Ratio (%)		Thickness Ratio of EPS Board
			(mm³)	ρ_b	ρ_h=ρ_v	
DW1.0-5	1.0	X	1500 × 1500 × 150			
DW1.5-5	1.5	X	2250 × 1500 × 150			
DW2.0-5	2.0	X	3000 × 1500 × 150	0.84 Φ8@100	0.30 Φ6.5@150	0.33
CW1.0-5	1.0	+	1500 × 1500 × 150			
CW1.5-5	1.5	+	2250 × 1500 × 150			
CW2.0-5	2.0	+	3000 × 1500 × 150			

Note: X = diagonal bracing; + = cross bracing. ρ_b = RR of concealed bracing; ρ_h = RR of horizontal web reinforcement; ρ_v = RR of vertical web reinforcement; Φ = hot-rolled plain steel bars.

2.2. Test Program

All specimens were fabricated at the scale of approximately 0.75 and were tested under vertical load and lateral cyclic load (Figure 3). The ALR applying to the loading beam herein was a constant of 0.1. The axial load was first applied to the rigid beam and remained stable throughout the test. Then, the displacement-controlled lateral load was uniformly increased to a drift ratio of Δ_y = 0.33% in five incremental levels (loaded once for every loading level). Afterward, the lateral load increased by 0.33% of the drift ratio (loaded twice for every loading level), and the test was finally terminated when the lateral resistances of the specimen decreased by 15% or wall failure occurred. During the experimental process, the lateral displacement at the top of the wall and the shear resistances at the base beam of the wall were recorded.

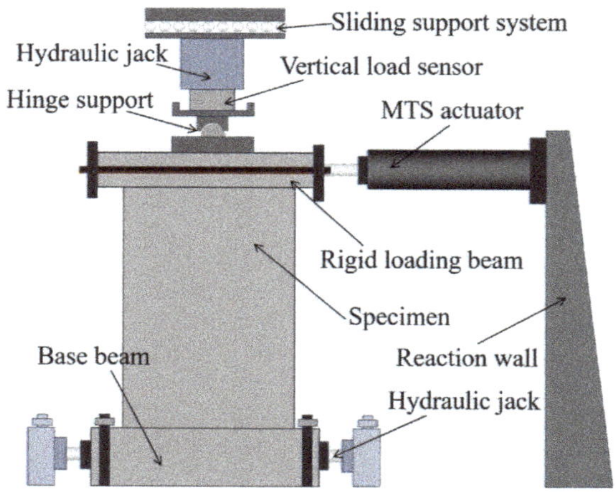

Figure 3. Schematics of test setup.

3. Numerical Simulation Methods and Validation

As mentioned above, it is meaningful to quantify the influence of thickness ratio on the hysteretic response of PBCS walls. However, because of the high cost, as well as the limited number of specimens, the effect of thickness ratio is not fully discussed. This section presents the FE models of PBCS walls based on the multi-layer shell element [42], and then, the numerical results are verified with experimental data. On this basis, the FE models of PBCS walls with consideration of parameters, such as the bracing type, SSR and especially the thickness ratio of EPS board, are established, and their hysteretic behaviors are obtained through quasi-static loading.

3.1. Description and Establishment of FE Model of PBCS Walls

The multi-layer shell element discretizes the FE model of the wall panel into several layers with specified thicknesses and different material properties [42]. This approach is adopted for modeling PBCS walls, since it can take into account the thickness and shape of the EPS board. The FE models herein consist of the modeling of longitudinal reinforcements and multi-layer shell elements, both of which work together and share common nodes. The multi-layer shell elements are composed of numerous layers of cover and core concrete, as well as stirrup layers (Figure 4). The truss element is utilized to simulate the behaviors of longitudinal reinforcements. In particular, the stirrup layer of diagonal braces is established with the angle between the diagonal brace and the concealed boundary column. The horizontal and vertical web reinforcement layers are set at 0° and 90°, respectively. Additionally, the material properties are valued according to experimental results. Since the density of the EPS board (16 kg/m^3) is much lower than that of the concrete (2500 kg/m^3), it is suitable to simulate the EPS board as a hollow core. It should be noted that the constitutive models for concrete and steel adopted in this paper are PlaneStressUserMaterial [42] and Reinforcing Steel Material [43], respectively. All numerical simulation analyses were conducted on the OpenSEES platform (V2.5.0) [44].

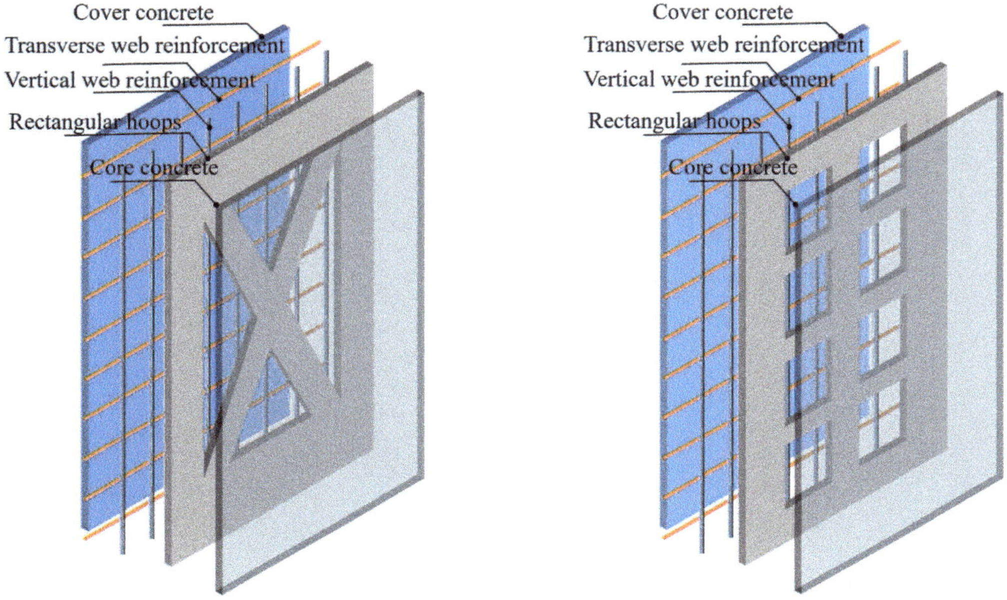

(**a**) Diagonal PBCS wall. (**b**) Cross PBCS wall.

Figure 4. Schematics of multi-layer shell element of PBCS walls (SSR = 2.0).

The thickness ratio of PBCS walls herein refers to the ratio of EPS thickness to wall panel thickness. As mentioned before, the thickness of the EPS board embedded in the shear walls would dramatically affect the structural weight and concrete use. On the premise of code compliance of the cover concrete, the thickness of the EPS board inside ranges from 10 mm to 100 mm, with an even interval of 10 mm, which corresponds to thickness ratios varying between 0.067 and 0.667. The concrete layer thickness in the multi-layer shell element of PBCS walls corresponding to the embedding EPS areas is changed to achieve the different thickness ratios. Overall, 60 FE models featuring various thickness ratios are established, coupling the parameters of SSRs and bracing types. The hysteresis responses of the models are quantitatively analyzed and discussed.

3.2. Validation of Numerical Models of PBCS Walls

Previous research has shown that the hysteresis behaviors, including energy dissipation capacity and lateral deformation capacity, of PBCS walls can be properly predicted by the multi-layer shell elements. Stiffness degradation, as an important factor of the hysteretic performance of shear walls, is taken herein to validate the accuracy of numerical models. Secant stiffness can be obtained by Equation (1).

$$K_i = \frac{|+F_i| + |-F_i|}{|+\Delta_i| + |-\Delta_i|} \tag{1}$$

where K_i, F_i, Δ_i denote the secant stiffness, lateral shear resistances and lateral top displacement of the ith cycle in both reverse loading directions.

Figure 5 presents the comparisons between numerical and experimental results on the stiffness degradation of PBCS walls. It is found that the FE model can depict the stiffness degradation of PBCS walls accurately. Nevertheless, the initial stiffness of the numerical model is generally greater than the experimental results. Kolozvari et al. [45] reviewed the FE modeling approaches for RC shear walls and found that all of the models overestimated the initial stiffness by even three times that of the experimental results. This was attributed to the presence of initial defects in the test specimens and micro-cracking in the concrete, which could not be properly simulated in FE modeling. However, stiffness degradation was less affected during the loading stage. As shown in Figure 5, the stiffness degradation curves matched well with experimental results after reaching a drift ratio of 0.67%.

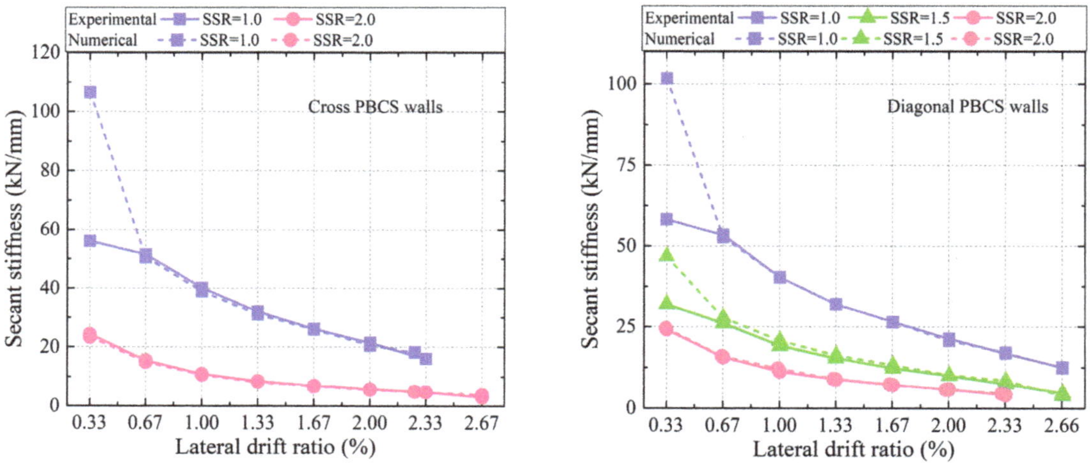

(**a**) Cross PBCS walls. (**b**) Diagonal PBCS walls.

Figure 5. Verification of the numerical models.

4. Results and Discussion

4.1. Concrete Use Reduction

By increasing the thickness ratio of the EPS board, the structural weight of the PBCS wall can be efficiently reduced; the concrete use of the structure is also decreased. Figure 6 illustrates the impact of thickness ratio on weight reduction in PBCS walls. It is evident that the volume of the embedded EPS board increases significantly with the increase in SSR and thickness ratio, resulting in lower use of concrete. Furthermore, the volumetric ratio of the EPS board to the entire wall is calculated to obtain the normalized curves of weight reduction (Figure 6b). A near 20% reduction in concrete use is observed for the cross PBCS walls, which is 11.67% higher than the maximum value of the diagonal PBCS walls. Compared to conventional shear walls, PBCS walls have certain insulation functions with additional filling of the walls with EPS panels. Meanwhile, the amount of concrete used in PBCS walls can be remarkably reduced, which is beneficial for energy conservation and emission reduction. The integrity of PBCS walls outperforms that of sandwich PC walls by embedding EPS boards between the bracings.

4.2. Skeleton Curves

Skeleton curves can be acquired by connecting the peak points of each loading level in the hysteretic curves. A total of 60 skeleton curves are shown in Figure 7, according to different SSRs and bracing types. Each sub-image illustrates the skeleton curves of a PBCS wall with a specific SSR and bracing type, under varying thickness ratios. Additionally, Figure 7a–f scale the skeleton curve at a drift ratio of 0.67% to provide a more detailed view of the impact of thickness ratio on the wall's lateral shear resistances.

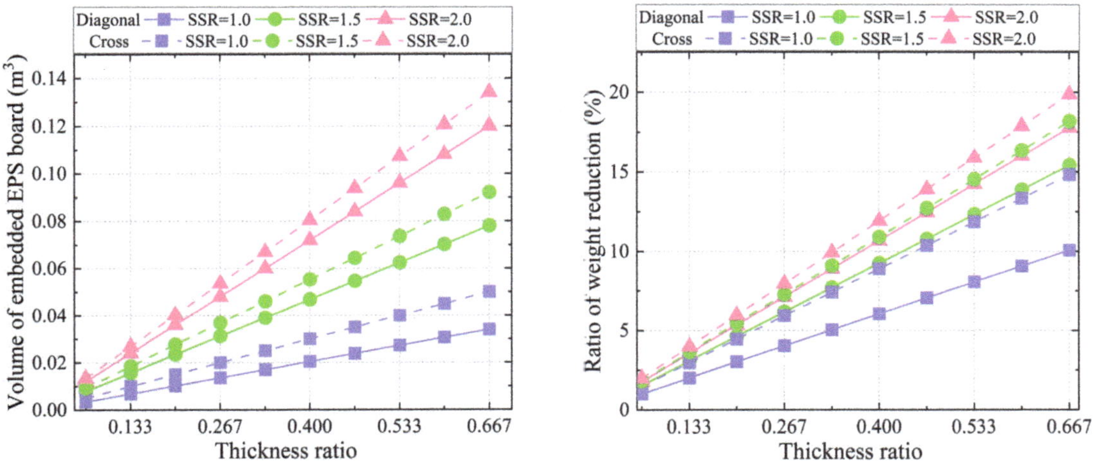

(**a**) Volume of embedded EPS boards. (**b**) Normalized results of weight reduction.

Figure 6. Effects of thickness ratio on concrete use reduction.

It is found that all numerical models can satisfy the Chinese code GB 50011-2010 [46] when their lateral drift ratios reach 1/1000 as well as 1/120 (i.e., 0.83%), which represent the maximum inter-story drift angle for RC shear wall structures subjected to frequent and infrequent earthquakes, respectively. It is evident in Figure 7 that the shear capacity and lateral stiffness of PBCS walls decrease at all loading stages as the thickness ratio of the EPS board increases. This is because the use of concrete, which is the primary factor for bearing and transmitting compressive stress in the wall, is directly weakened. On the contrary, an increase in the thickness ratio leads to a slight enhancement in the ultimate displacement of the PBCS wall (Figure 7e).

Moreover, by comparing the skeleton curves of PBCS walls with different SSRs, it is concluded that the lateral elastic stiffness of PBCS walls with low SSRs experienced a more significant variation than that of walls with a high SSR (Figure 7a,b). This is in part because shear walls with low SSRs are more prone to shear failure characterized by concrete crushing. Additionally, the increase in the thickness ratio of the EPS board further weakens the bearing capacity provided by concrete. In addition, it is suggested in Figure 7a,e that the thickness ratio of the cross PBCS wall with a low SSR (SSR = 1.0) has a more noticeable impact on its hysteresis response relative to the other cases. However, the hysteresis response of the diagonal PBCS wall is more sensitive to the escalation in thickness ratio at a high SSR (SSR = 2.0).

The findings of the hysteresis analyses indicate that the composite actions for cross PBCS shear walls with a high SSR of 2.0 or diagonal PBCS walls with a low SSR of 1.0 can remain stable by increasing the thickness ratio of the EPS board. Specifically, this approach not only reduces the consumption of concrete and the overall weight of the structure, but it also ensures that the cyclic response of the PBCS wall will not be significantly affected by the increased thickness ratio.

4.3. Lateral Bearing Capacity

Figure 8 exhibits the lateral bearing capacity of PBCS walls with different SSRs, bracing types and thickness ratios of the EPS board. As shown in the figure, there is a gradual decrease in peak forces as the thickness ratio increases for both diagonal and cross layouts of PBCS walls. This is due to the reduction in concrete use. Nevertheless, a maximum decline of only 4.8% and 7.9% in lateral shear resistances is observed for the diagonal and cross PBCS walls, respectively, with a low SSR of 1.0. Additionally, it can be concluded from Figure 8 that the SSR outweighs the bracing type and thickness ratio in affecting the shear bearing capacity of PBCS walls.

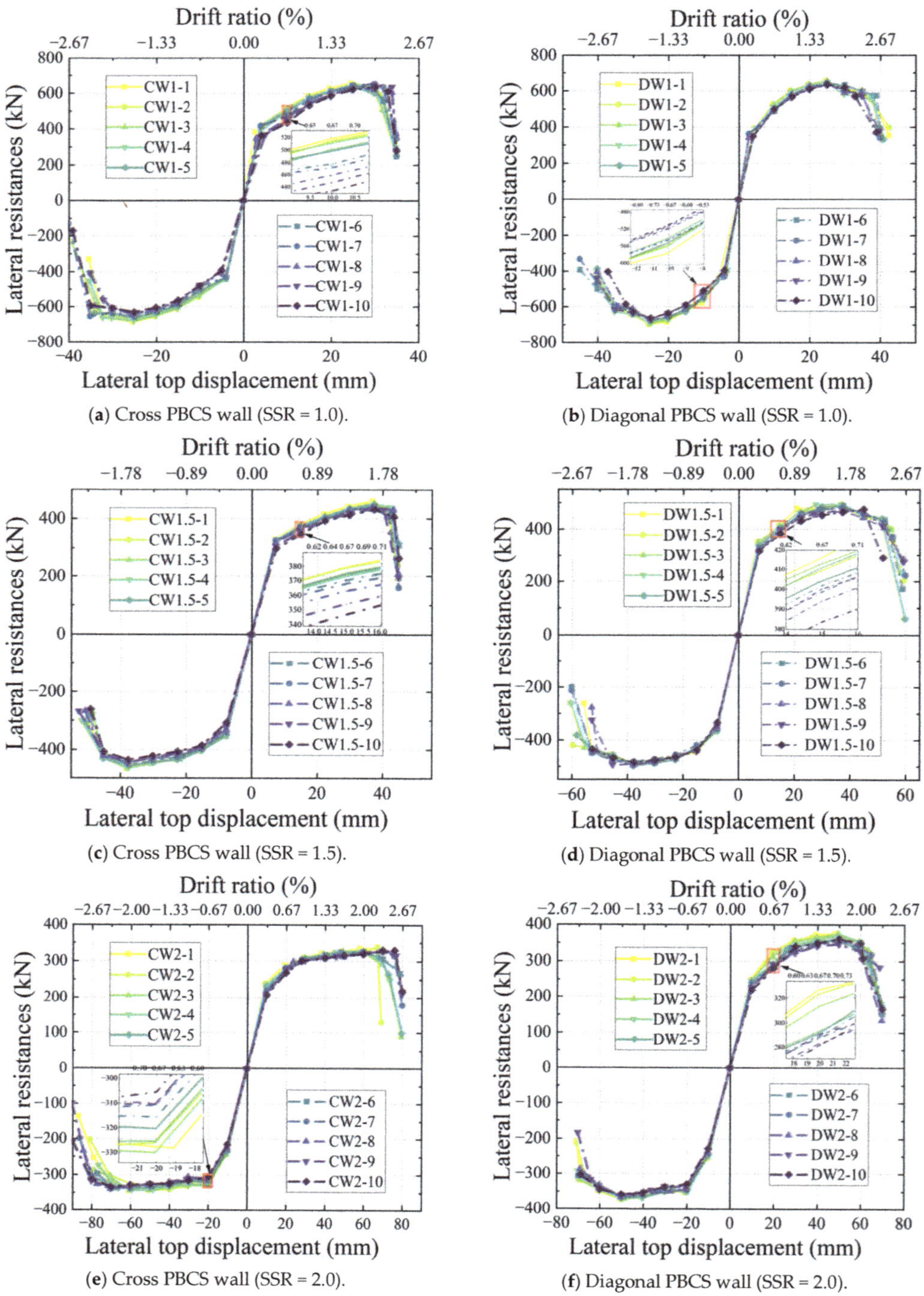

(**a**) Cross PBCS wall (SSR = 1.0).

(**b**) Diagonal PBCS wall (SSR = 1.0).

(**c**) Cross PBCS wall (SSR = 1.5).

(**d**) Diagonal PBCS wall (SSR = 1.5).

(**e**) Cross PBCS wall (SSR = 2.0).

(**f**) Diagonal PBCS wall (SSR = 2.0).

Figure 7. Skeleton curves of PBCS walls.

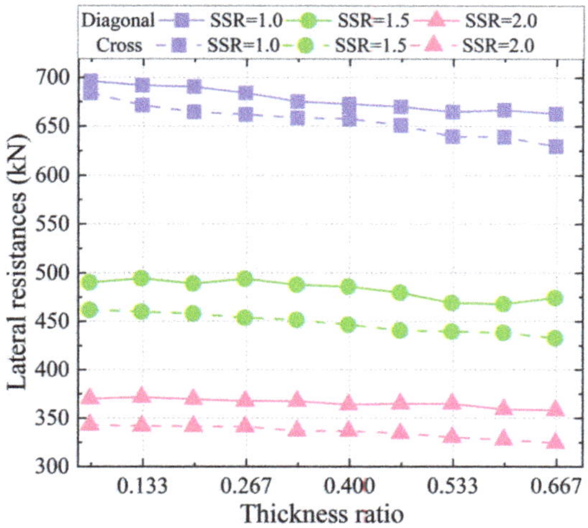

Figure 8. Peak forces of PBCS walls with different thicknesses of EPS board.

The effect of the thickness ratio on the bearing capacity of PBCS walls demonstrates an approximately linear trend. Thus, fitting surfaces are performed to predict the peak shear resistances of PBCS walls regarding the SSR and thickness ratio. The results are listed as follows:

$$P_D = 992 - 311.96R_S - 41.05R_T \tag{2}$$

$$P_C = 980 - 319.94R_S - 53.54R_T \tag{3}$$

where P_D and P_C are the bearing capacities of the diagonal and cross PBCS walls (expressed in kN), respectively; R_S denotes the SSR; and R_T represents the thickness ratio of the EPS board (in percent). The R-squares of the fitting results are 0.977 and 0.969, respectively. The fitting equations here are shown to be applicable for accurately predicting the bearing capacity of PBCS walls. Due to the limited sample size, this formulation is only applicable when other parameters of PBCS walls are kept constant with the experimental setups.

4.4. Stiffness Degradation

To facilitate comparison, the stiffness–drift ratio curves of each loading level were derived by dividing the lateral displacement by the corresponding height of the wall, as shown in Figure 9. For simplicity, the lines presented in Figure 9 vary from light colors to dark colors, which correspond to the thickness ratio of the EPS board from 0.067 to 0.667. It can be observed that before the drift ratio reaches 0.67%, the stiffness in all cases decreases rapidly. During this stage, the maximum stiffness degradation occurs at SSR = 1.0, with stiffness of the cross PBCS wall and diagonal PBCS wall decreasing by 54.9% and 49.5%, respectively (compared to initial stiffness). The deterioration in stiffness during this stage is mainly attributed to the large number of concrete cracks generated in the early stage of loading. Subsequently, the lateral stiffness of the wall gradually decreases, with stiffness of the cross PBCS wall and diagonal PBCS wall descending, respectively, by 31.1% and 34.6% (compared to initial stiffness), from a drift ratio of 0.67%–2.0%. Stiffness degradation during this stage is mainly due to the further cracking of concrete and the ability of steel bars to provide resistance even after entering the plastic stage.

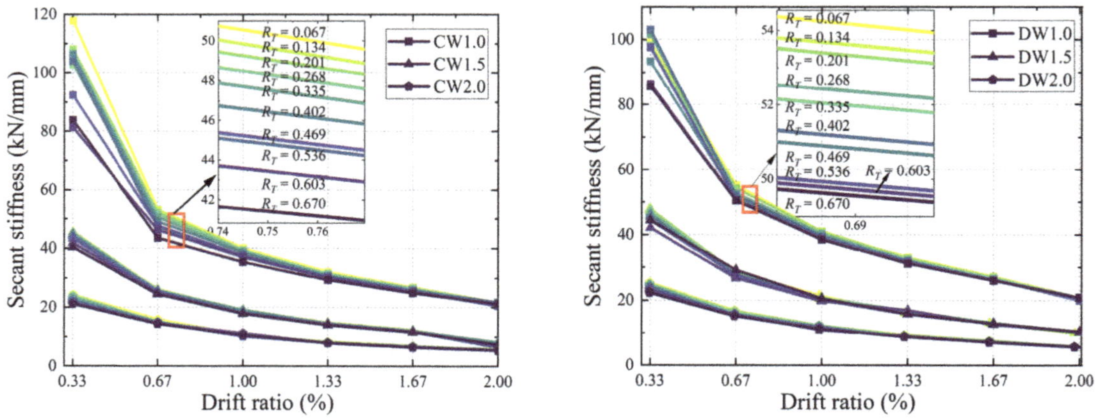

(**a**) Cross PBCS walls. (**b**) Diagonal PBCS walls.

Figure 9. Stiffness degradation of PBCS walls.

Furthermore, Figure 9 illustrates that the secant stiffness of the wall witnesses an evident drop with the increase in the EPS board thickness ratio when the horizontal displacement arrives at the same level, before the wall reaches its peak bearing capacity. However, after the drift ratio exceeds 1.33%, the impact of the EPS board thickness ratio on stiffness becomes negligible. This is due to the fact that the wall panel is full of concrete cracking, and stress redistribution occurs when the peak force stage shows up, rendering the EPS board ineffective as a hollow core. Additionally, the stiffness of PBCS walls with larger SSRs is slightly affected by the EPS board thickness ratio during the entire loading process. This can be attributed to the bending failure mode, which usually occurs in walls with higher SSRs, characterized by concrete crushing at the wall bottom and yielding longitudinal bars in the boundary columns. The damage to the upper and middle areas of the wall, where the EPS board is embedded, is small in extent, resulting in minor impact of the EPS board on stiffness.

To comprehensively investigate the influences of embedded EPS boards on the stiffness degradation of PBCS walls, the stiffness degradation curves of cases with similar amounts of EPS embedded are shown in Figure 10. When the thickness ratios of cross and diagonal PBCS walls are 0.467 and 0.533, respectively, the EPS volumes of the two types of PBCS walls are similar (with the same SSR). This means that in order to embed the same volume of the EPS board, the diagonal PBCS wall requires a thicker EPS board than the cross PBCS wall. Figure 10 shows that when the filling volumes are the same, the stiffness degradation process of the two types of walls is basically the same. The diagonal PBCS wall exhibits slightly greater stiffness than the cross PBCS wall during the loading process. This can be attributed to the fact that the diagonal bracing layout has higher stiffness compared to the cross bracing layout.

4.5. Displacement Ductility

Ductility of a shear wall is a crucial factor, which reflects its deformation capacity. It is typically represented by the ratio of ultimate displacement ($\Delta_{0.85p}$) to yielding displacement (Δ_y) of the wall. The yielding displacement of the wall is determined using the method proposed by Park et al. [47], while the ultimate displacement is obtained when the peak strength of the wall decreases by 15%, as illustrated in Figure 11. In this approach, α is the load coefficient, which is taken as 0.75.

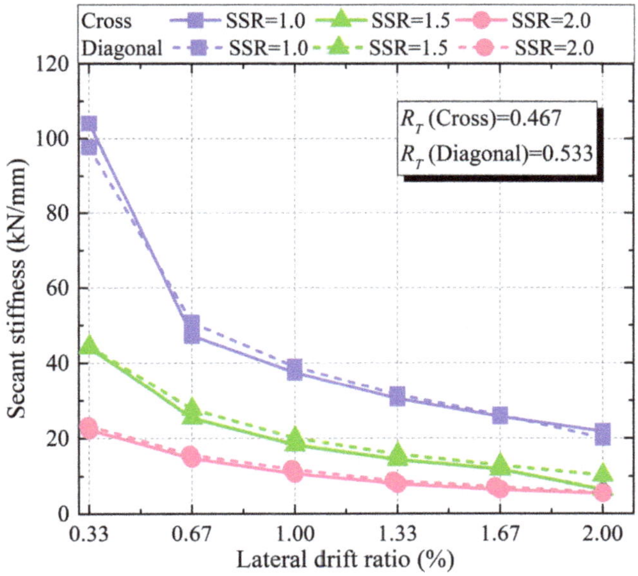

Figure 10. Stiffness degradation of PBCS walls with similar amounts of EPS.

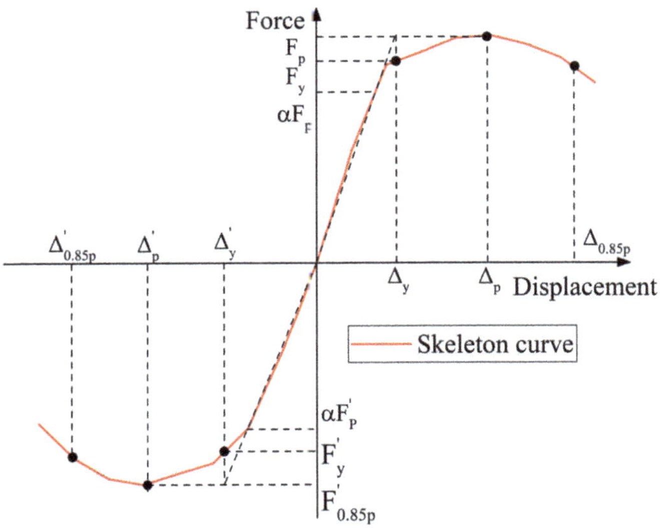

Figure 11. Feature points based on skeleton curve.

Figure 12 illustrates the yielding displacement of PBCS walls with different bracing types. It is evident that the yielding displacement in all cases generally shows an upward trend when the EPS board thickness ratio rises. Shear walls typically reach yielding resistances (F_y) when the longitudinal reinforcing bar yield is reached. This depends considerably on the sectional area of longitudinal bars in the boundary columns, which is a constant in this study, and thus results in a stable F_y. Meanwhile, the stiffness of the wall tails off with the thickness ratio increases. In view of this, the corresponding yielding displacement of PBCS walls is increased to achieve the yielding resistances identified.

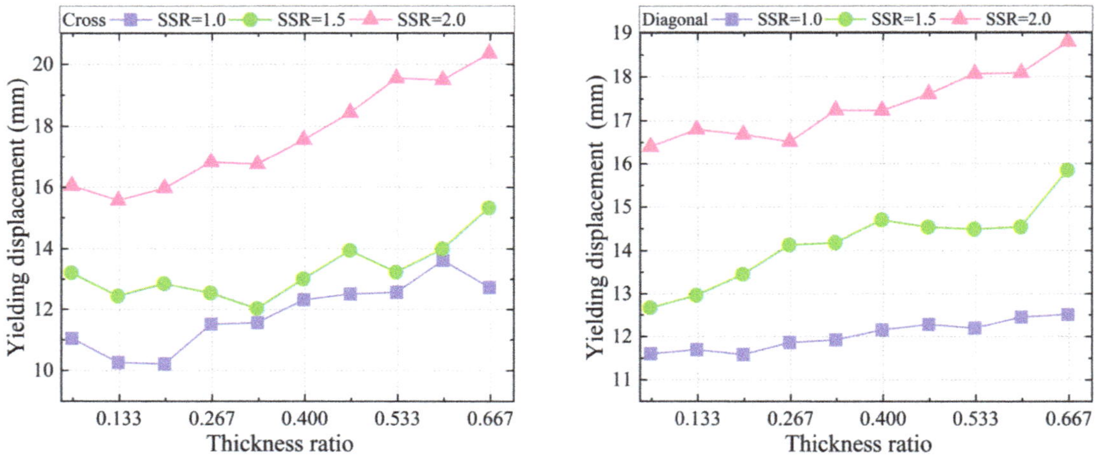

(**a**) Cross PBCS walls. (**b**) Diagonal PBCS walls.

Figure 12. Yielding displacement of PBCS walls.

The ductility capacities of PBCS walls with different bracing types are presented in Figure 13. It can be seen that the ductility of the PBCS wall descends evidently with an increase in the thickness ratio. This can be explained by a previous study [41], which demonstrated that PBCS walls with high ALR are susceptible to low ductility capacities. Specifically, the increase in the thickness of the EPS board leads to a reduction in the net cross-sectional area of the wall. As a consequence, the ALR is improved in spite of the axial loads remaining constant. In addition, a horizontal line $\mu = 3.0$ is presented in the figure to represent the code requirement that the ductility of the shear wall should not be lower than 3.0 [46]. It can be observed that the ductility of all diagonal PBCS walls can nearly meet the code requirements as the EPS board thickness ratio changes. In particular, the diagonal PBCS wall with SSR = 1.0 should be designed with a thickness ratio below 0.6 to satisfy the requirements of displacement ductility greater than 3.0. In addition, the ductility of the cross PBCS wall generally falls below the code requirements when the SSR is 1.0, indicating that the design schemes for cross PBCS walls with SSR = 1.0 or lower should be avoided.

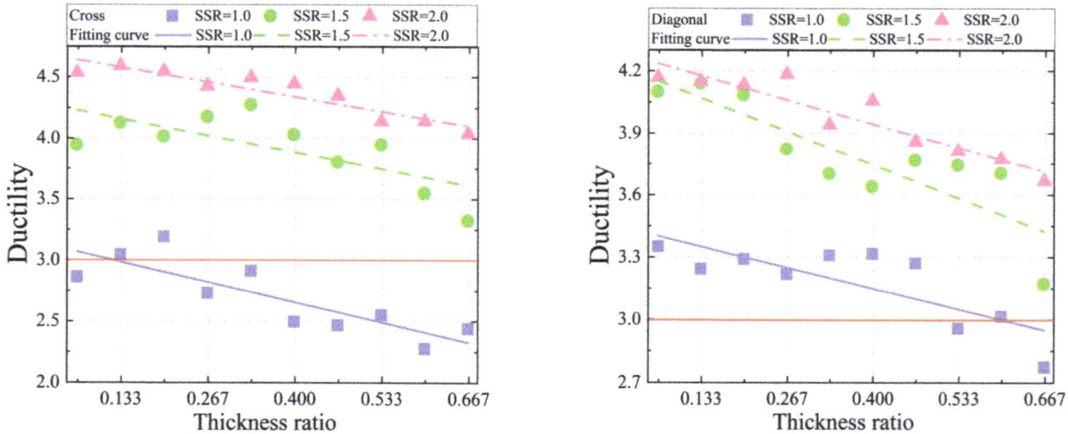

(**a**) Cross PBCS walls. (**b**) Diagonal PBCS walls.

Figure 13. Ductility of PBCS walls.

4.6. Energy Dissipation

The accumulated energy dissipation refers to the energy dissipated from the beginning to the *i*th cycle of the loading scheme. Figure 14 demonstrates the accumulated energy dissipation curves for PBCS walls with different thickness ratios during the loading process. As the lateral drift ratio increases, the cumulative energy dissipation in each case exhibits a substantial increase. It is apparent that the thickness ratio of the EPS board does not affect the accumulated energy dissipation of the specimen until the eighth cycle, i.e., before the drift ratio reaches 0.83%. At this stage, the energy dissipation of the wall mainly occurs with the cracking of concrete on both sides of the boundary columns and deformation of the steel bars. The panel zone with the EPS board embedded remains nearly undamaged.

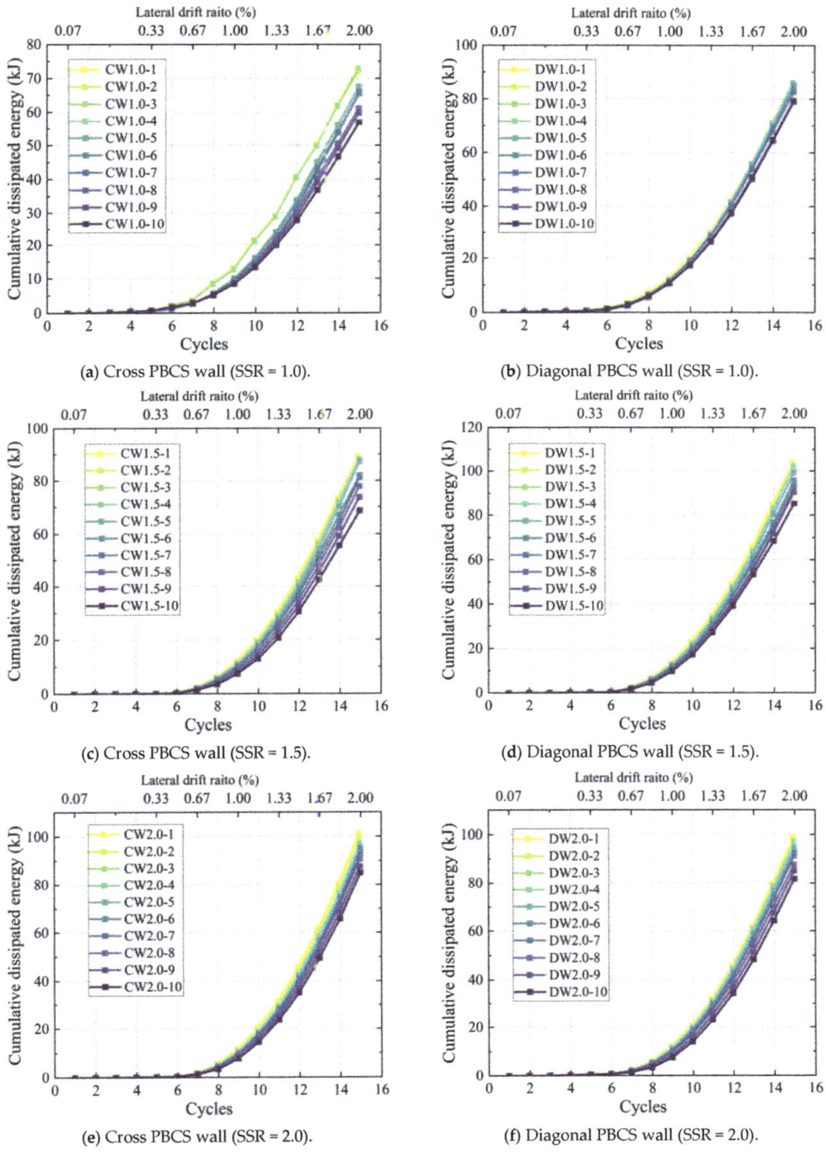

(**a**) Cross PBCS wall (SSR = 1.0). (**b**) Diagonal PBCS wall (SSR = 1.0).

(**c**) Cross PBCS wall (SSR = 1.5). (**d**) Diagonal PBCS wall (SSR = 1.5).

(**e**) Cross PBCS wall (SSR = 2.0). (**f**) Diagonal PBCS wall (SSR = 2.0).

Figure 14. Cumulative dissipated energy of PBCS walls.

As the loading process progresses, the impact of the thickness ratio of EPS boards on the energy dissipation of PBCS walls becomes increasingly apparent. Concrete cracking extends toward the area filled with EPS boards and widens as the lateral drift increases. The greater the thickness ratio of the EPS board, the lower the resistance provided by the concrete in this area, resulting in a reduced energy dissipation capacity.

In addition, the energy dissipation of cross PBCS walls with SSRs of 1.0 and 1.5 is more sensitive to the variation in thickness ratio relative to diagonal PBCS walls (Figure 14a–d). The cross PBCS wall with SSR = 1.0 saw a fall of 20.7% (from 71.8 kJ to 57 kJ) in accumulated dissipated energy, while a decrease of only 7.8% (from 86.3 kJ to 79.5 kJ) was observed in the diagonal PBCS wall. PBCS walls with low SSRs are especially vulnerable to shear failure characterized by diagonal cracks, whereas the diagonal braces in PBCS walls can effectively slow down the development of these diagonal cracks. Therefore, diagonal PBCS walls with low SSRs behave better in terms of energy dissipation.

However, as the thickness ratio rises, diagonal PBCS walls with SSRs = 2.0 experience a more significant decline in energy dissipation compared to cross PBCS walls. Generally, walls with SSRs = 2.0 are more susceptible to a bending failure, with horizontal cracks appearing in the lower part of the wall, where cross braces can play a more effective role than diagonal ones. Therefore, for PBCS walls with SSRs of 1.0 and 1.5, diagonal braces with high thickness ratio could be adopted to achieve comparable energy dissipation capacity. For PBCS walls with SSRs = 2.0, it is more advisable to employ cross PBCS walls with a higher thickness ratio than diagonal ones.

5. Conclusions and Recommendations

In this paper, the effects of the thickness ratio of the EPS board on the cyclic behaviors of PBCS walls with different SSRs and bracing types were comprehensively investigated. Numerical models were developed and verified against experimental data in terms of stiffness degradation capacity. A total of 60 numerical cases were established with parameters such as the thickness ratio, SSR and bracing types. The hysteretic behaviors of PBCS walls were discussed with respect to skeleton curves, lateral bearing capacity, stiffness degradation, displacement ductility and energy dissipation. Based on the ductility requirements, the thickness ratio limits of EPS boards in PBCS walls with different bracing types were addressed. The main conclusions and recommendations can be summarized as follows:

1. The stiffness degradation capacity of PBCS walls can be precisely captured by numerical models based on the multi-layer shell element. Moreover, increasing the thickness ratio of EPS boards in PBCS walls can significantly reduce the amount of concrete used. Cross PBCS walls witnessed a roughly 20% reduction in concrete use when the thickness ratio rose from 0.067 to 0.667, which was 11.67% higher than what the diagonal PBCS walls achieved.
2. Based on the findings, the recommendations for determining the thickness of EPS panels are put forward for practical design of PBCS walls. When the thickness ratio of the EPS board increased, the lateral bearing capacity, secant stiffness and displacement ductility of the PBCS wall experienced a consistent decline, while the yielding displacement had a gradual upward trend. Specifically, the lateral bearing capacity of diagonal and cross PBCS walls experienced declines of 4.8% and 7.9%, respectively. In addition, the equations developed for predicting the bearing capacity of PBCS walls were demonstrated to be sufficiently accurate. Furthermore, to ensure the ductility is greater than 3.0, the diagonal PBCS wall with SSR = 1.0 should be designed with a thickness ratio below 0.6, and a cross PBCS wall with SSR = 1.0 should be avoided.
3. The accumulated dissipated energy of cross PBCS walls (SSR = 1.0) and diagonal PBCS walls (SSR = 1.0) saw decreases of 20.7% and 7.8% with the EPS board thickness ascending, respectively. A diagonal PBCS wall with high thickness ratio of the EPS board could be adopted to achieve comparable energy dissipation capacity when the

SSR is below 1.5. For the SSR of 2.0, cross PBCS walls are more suitable than diagonal ones, as they dissipate more energy.

4. Owing to the limited scope of SSR involved in this study, the relevant suggestions obtained herein are applicable to PBCS walls with frequently configured SSR, ranging from 1.0 to 2.0. Like other RC shear wall modeling approaches, an overestimation of initial stiffness was also observed in the FE models of PBCS walls based on the multi-layer shell element. This is a promising study for optimizing the deficiency. A quantitative analysis of the influence of the EPS board thickness ratio on the thermal insulation performance of PBCS walls would be a valuable development in future studies. Additionally, in order to achieve better weight reduction effects for PBCS walls as well as further reduce concrete use, the adoption of new lightweight and high strength concrete materials and embedded thermal insulation layer in PBCS walls is also necessary.

Author Contributions: Conceptualization, H.L.; Methodology, Y.T.; Validation, Y.T.; Formal analysis, Y.T. and H.L.; Data curation, Y.T.; Writing—original draft, Y.T.; Writing—review & editing, H.L.; Supervision, H.L.; Project administration, H.L.; Funding acquisition, H.L. All authors have read and agreed to the published version of the manuscript.

Funding: This research is financially supported by the National Natural Science Foundation of China (No. 51738007).

Data Availability Statement: The data that support the findings of this study are available from the corresponding author upon reasonable request.

Conflicts of Interest: The authors declare no conflict of interest.

References

1. Dong, Y.H.; Jaillon, L.; Chu, P.; Poon, C.S. Comparing carbon emissions of precast and cast-in-situ construction methods—A case study of high-rise private building. *Constr. Build. Mater.* **2015**, *99*, 39–53. [CrossRef]
2. Kurama, Y.C.; Sritharan, S.; Fleischman, R.B.; Restrepo, J.I.; Henry, R.S.; Cleland, N.M.; Ghosh, S.K.; Bonelli, P. Seismic-Resistant Precast Concrete Structures: State of the Art. *J. Struct. Eng.* **2018**, *144*, 03118001. [CrossRef]
3. Wibowo, A.; Wijatmiko, I.; Nainggolan, C.R. Cyclic Behaviour of Expanded Polystyrene (EPS) Sandwich Reinforced Concrete Walls. *Adv. Mater. Sci. Eng.* **2018**, *2018*, 7214236. [CrossRef]
4. O'Hegarty, R.; Kinnane, O. Review of precast concrete sandwich panels and their innovations. *Constr. Build. Mater.* **2020**, *233*, 117145. [CrossRef]
5. Lombardi, R.; Jünemann, R.; Lopez, M. Experimental assessment of the behavior of expanded glass lightweight reinforced concrete walls. *J. Build. Eng.* **2022**, *49*, 104043. [CrossRef]
6. Mousavi, S.A.; Zahrai, S.M.; Bahrami-Rad, A. Quasi-static cyclic tests on super-lightweight EPS concrete shear walls. *Eng. Struct.* **2014**, *65*, 62–75. [CrossRef]
7. Pakizeh, M.R.; Parastesh, H.; Hajirasouliha, I.; Farahbod, F. Seismic performance of CFS shear wall systems filled with polystyrene lightweight concrete: Experimental investigation and design methodology. *Steel Compos. Struct.* **2023**, *46*, 497.
8. Xu, Z.; Chen, Z.; Yang, S. Effect of a new type of high-strength lightweight foamed concrete on seismic performance of cold-formed steel shear walls. *Constr. Build. Mater.* **2018**, *181*, 287–300. [CrossRef]
9. Xu, Z.; Chen, Z.; Yang, S. Seismic behavior of cold-formed steel high-strength foamed concrete shear walls with straw boards. *Thin-Walled Struct.* **2018**, *124*, 350–365. [CrossRef]
10. Xu, Z.; Chen, Z.; Dong, X.; Zuo, Y. Experimental Study on Seismic Behavior of Lightweight Concrete-Filled Cold-Formed Steel Shear Walls Strengthened Using Horizontal Reinforcement. *J. Earthq. Eng.* **2023**, *27*, 4126–4160. [CrossRef]
11. Ximei, Z.; Jiayu, Y.; Can, C. Seismic performance and flexible connection optimization of prefabricated integrated short-leg shear wall filled with ceramsite concrete. *Constr. Build. Mater.* **2021**, *311*, 125224. [CrossRef]
12. Chen, F.; Li, Z.; He, M.; Wang, Y.; Shu, Z.; He, G. Seismic performance of self-centering steel-timber hybrid shear wall structures. *J. Build. Eng.* **2021**, *43*, 102530. [CrossRef]
13. Kuai, L.; Ormarsson, S.; Vessby, J.; Maharjan, R. A numerical and experimental investigation of non-linear deformation behaviours in light-frame timber walls. *Eng. Struct.* **2022**, *252*, 113599. [CrossRef]
14. Orlowski, K.; Baduge, S.K.; Mendis, P. Prefabricated Composite Steel-Timber Stiffened Wall Systems with Post-Tensioning: Structural Analysis and Experimental Investigation under Vertical Axial Load. *J. Struct. Eng.* **2021**, *147*, 04020325. [CrossRef]
15. Wang, R.; Wei, S.Q.; Li, Z.; Xiao, Y. Performance of connection system used in lightweight glubam shear wall. *Constr. Build. Mater.* **2019**, *206*, 419–431. [CrossRef]

16. Darzi, S.; Karampour, H.; Bailleres, H.; Gilbert, B.P.; Fernando, D. Load bearing sandwich timber walls with plywood faces and bamboo core. *Structures* **2020**, *27*, 2437–2450. [CrossRef]
17. Hamid, N.H.; Mander, J.B. Lateral Seismic Performance of Multipanel Precast Hollowcore Walls. *J. Struct. Eng.* **2010**, *136*, 795–804. [CrossRef]
18. Dal Lago, B.; Muhaxheri, M.; Ferrara, L. Numerical and experimental analysis of an innovative lightweight precast concrete wall. *Eng. Struct.* **2017**, *137*, 204–222. [CrossRef]
19. Wang, W.; Wang, X. Experimental and numerical investigations on concentrated-hollow RC shear walls. *Eng. Struct.* **2021**, *242*, 112570. [CrossRef]
20. Pessiki, S.; Mlynarczyk, A. Experimental evaluation of the composite behavior of precast concrete sandwich wall panels. *PCI J.* **2003**, *48*, 54–71. [CrossRef]
21. Fernando, P.L.N.; Jayasinghe, M.T.R.; Jayasinghe, C. Structural feasibility of Expanded Polystyrene (EPS) based lightweight concrete sandwich wall panels. *Constr. Build. Mater.* **2017**, *139*, 45–51. [CrossRef]
22. Sevil Yaman, T.; Lucier, G. Shear Transfer Mechanism between CFRP Grid and EPS Rigid Foam Insulation of Precast Concrete Sandwich Panels. *Buildings* **2023**, *13*, 928. [CrossRef]
23. Barbosa, K.; Silva, W.T.M.; Silva, R.; Vital, W.; Bezerra, L.M. Experimental Investigation of Axially Loaded Precast Sandwich Panels. *Buildings* **2023**, *13*, 1993. [CrossRef]
24. Lu, Y.; Chen, W.; Xiong, F.; Yan, H.; Ge, Q.; Zhao, F. Seismic Performance of a Full-Scale Two-Story Bolt-Connected Precast Concrete Composite Wall Panel Building Tested on a Shake Table. *J. Struct. Eng.* **2021**, *147*, 04021209. [CrossRef]
25. He, J.-X.; Xu, Z.-D.; Zhang, L.-Y.; Lin, Z.-H.; Hu, Z.-W.; Li, Q.-Q.; Dong, Y.-R. Shaking table tests and seismic assessment of a full-scale precast concrete sandwich wall panel structure with bolt connections. *Eng. Struct.* **2023**, *278*, 115543. [CrossRef]
26. Li, H.-N.; Tang, Y.-C.; Li, C.; Wang, L.-M. Experimental and numerical investigations on seismic behavior of hybrid braced precast concrete shear walls. *Eng. Struct.* **2019**, *198*, 109560. [CrossRef]
27. Yu, S.; Liu, Y.; Wang, D.; Ma, C.; Liu, J. Theoretical, experimental and numerical study on the influence of connectors on the thermal performance of precast concrete sandwich walls. *J. Build. Eng.* **2022**, *57*, 104886. [CrossRef]
28. Liu, P.; Gong, Y.F.; Tian, G.H.; Miao, Z.K. Preparation and experimental study on the thermal characteristics of lightweight prefabricated nano-silica aerogel foam concrete wallboards. *Constr. Build. Mater.* **2021**, *272*, 121895. [CrossRef]
29. Xu, G.; Li, A. Seismic performance of a new type precast concrete sandwich wall based on experimental and numerical investigation. *Soil Dyn. Earthq. Eng.* **2019**, *122*, 116–131. [CrossRef]
30. Kumar, S.; Chen, B.; Xu, Y.; Dai, J.-G. Structural behavior of FRP grid reinforced geopolymer concrete sandwich wall panels subjected to concentric axial loading. *Compos. Struct.* **2021**, *270*, 114117. [CrossRef]
31. Kilar, V.; Koren, D.; Bokan-Bosiljkov, V. Evaluation of the performance of extruded polystyrene boards—Implications for their application in earthquake engineering. *Polym. Test.* **2014**, *40*, 234–244. [CrossRef]
32. Niu, F.; Jiang, H.; Su, W.; Jiang, W.; He, J. Performance degradation of polymer material under freeze-thaw cycles: A case study of extruded polystyrene board. *Polym. Test.* **2021**, *96*, 107067. [CrossRef]
33. Gombeda, M.J.; Naito, C.J.; Quiel, S.E. Development and performance of a ductile shear tie for precast concrete insulated wall panels. *J. Build. Eng.* **2020**, *28*, 101084. [CrossRef]
34. Gombeda, M.J.; Naito, C.J.; Quiel, S.E. Flexural performance of precast concrete insulated wall panels with various configurations of ductile shear ties. *J. Build. Eng.* **2021**, *33*, 101574. [CrossRef]
35. Choi, I.; Kim, J.; Kim, D.; Park, J. Effects of grid-type shear connector arrangements used for insulated concrete sandwich wall panels with a low aspect ratio. *J. Build. Eng.* **2022**, *46*, 103754. [CrossRef]
36. Naito, C.; Hoemann, J.; Beacraft, M.; Bewick, B. Performance and Characterization of Shear Ties for Use in Insulated Precast Concrete Sandwich Wall Panels. *J. Struct. Eng.* **2012**, *138*, 52–61. [CrossRef]
37. Daniel Ronald Joseph, J.; Prabakar, J.; Alagusundaramoorthy, P. Experimental studies on through-thickness shear behavior of EPS based precast concrete sandwich panels with truss shear connectors. *Compos. Part B-Eng.* **2019**, *166*, 446–456. [CrossRef]
38. Choi, W.; Jang, S.-J.; Yun, H.-D. Design properties of insulated precast concrete sandwich panels with composite shear connectors. *Compos. Part B-Eng.* **2019**, *157*, 36–42. [CrossRef]
39. Xu, G.; Li, A. Experimental and numerical studies on the lateral performance of concrete sandwich walls. *Struct. Des. Tall Spec. Build.* **2020**, *29*, e1715. [CrossRef]
40. Lameiras, R.; Barros, J.A.O.; Valente, I.B.; Poletti, E.; Azevedo, M.; Azenha, M. Seismic behaviour of precast sandwich wall panels of steel fibre reinforced concrete layers and fibre reinforced polymer connectors. *Eng. Struct.* **2021**, *237*, 112149. [CrossRef]
41. Tang, Y.-C.; Li, H.-N.; Li, C. Parametric Studies on Seismic Performance of New Precast Braced Concrete Shear Walls under Cyclic Loading. *J. Struct. Eng.* **2023**, *149*, 04023061. [CrossRef]
42. Lu, X.; Xie, L.; Guan, H.; Huang, Y.; Lu, X. A shear wall element for nonlinear seismic analysis of super-tall buildings using OpenSees. *Finite Elem. Anal. Des.* **2015**, *98*, 14–25. [CrossRef]
43. Kunnath, S.K.; Heo, Y.; Mohle, J.F. Nonlinear Uniaxial Material Model for Reinforcing Steel Bars. *J. Struct. Eng.* **2009**, *135*, 335–343. [CrossRef]
44. Mazzoni, S.; McKenna, F.; Scott, M.H.; Fenves, G.L. OpenSees command language manual. *Pac. Earthq. Eng. Res. (PEER) Cent.* **2006**, *264*, 137–158.

45. Kolozvari, K.; Biscombe, L.; Dashti, F.; Dhakal, R.P.; Gogus, A.; Gullu, M.F.; Henry, R.S.; Massone, L.M.; Orakcal, K.; Rojas, F.; et al. State-of-the-art in nonlinear finite element modeling of isolated planar reinforced concrete walls. *Eng. Struct.* **2019**, *194*, 46–65. [CrossRef]
46. *GB 50011-2010*; Ministry of Housing and Urban-Rural Development, PRC. Code for Seismic Design of Buildings. China Construction Industry Press: Beijing, China, 2016. (In Chinese)
47. Park, R.; Priestley, M.J.N.; Gill, W.D. Ductility of Square-Confined Concrete Columns. *J. Struct. Div.* **1982**, *108*, 929–950. [CrossRef]

buildings

MDPI

Article

Hysteretic Model for RC Columns Based on Effective Hysteretic Energy Dissipation with Positive and Negative Directions

Ge Zhang [1,2], Baitao Sun [1,*] and Wen Bai [1]

[1] Institute of Engineering Mechanics, China Earthquake Administration, Harbin 150000, China; zhangge617@163.com (G.Z.); baiwen@iem.ac.cn (W.B.)
[2] School of Automation, Harbin University of Science and Technology, Harbin 150000, China
* Correspondence: sunbt@iem.ac.cn

Abstract: Accurately simulating the nonlinear response of reinforced concrete (RC) columns under cyclic loading is crucial in performance-based seismic design for building structures, especially regarding strength degradation. This paper presents the description, calibration and simulation of the hysteretic model for RC columns based on effective hysteretic energy dissipation with positive and negative directions. During the analysis of previous experimental data, the relationship between hysteresis energy dissipation, maximum displacement, and the effects of positive and negative loading directions on strength degradation has been summarized. The proposed method for determining the yield strength of the hysteresis loop is based on the farthest point method. Calibration of the hysteretic models' existing RC columns' experimental data demonstrates that the proposed model can simulate the main characteristics that influence deterioration.

Keywords: RC columns; strength deterioration; asymmetric loading; effective hysteresis energy dissipation

Citation: Zhang, G.; Sun, B.; Bai, W. Hysteretic Model for RC Columns Based on Effective Hysteretic Energy Dissipation with Positive and Negative Directions. *Buildings* **2023**, *13*, 1140. https://doi.org/10.3390/buildings13051140

Academic Editors: Chao Li, Weiping Wen, Jian Zhong, Xiaowei Wang and Suiwen Wu

Received: 23 March 2023
Revised: 22 April 2023
Accepted: 23 April 2023
Published: 25 April 2023

1. Introduction

It is essential for performance-based seismic design to reasonably evaluate the performance of structures under seismic loading [1]. Reinforced concrete (RC) columns are the critical members bearing vertical loads and resisting horizontal loads in the actual structure. The phenomena in the experiments and the earthquake damage indicate that the seismic performance of the structure is related to the loading histories [2]. With the increasing duration and strength of strong ground motions, the strength of the columns decreases more significantly when the earthquake occurs. Therefore, the strength degradation of RC columns should be reasonably evaluated when analyzing the seismic performance of the structure.

Strength degradation in the cyclic loading refers to the phenomenon that the strength gradually decreases each time as the number of cycles increases, while the load reaches the same displacement [3]. In most seismic studies, hysteresis models that include strength degradation are developed and form a non-deteriorating backbone curve and hysteresis rules [4]. Park [5] and Miramontes [6] each proposed hysteresis rules for reloading paths pointing to points with larger displacement values. Ozcebe [7] and Sivaselvan [8] developed rules for reloading paths that point to strength-discounting points with the same displacement values. In traditional RC column hysteresis tests, the force–displacement curves at the column end sections are primarily obtained. Chalioris and his research group [9] ingeniously arranged piezoelectric sensors at the beam–column joints to trace the structural damage progression with high precision and fidelity. Capitalizing on the obtained data, they simulated the cyclic lateral behavior of carbon fiber-reinforced polymer bars as a longitudinal reinforcement in the beam using the finite element analysis software ABAQUS [10]. Regarding the simulation of other structural members, Thomoglou et al. [11] implemented an intricate micro-nonlinear finite element model in ANSYS to simulate and validate the shear behavior of unreinforced masonry walls containing units, mortar, and fiber-reinforced composites with their complex interactions.

The loading path is a critical factor that affects the strength degradation of RC columns. It primarily depends on two key indicators, namely maximum displacement and hysteresis energy dissipation. In most studies on the effect of loading path on strength degradation, the strength degradation rate proposed by Roufaiel [12], Chung [13] and Yousseef [14] was only related to the maximum displacement. Kunnath [15], Rahnama [16], Sucuoglu [17] and Lee [18] summarized degradation models associated with hysteretic energy dissipation. Qu [19] presented the concept of effective hysteretic energy dissipation, which applied the two factors of maximum displacement and hysteretic energy dissipation together to the hysteretic model of strength degradation. Huang [20] proposed a neural network-based hysteresis model to characterize the relationship between lateral force and displacement of RC columns with different properties.

When RC columns are subjected to seismic loading, both the steel rebar and concrete exhibit complex nonlinear properties. However, previous studies mostly utilized symmetric hysteresis experiments to simulate the response of RC columns under seismic loading, resulting in distortion of the simulation under asymmetric hysteresis. Previous studies have also utilized a complete hysteresis loop as the basis for their analysis, which has been shown to result in a failure of hysteresis models to accurately reflect the strength degradation when loading the opposite side after experiencing large deformation on one side. The simulation model proposed in this paper fully accounts for the asymmetry in horizontal loading procedures. It can predict the response of RC columns under seismic loading with superior accuracy. This model provides a foundation for analyzing and evaluating the seismic capacity of structures in the future.

2. The Proposed Hysteretic Model

2.1. Backbone Curve

The backbone curve represents the maximum strength that a structural member can develop at a given level of deformation, resulting in an effective limit to the strength of a member in force–deformation space [21]. The developed model includes four distinct sections: the elastic branch, hardening branch, softening branch, and residual branch, as illustrated in Figure 1.

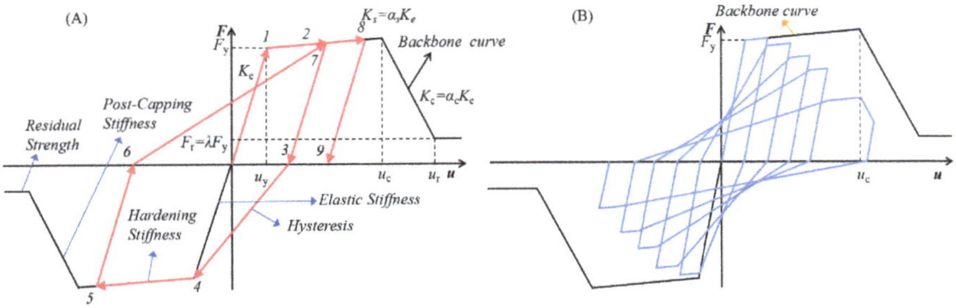

Figure 1. The proposed hysteretic model. (**A**) Backbone curve and hysteretic rule. (**B**) Strength degradation.

The backbone curve is a quad-linear idealization that is defined by several parameters, including the yield strength F_y, the residual strength F_r, the yield deformation u_y, the strain-hardening stiffness K_s, the cap deformation u_c, and the residual deformation u_r. In the elastic branch, the elastic stiffness is calculated, $K_e = F_y/u_y$, which can determine the cap strength, F_c. During the hardening stage, the strain-hardening stiffness is calculated, $K_s = \alpha_s K_e$. The ratio of the cap deformation to the yield deformation can be described as the ductility capacity, (u_c/u_y). The softening stage is defined by the post-capping stiffness, $K_c = \alpha_c K_e$. The residual strength can be defined in the model, $F_r = \lambda F_y$. If the deformation exceeds the residual displacement, the load capacity of the member is not less than the residual strength.

The above parameters of the backbone curves can be obtained from load–deformation data from the experiments or theories of previous studies [22–26].

2.2. Hysteretic Rule

The hysteretic rule defines the load–deformation path under cyclic loading [27]. The hysteretic rule for the developed model is shown in Figure 1A, where the load path points to the maximum displacement of the previous cycle, based on Clough's theory [28]. This paper focuses on the strength degradation theory; therefore, the stiffness in the post-capping and residual stages is assumed to be the same as the backbone curve.

2.3. Cyclic Strength Degradation

Figure 1B shows the cyclic strength degradation, which controls the amount of the hysteretic curve that moves to the x coordinate axis with the cyclic deformation.

2.3.1. Calculation of Strength Degradation Value

In previous studies [4,19], it requires much time and effort to obtain the strength degradation value from the experimental hysteresis data. This paper introduces a new calculation method based on the farthest point method [29], as shown in Figure 2. The point on the curve farthest from the line connecting the intersection of the origin and the cap point is the yield point. As shown in the hysteretic model presented in this paper (Figure 1A), the RC column yields at point 1 within the initial hysteresis loop and reaches the cap response at point 2. Point 1 on the hysteresis curve represents the point that is furthest from the line connecting the origin and point 2. If there is more than one point, the yield point can be obtained by averaging the load values of these points and corresponding to the curve. The method has the advantages of easy operation and accurate calculation by programming, which can avoid the calculation error caused by manual value.

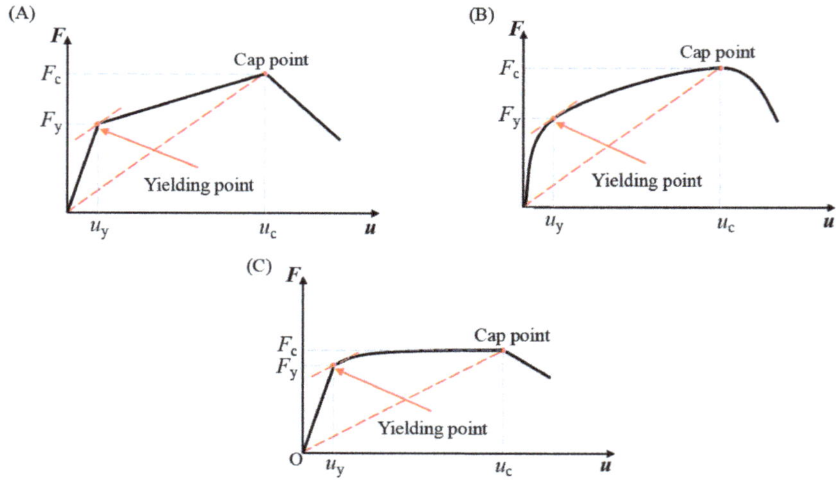

Figure 2. Farthest point method used to determine yield point. (**A**) Bilinear curve. (**B**) Without obvious turning curve. (**C**) Traditional elastic–plastic curve.

The yield point can be obtained form the backbone curve based on the Equation (1).

$$(F_{ys}, u_{ys}) = \max_{(F_{ys}, u_{ys})=(F,u)} d = \frac{|F_c \cdot u - u_c \cdot F|}{\sqrt{F_c{}^2 + u_c{}^2}} \quad (1)$$

where (F,u) is the coordinate of any point on the backbone curve of the component; (F_{ys}, u_{ys}) is the coordinate of the yield point determined by the farthest point method; (F_c, u_c) is the coordinate of the cap point; and $0 \le u \le u_p$.

Since the reloaded hysteresis loop does not pass through the origin, the calculation is derived as Equation (2). Since the columns are axisymmetric, it is assumed that the yield strength in the negative direction is equal to the yield strength in the positive direction.

$$(F_{ys}, u_{ys}) = \max_{(F_{ys},u_{ys})=(F,u)} d = \frac{|(u_c - u_o)F - (F_c - F_o)u + u_oF_c - u_cF_o|}{\sqrt{(u_c - u_o)^2 + (F_c - F_o)^2}} \tag{2}$$

where (F_o, u_o) is the intersection coordinate of the hysteresis loop and the coordinate axis, and $0 \le u \le u_p$. The positive and negative directions discussed in this paper are the directions of force loading on RC columns. Therefore, $F_o = 0$ and Equation (2) can be simplified to Equation (3), which is suitable for positive and negative directions. In some experiments, the hysteresis loop loading did not exceed the cap point; thus, these hysteresis loops could not be calculated.

$$(F_{ys}, u_{ys}) = \max_{(F_{ys},u_{ys})=(F,u)} d = \frac{|(u_c - u_o)F - F_cu + u_oF_c|}{\sqrt{(u_c - u_o)^2 + F_c^2}} \tag{3}$$

Since the columns are axisymmetric, it is assumed that the yield strength in the negative direction is equal to the yield strength in the positive direction, as shown in Figure 3.

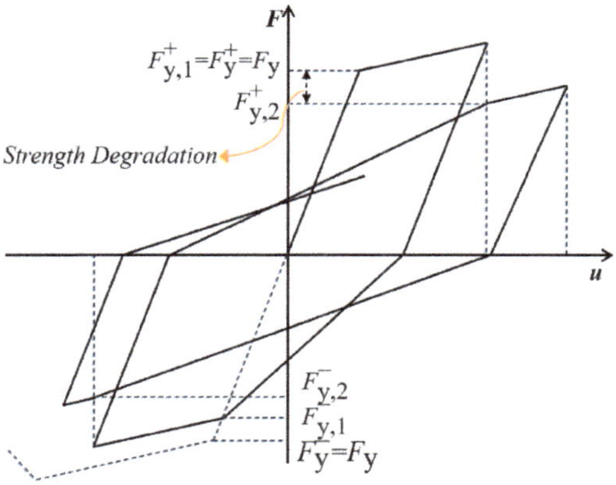

Figure 3. Strength deterioration.

2.3.2. Strength Deterioration Rule

The authors' previous [30] study showed that the degradation mechanism of the RC columns is usually considered to be the bond slip between the reinforcement and the concrete and the spalling of the concrete protective layer, as shown in Figure 4. Positive and negative sequential loading is the most common loading scheme in studying the cyclic behavior of structural members. At the same time, displacement loading with monotonically increasing amplitude is the most widely used loading path. However, the deformation of the column in the positive and negative directions is usually asymmetrical under the action of the ground motion.

Figure 4. Experimental phenomenon of RC column. (**A**) Mortar surface of RC column. (**B**) Concrete surface of RC column.

According to Takemura and Kawashima [31], two identical batch columns with the same design parameters were applied with different hysteresis loads. The specimen was first applied with a large displacement in the positive direction, and then, the same displacement was applied in the negative direction, as shown in Figure 5A. The other specimen was loaded with a large displacement in the positive direction only, and then, the displacement was gradually reduced, as shown in Figure 5B. Figure 6A shows the difference between the positive yield strength of 150 kN and the negative yield strength of 110 kN. Comparing Figure 6A,B, with only positive loading, the strength of the second hysteresis loop is significantly greater than that of both positive and negative loading. Therefore, the column damage by positive or negative deformation causes the strength deterioration in both negative and positive directions.

Figure 5. Loading hysteresis of Takemura and Kawashima test. (**A**) Positive and negative hysteretic loading. (**B**) Positive hysteretic loading.

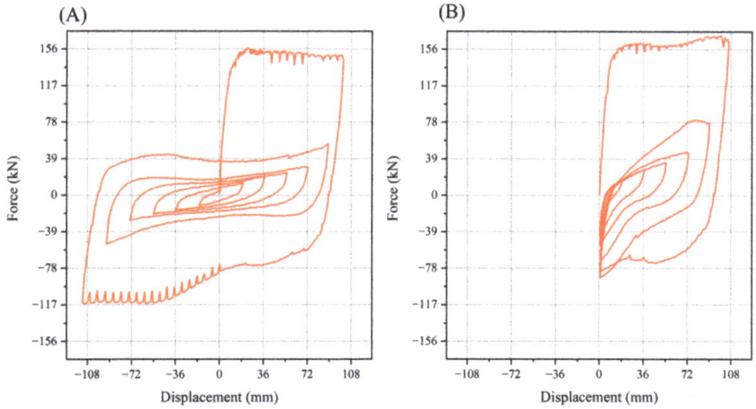

Figure 6. Force–displacement hysteresis curve of Takemura and Kawashima' test. (**A**) positive and negative hysteretic loading. (**B**) positive hysteretic loading.

On the basis of the above findings, the cyclical strength deterioration rates are proposed in this paper. The cyclic deterioration in excursion i is defined by the strength degradation parameter $\beta_i^{+/-}$, considering positive and negative directional effects. The strength degradation is described by the following Equation (4).

$$F_{y,i}^{+/-} = \left(1 - \beta_i^{+/-}\right) F_y \tag{4}$$

where $F_{y,i}^{+/-}$ is the deteriorated yield strength in negative and positive directions of ith circle. $\beta_i^{+/-}$ and $F_{y,i}^{+/-}$ are each the time that the updated inelastic path crosses the horizontal axis. In the proposed hysteresis model, $\beta_i^{+/-}$ must be within the limits $0 < \beta_i^{+/-} \leq 1 - \lambda$, due to the residual branch. If $\beta_i^{+/-}$ is outside these limits ($\beta_i^{+/-} > 1 - \lambda$), RC columns are advanced to the residual strength stage. The details of $\beta_i^{+/-}$ will be presented in subsequent sections.

3. Effective Hysteretic Energy Dissipation with Positive and Negative Directions

Liu [32] conducted a series of constant amplitude cyclic loading tests aimed at investigating the relationship between load displacement amplitude and low cycle fatigue life of RC members. The findings revealed that under larger displacement loading, even a relatively small amount of hysteretic energy dissipation could cause failure of the RC members. In light of these results and Section 2.3, Equation (5) was proposed to account for effective hysteretic energy dissipation in both positive and negative directions, taking into consideration the loading displacement, hysteresis energy dissipation, as well as the positive and negative directions.

$$E_{e,i}^{+/-} = \sum_{j=1}^{i} \left(a \cdot E_j^{+/-} \left(\frac{u_j^{+/-}}{u_c} \right)^c + b \cdot E_j^{-/+} \left(\frac{u_j^{-/+}}{u_c} \right)^c \right) \tag{5}$$

where $E_{e,i}^{+/-}$ is the effective hysteretic energy dissipation up to the ith loading cycle in positive and negative directions; $u_j^{+/-}$ refers to the deformation amplitude of the jth cycle in positive and negative directions; u_c is the cap deformation; $E_j^{-/+}$ represents the hysteretic energy dissipation up to the ith loading cycle in positive and negative directions. a, b and c are the model coefficients, and $a + b = 2$.

The strength degradation parameter $\beta_i^{+/-}$ is suggested in Equation (6).

$$\beta_i^{+/-} = \frac{E_{\mathrm{e},i}^{+/-}}{kF_{\mathrm{y}}u_{\mathrm{c}}} \tag{6}$$

where k is the parameter determined from the test.

Equation (6) is substituted into Equation (4). Therefore, the yield strength of each hysteresis loop is expressed by Equation (7).

$$F_{\mathrm{y},i}^{+/-} = \left(1 - \frac{E_{\mathrm{e},i}^{+/-}}{kF_{\mathrm{y}}u_{\mathrm{c}}}\right)F_{\mathrm{y}} \tag{7}$$

In this model, the absolute value of the yield strength in negative and positive directions of ith circle $F_{\mathrm{y},i}^{+/-}$ is as the value of the effective hysteretic energy dissipation $E_{\mathrm{e}}^{+/-}$ increases.

4. The Influence Degree of Loading Directions on Strength Deterioration

4.1. Rectangular RC Column Dataset

This study aimed to determine the relative influence of positive and negative loading directions on the strength deterioration of RC columns. For this purpose, 98 RC columns with rectangular cross-sections were selected from the PEER Structural Performance Database and the authors' previous studies [33]. The stacked histograms in Figure 7 show the distribution of the design parameters of the RC columns, which include: (1) compressive strength of concrete $(f_{\mathrm{c}}(MPa))$, (2) yield strength of longitudinal reinforcement $(f_{\mathrm{yl}}(MPa))$, (3) yield strength of transverse reinforcement $(f_{\mathrm{yt}}(MPa))$, (4) axial compression ratio $(n = N/(Af_{\mathrm{c}}))$, (5) cross-sectional area $(A(mm^2))$, (6) length $(L(mm))$, (7) longitudinal reinforcement ratio (ρ_{l}), and (8) Vol transverse reinforcement ratio (ρ_{v}). Different colors indicate distinct limit zones, and these restrictions are already displayed in Figure 7. The parameters of the database can be found in Appendix A.

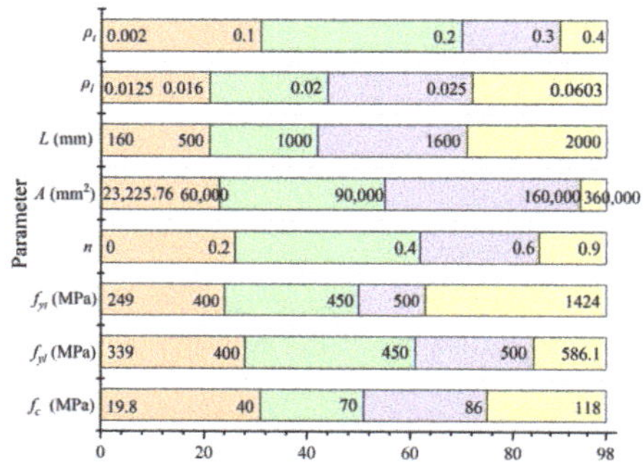

Figure 7. Stacked histograms of the design parameters of the RC columns.

4.2. Proposed Regression Coefficients

In this sub-section, the rectangular RC column dataset will be analyzed in order to determine the impact of effective hysteretic energy dissipation with positive and negative directions on the strength degradation of RC columns. This means that by fitting the hysteresis curves within the dataset, the most suitable values for the three coefficients (a, b, and c) in Equation (5) can be determined.

The value a represents the magnitude of effective hysteresis energy dissipation during one half of the hysteresis loop, and b represents the magnitude of effective hysteresis energy dissipation during the other half. It is assumed that a takes on values between 1.02 and

2, with an interval of 0.02, and b is defined as $2 - a$. This means that b takes on values between 0 and 0.98, with the same interval of 0.02. The variable c is used to quantify the impact of the ratio between the maximum displacement of a half hysteresis loop and the cap displacement on the strength degradation. The value of c is assumed to range from 0.01 to 4, with increments of 0.01. The researchers used a statistical method called least squares linear fit [34] to process Equation (5). This method is commonly used in data analysis to find the best-fitting straight line through a set of data points. The goal is to minimize the sum of the squared differences between the observed values and the corresponding fitted values on the line. R-squared (R^2) is used to evaluate the performance of a regression model in the paper, as shown in Equation (8). R^2 is typically expressed as a percentage between 0% and 100%. The higher the R^2 value, the better the fit of the regression model [35].

$$R^2 = 1 - \frac{\sum (y_i - f_i)^2}{\sum (y_i - y_{\text{mean}})^2} \tag{8}$$

where y_i represents the actual value of the ith observation, f_i represents the predicted value of the ith observation, and y_{mean} represents the mean value of all observations.

The average R^2 of the 98 reinforced concrete columns is depicted in Figure 8. When $a = 1.12$, $b = 2 - a = 0.88$, and $c = 1.85$, the mean value of R^2 is maximum, indicating a better fit. Figure 9 shows the statistical information for R^2, when $a = 1.12$, $b = 0.88$, and $c = 1.85$. After performing the calculation, we can transform Equation (7) into Equation (9). The degradation in strength caused by the effective hysteresis energy in the same direction is 1.27 times greater than the effect of the effective hysteresis energy in the opposite direction. This finding aligns with and corroborates the effects of bidirectional hysteretic energy dissipation and strength deterioration elucidated in Section 2.3.2.

$$F_{y,i}^{+/-} = \left(1 - \frac{\sum_{j=1}^{i} \left(1.12 \cdot E_j^{+/-} \left(\frac{u_j^{+/-}}{u_c} \right)^{1.85} + 0.88 \cdot E_j^{-/+} \left(\frac{u_j^{-/+}}{u_c} \right)^{1.85} \right)^{+/-}}{k F_y u_c} \right) F_y \tag{9}$$

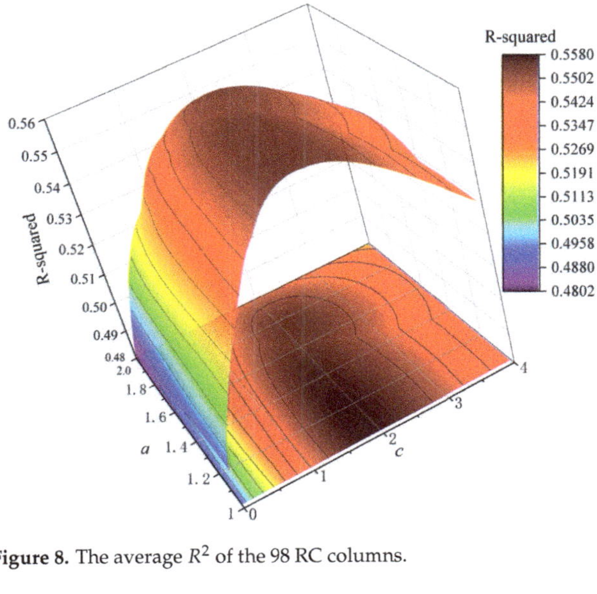

Figure 8. The average R^2 of the 98 RC columns.

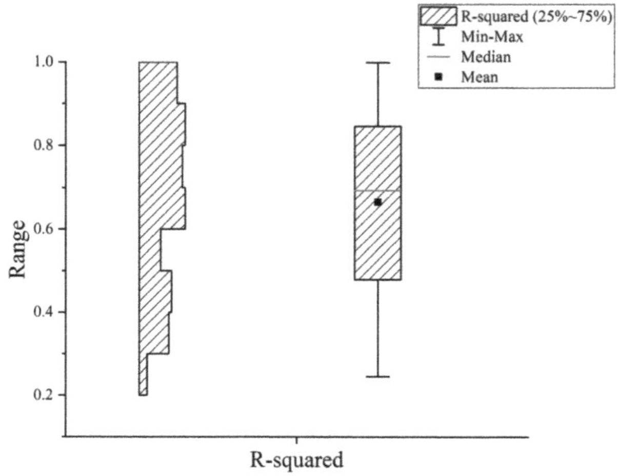

Figure 9. The statistical information for R^2.

5. Modeling of Strength Deterioration of RC Columns

The fiber beam element in ABAQUS (a software suite for finite element analysis and computer-aided engineering) is able to simulate the applicability of bending and axial force interactions for different cross-sectional profiles and reinforcement modeling members. By creating fiber bundles within the cross-section of the beam, the complex stress–strain behavior of RC columns can be captured more accurately, thus improving the accuracy of the model [36].

The strength degradation of the RC columns is a macroscopic behavior, which can be observed through the reduction of load-carrying capacity over time due to various factors. The mechanisms that contribute to this degradation are complex and involve multiple factors, including bond slip between the steel rebar and concrete and spalling of the concrete protective layer.

The constitutive relationship of the steel rebar plays a crucial role in the behavior of RC columns subjected to cyclic loading. Youssef [14] demonstrated that incorporating the degradation of the load-carrying capacity of the member into the constitutive relationships of the steel rebar can effectively simulate the load-carrying capacity degradation of RC columns under incremental loading. Based on the above conclusions, the forces and displacements in Equation (9) are replaced by the stresses and strains of the steel rebar, as shown in Equation (10).

$$\sigma_{y,i}^{+/-} = \left(1 - \frac{\sum_{j=1}^{i} \left(1.12 \cdot E_j^{+/-} \left(\frac{\varepsilon_j^{+/-}}{\varepsilon_c} \right)^{1.85} + 0.88 \cdot E_j^{-/+} \left(\frac{\sigma_j^{-/+}}{\sigma_c} \right)^{1.85} \right)^{+/-}}{k \sigma_y \varepsilon_c} \right) \sigma_y \quad (10)$$

where σ_y is the yield stress of the steel rebar; $\sigma_{y,i}^{+/-}$ is the yield stress of the steel rebar in negative and positive directions; ε_c can be taken as the strain of the steel rebar when the member is monotonically loaded to the cap deformation u_c.

There are more factors affecting the value of ε_c. After simulation, the proposed value of ε_c is shown in Equation (11).

$$\varepsilon_c = 0.15 \lambda_V / n \quad (11)$$

where $\lambda_V = \rho_v f_{yt} / f_c$ is the stirrup characteristic value; ρ_v is the vol transverse reinforcement ratio; f_{yt} is the yield strength of transverse reinforcement; f_c is the yield strength of longitudinal reinforcement. $n = F/(Af_c)$ is the axial compression ratio, which takes the value 0.1. If it is less than 0.1, N is the axial compression, and A is the cross-sectional area.

The hysteretic models have been calibrated by force–displacement data from 12 experiments on RC columns with different design parameters and loading histories, based on the general purpose program ABAQUS. When accounting for strength deterioration, Equation (10) will be used to update the yield strength of the steel rebar during the analysis.

For the simulation of eight RC columns with different design parameters, the model results and experimental results are displayed in Figures 10–13. The majority of the experimental results were effectively simulated, including the presentation of strength degradation. Additionally, information such as the axial compression ratio and stirrup characteristic values was added to the figures for a better understanding of the model's applicability. In cases where the axial compression ratio is less than 0.5, the model simulation performs exceptionally well. This indicates that when the axial compression ratio is low, the analysis method can effectively predict the performance of RC columns. However, some discrepancies were observed in specific cases.

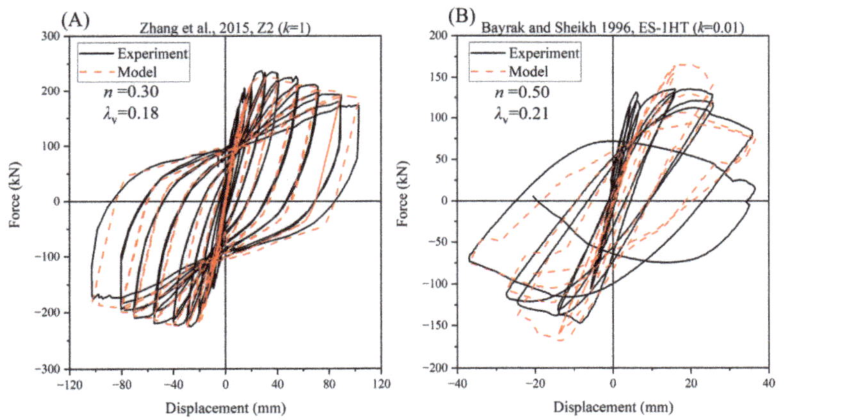

Figure 10. Comparison between the analysis and experimental results of Zhang [33] and Bayrak [37].

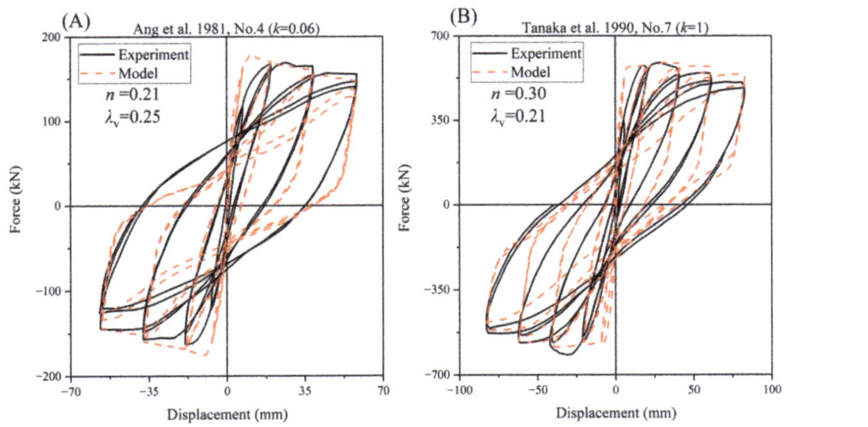

Figure 11. Comparison between the analysis and experimental results of Ang [38] and Tanaka [39].

For instance, in Figure 10B (Bayrak and Sheikh, 1996, ES-1HT), when high-strength concrete with a compressive strength of 72.1 MPa was used for the RC columns, the analytical model tended to underestimate the actual strength of the columns. This could be attributed to the limitations of the model in capturing the behavior of high-strength concrete or due to simplifications made in the model assumptions.

Regarding the specimens of Watson et al. 1989, No. 7 ($n = 0.7$) and Zhou et al. 1987, No. 223-09 ($n = 0.9$), where the axial pressure was relatively large, discrepancies were observed in the compressive-bending load capacity of the simulated members compared to the experimental results. These discrepancies could be attributed to the model's limitations in simulating the complex interaction between axial pressure, bending, and confinement effects.

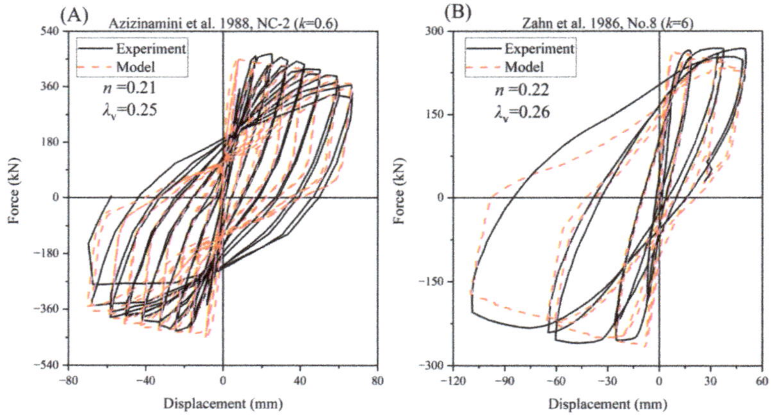

Figure 12. Comparison between the analysis and experimental results of Azizinamini [40] and Zahn [41].

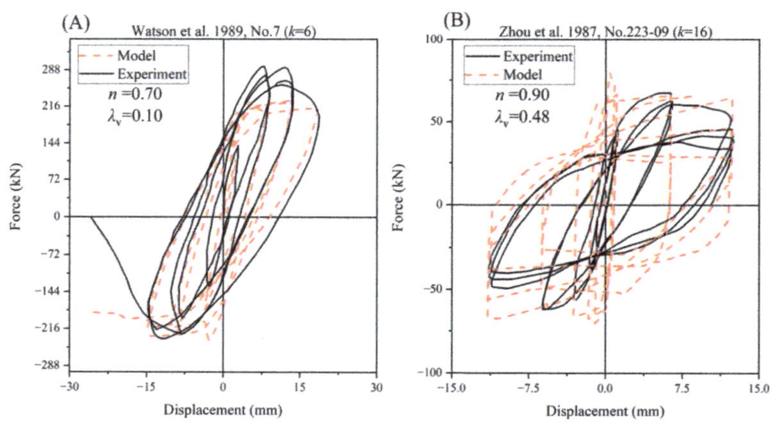

Figure 13. Comparison between the analysis and experimental results of Watson [42] and Zhou [43].

The analytical model also demonstrates good simulation results for non-square rectangular columns, as evidenced by the specimens No. 40.033a and No. 40.033, as shown in Figure 14. These specimens had dimensions of 152 mm in width and 305 mm in depth. The accurate simulation of non-square rectangular columns is crucial, as it highlights the model's versatility and applicability to a wider range of column geometries.

Figures 14 and 15 effectively simulate the hysteresis performance of the column under different loading paths. These figures demonstrate the role of effective hysteretic energy dissipation in both positive and negative directions on the column's strength degradation. The ability to simulate hysteresis performance and energy dissipation is essential in understanding the behavior of RC columns under cyclic loading, which is commonly encountered in situations such as earthquakes.

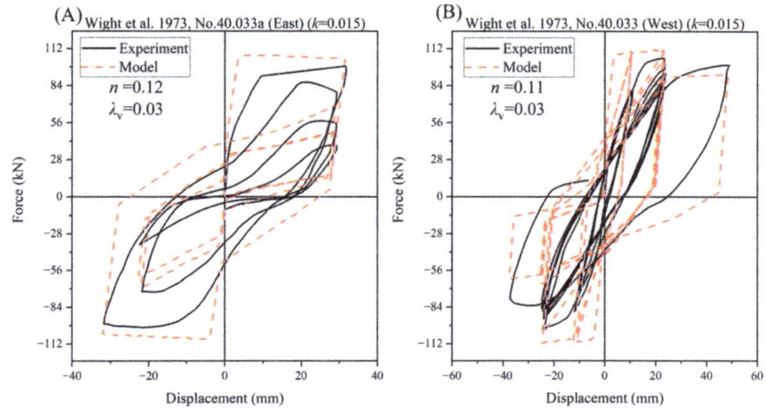

Figure 14. Comparison between the analysis and experimental results of Wight [44].

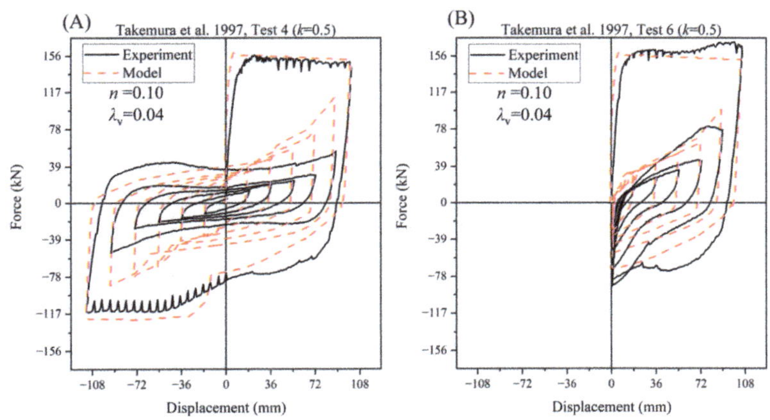

Figure 15. Comparison between the analysis and experimental results of Takemura [31].

Despite these discrepancies, the analytical model generally provided effective simulations in the majority of experimental results. However, it is crucial to address these limitations and improve the model's accuracy to provide a better understanding of the behavior of RC columns with different design parameters and under various loading conditions.

6. Discussion

1. In previous studies, the method of yield strengths ΔF_{yi} reduction of the backbone curve derived from the force reduction at the reloading point ΔF_{oi} was proposed by Qu [19], as shown in the Equation (12). However, that method has the disadvantage of requiring a lot of manual processing of experimental data. The furthest point method has a wider range of practical applications and is more convenient for converting into calculations, resulting in significantly reduced data-processing time.

$$\Delta F_{yi} = \frac{\Delta F_{oi}}{1 - \alpha_s} \tag{12}$$

2. Compared with the model of Qu [19], which only considers hysteretic energy dissipation and maximum historical displacement, the model proposed in this study considers the effects of positive and negative directions, which simulates the hysteretic behavior of the member more accurately and improves the simulation accuracy. Hysteretic energy is provided in the form of effective hysteretic energy, which comprehensively

considers the effects of both directions and avoids the hysteretic energy in a single direction from dominating the simulation results. When the displacement in the first hysteresis loop is large, the model that considers the effects of both directions can produce more accurate simulation results. It is difficult to accurately simulate this situation by relying solely on the hysteretic energy of the symmetrical loading.

3. Figure 14 demonstrates the practical application of the proposed strength degradation model for non-square rectangular columns. The model can potentially be applied to beam simulations, depending on the force distribution in the beam and the column.

4. The model may produce some errors when simulating high-strength concrete and high axial compression ratios. Future studies could improve the accuracy of the simulation by taking into account the effect of transverse reinforcement on the concrete constitutive relationships.

5. During the simulation, the coefficients k differed significantly for different specimens of different design parameters. In future research, a neural network algorithm can be applied to find the relationship between the design parameters and the coefficients k.

7. Conclusions

This paper presents a hysteretic model for simulating the seismic response of RC columns incorporating loading direction and hysteretic energy dissipation. The model provides the following novel insights:

1. A new approach based on the farthest point method for accurately identifying yield points in bidirectional hysteresis loops. This technique can be readily implemented for computational efficiency.

2. An innovative hysteretic energy dissipation (effective hysteretic energy dissipation with positive and negative directions) model that considers loading orientation and maximum displacement, in addition to cumulative hysteretic energy. This formulation more realistically captures the damage caused by excursions in positive and negative directions and at varying displacement amplitudes.

3. Methodologies for introducing strength degradation in fiber beam element models to enable precise reproduction of the hysteretic behavior of RC columns, especially under cyclic loading.

The proposed model enhances simulation capabilities for structural response under asymmetric seismic ground motions. By incorporating salient features influencing damage and strength deterioration in RC members, the model can improve predictions of structural performance during severe earthquakes, enabling more reliable seismic design and assessment.

Author Contributions: Conceptualization, B.S. and G.Z.; methodology, B.S. and G.Z.; software, G.Z. and W.B.; validation, G.Z.; formal analysis, B.S., G.Z. and W.B.; investigation, G.Z.; resources, B.S.; data curation, G.Z.; writing—original draft preparation, G.Z.; writing—review and editing, G.Z.; visualization, G.Z.; supervision, B.S.; project administration, B.S.; funding acquisition, B.S. All authors have read and agreed to the published version of the manuscript.

Funding: This research was funded by National Natural Science Foundation of China (U2239252), the Scientific Research Fund of the Institute of Engineering Mechanics, the China Earthquake Administration (grant no. 2019EEEV0103) and the National Key R&D Program of China (grant no. 2019YFC1509301).

Institutional Review Board Statement: Not applicable.

Informed Consent Statement: Not applicable.

Data Availability Statement: The models and codes that support the findings of this study are available from the corresponding author upon reasonable request.

Acknowledgments: The work presented in this paper was supported by National Natural Science Foundation of China (U2239252), the Scientific Research Fund of the Institute of Engineering Mechanics, the China Earthquake Administration (grant no. 2019EEEV0103) and the National Key R&D

Program of China (grant no. 2019YFC1509301), and the Program for Innovative Research Team in China Earthquake Administration.

Conflicts of Interest: The authors declare that they have no known competing financial interests or personal relationships that could have appeared to influence the work reported in this paper.

Abbreviations

The following abbreviations are used in this manuscript:

RC	Reinforced concrete
A	Cross-sectional area
E_e	Effective hysteretic energy dissipation
E	Hysteretic energy dissipation
N	Axial compression
F	Horizontal loading
F_c	Cap strength
F_r	Residual strength
F_y	Yield strength
K_c	Post-capping stiffness
K_e	Elastic stiffness
K_s	Strain-hardening stiffness
L	Length
n	Axial compression ratio
α_s	Strain-hardening stiffness ratio
α_c	Post-capping stiffness ratio
λ	Residual strength ratio
λ_v	Stirrup characteristic value
ρ_l	Longitudinal reinforcement ratio
ρ_v	Vol transverse reinforcement ratio
u_c	Cap deformation
u_c/u_y	Ductility capacity
u	Deformation
u_r	Residual deformation
u_y	Yield deformation
a, b, c	Model coefficients
k	Parameter determined from the test
f_c	Concrete compressive strength
f_{yl}	Yield strength of longitudinal reinforcement
f_{yt}	Yield strength of transverse reinforcement
σ	Stress of steel rebar
σ_y	Yield Stress of steel rebar
ε	Strain of steel rebar
ε_c	Cap Strain of steel rebar

Appendix A

Table A1 summarize the rectangular RC column database and provide input parameters used for model calibration. These tables are useful for understanding the key characteristics of the database.

Table A1. RC column database parameters.

#	ID	f_c (MPa)	f_{yl} (MPa)	f_{yt} (MPa)	n	$A(\text{mm}^2)$	$L(\text{mm})$	ρ_l	ρ_v
1	Gill et al. 1979, No.2 [45]	41.4	375	316	0.21	302,500	1200	0.0179	0.0230
2	Ang et al. 1981, No. 3 [38]	23.6	427	320	0.38	160,000	1600	0.0151	0.0280
3	Ang et al. 1981, No. 4 [38]	25	427	280	0.21	160,000	1600	0.0151	0.0220
4	Soesianawati et al. 1986, No. 1 [46]	46.5	446	364	0.1	160,000	1600	0.0151	0.0090
5	Soesianawati et al. 1986, No. 2 [46]	44	446	360	0.3	160,000	1600	0.0151	0.0120
6	Soesianawati et al. 1986, No. 3 [46]	44	446	364	0.3	160,000	1600	0.0151	0.0080
7	Soesianawati et al. 1986, No. 4 [46]	40	446	255	0.3	160,000	1600	0.0151	0.0060
8	Zahn et al. 1986, No. 7 [41]	28.3	440	466	0.22	160,000	1600	0.0151	0.0160
9	Zahn et al. 1986, No. 8 [41]	40.1	440	466	0.39	160,000	1600	0.0151	0.0200
10	Watson et al. 1989, No. 5 [42]	41	474	372	0.5	160,000	1600	0.0151	0.0070
11	Watson et al. 1989, No. 6 [42]	40	474	388	0.5	160,000	1600	0.0151	0.0030
12	Watson et al. 1989, No. 7 [42]	42	474	308	0.7	160,000	1600	0.0151	0.0130
13	Watson et al. 1989, No. 8 [42]	39	474	372	0.7	160,000	1600	0.0151	0.0070
14	Watson et al. 1989, No. 9 [42]	40	474	308	0.7	160,000	1600	0.0151	0.0230
15	Tanaka et al. 1990, No. 7 [39]	32.1	511	325	0.3	302,500	1650	0.0125	0.0210
16	Ohno et al. 1984, L2 [47]	24.8	362	325	0.03	160,000	1600	0.0142	0.0030
17	Zhou et al. 1987, No. 104-08 [43]	19.8	341	559	0.8	25600	160	0.0222	0.0070
18	Zhou et al. 1987, No. 214-08 [43]	21.1	341	559	0.8	25,600	320	0.0222	0.0070
19	Zhou et al. 1987, No. 223-09 [43]	21.1	341	559	0.9	25,600	320	0.0222	0.0180
20	Zhou et al. 1987, No. 302-07 [43]	28.8	341	559	0.7	25,600	480	0.0222	0.0070
21	Kanda et al. 1988, 85STC-1 [48]	27.9	374	506	0.11	62,500	750	0.0162	0.0040
22	Kanda et al. 1988, 85STC-2 [48]	27.9	374	506	0.11	62,500	750	0.0162	0.0040
23	Kanda et al. 1988, 85STC-3 [48]	27.9	374	506	0.11	62,500	750	0.0162	0.0040
24	Arakawa et al. 1989, OA2 [49]	31.8	340	249	0.18	32,400	225	0.0313	0.0020
25	Muguruma et al. 1989, AL-1 [50]	85.7	399.6	328.4	0.4	40,000	500	0.0380	0.0160
26	Muguruma et al. 1989, AL-2 [50]	85.7	399.6	328.4	0.63	40,000	500	0.0380	0.0160
27	Mugumura et al. 1989, AH-2 [50]	85.7	399.6	792.3	0.63	40,000	500	0.0380	0.0160
28	Muguruma et al. 1989, BL-2 [50]	115.8	399.6	328.4	0.42	40,000	500	0.0380	0.0160
29	Muguruma et al. 1989, BH-2 [50]	115.8	399.6	792.3	0.42	40,000	500	0.0380	0.0160
30	Sakai et al. 1990, B1 [51]	99.5	379	774	0.35	62,500	500	0.0243	0.0050
31	Sakai et al. 1990, B2 [51]	99.5	379	774	0.35	62,500	500	0.0243	0.0070
32	Sakai et al. 1990, B4 [51]	99.5	379	1126	0.35	62,500	500	0.0243	0.0050
33	Sakai et al. 1990, B7 [51]	99.5	339	774	0.35	62,500	500	0.0181	0.0050
34	Amitsu et al. 1991, CB060C [52]	46.3	441	414	0.74	77,284	323	0.0412	0.0090
35	Atalay et al. 1975, No. 9 [53]	33.3	363	392	0.26	93,025	1676	0.0163	0.0150
36	Atalay et al. 1975, No. 11 [53]	31	363	373	0.28	93,025	1676	0.0163	0.0150
37	Atalay et al. 1975, No. 12 [53]	31.8	363	373	0.27	93,025	1676	0.0163	0.0090
38	Umehara et al. 1982, 2CUS [54]	42	441	414	0.27	94,300	455	0.0301	0.0030
39	Azizinamini et al. 1988, NC-2 [40]	39.3	439	454	0.21	208,849	1372	0.0194	0.0220
40	Azizinamini et al. 1988, NC-4 [40]	39.8	439	616	0.31	208,849	1372	0.0194	0.0130
41	Saatcioglu et al. 1989, U1 [55]	43.6	430	470	0	122,500	1000	0.0321	0.0090
42	Saatcioglu et al. 1989, U4 [55]	32	438	470	0.15	122,500	1000	0.0321	0.0250
43	Galeota et al. 1996, BA1 [56]	80	430	430	0.2	62,500	1140	0.0151	0.0180
44	Galeota et al. 1996, BA2 [56]	80	430	430	0.3	62,500	1140	0.0151	0.0180
45	Galeota et al. 1996, BA4 [56]	80	430	430	0.2	62,500	1140	0.0151	0.0180
46	Galeota et al. 1996, CA1 [56]	80	430	430	0.2	62,500	1140	0.0151	0.0370
47	Galeota et al. 1996, CA2 [56]	80	430	430	0.3	62,500	1140	0.0151	0.0370
48	Galeota et al. 1996, CA4 [56]	80	430	430	0.3	62,500	1140	0.0151	0.0370
49	Galeota et al. 1996, BB [56]	80	430	430	0.2	62,500	1140	0.0603	0.0180
50	Galeota et al. 1996, BB4 [56]	80	430	430	0.3	62,500	1140	0.0603	0.0180
51	Galeota et al. 1996, BB4B [56]	80	430	430	0.3	62,500	1140	0.0603	0.0180
52	Galeota et al. 1996, CB1 [56]	80	430	430	0.2	62,500	1140	0.0603	0.0370
53	Xiao et al. 1998, HC4-8L19-T10-0.1P [57]	76	510	510	0.1	64,516	508	0.0355	0.0370
54	Xiao et al. 1998, HC4-8L19-T10-0.2P [57]	76	510	510	0.2	64,516	508	0.0355	0.0370
55	Xiao et al. 1998, HC4-8L16-T10-0.1P [57]	86	510	510	0.1	64,516	508	0.0246	0.0370
56	Xiao et al. 1998, HC4-8L16-T10-0.2P [57]	86	510	510	0.19	64,516	508	0.0246	0.0370
57	Xiao et al. 1998, HC4-8L16-T6-0.1P [57]	86	510	449	0.1	64,516	508	0.0246	0.0160
58	Xiao et al. 1998, HC4-8L16-T6-0.2P [57]	86	510	449	0.19	64,516	508	0.0246	0.0160

Table A1. *Cont.*

#	ID	f_c (MPa)	f_{yl} (MPa)	f_{yt} (MPa)	n	$A(\mathrm{mm}^2)$	$L(\mathrm{mm})$	ρ_l	ρ_v
59	Sugano. 1996, UC10H [58]	118	393	1415	0.6	50,625	450	0.0186	0.0077
60	Sugano. 1996, UC15H [58]	118	393	1424	0.6	50,625	450	0.0186	0.0119
61	Sugano. 1996, UC20H [58]	118	393	1424	0.6	50,625	450	0.0186	0.0152
62	Sugano. 1996, UC15L [58]	118	393	1424	0.35	50,625	450	0.0186	0.0119
63	Sugano. 1996, UC20L [58]	118	393	1424	0.35	50,625	450	0.0186	0.0152
64	Bayrak et al. 1996, ES-1HT [37]	72.1	454	463	0.5	93,025	1842	0.0258	0.0320
65	Bayrak et al. 1996, AS-3HT [37]	71.8	454	542	0.5	93,025	1842	0.0258	0.0280
66	Bayrak et al. 1996, AS-4HT [37]	71.9	454	463	0.5	93,025	1842	0.0258	0.0510
67	Bayrak et al. 1996, AS-5HT [37]	101.8	454	463	0.45	93,025	1842	0.0258	0.0400
68	Bayrak et al. 1996, AS-6HT [37]	101.9	454	463	0.46	93,025	1842	0.0258	0.0670
69	Bayrak et al. 1996, ES-8HT [37]	102.2	454	463	0.47	93,025	1842	0.0258	0.0430
70	Saatcioglu et al. 1999, BG-2 [59]	34	455.6	570	0.43	122,500	1645	0.0195	0.0200
71	Saatcioglu et al. 1999, BG-9 [59]	34	427.8	580	0.46	122,500	1645	0.0328	0.0130
72	Saatcioglu et al. 1999, BG-10 [59]	34	427.8	570	0.46	122,500	1645	0.0328	0.0270
73	Matamoros et al. 1999, C10-05N [60]	69.6	586.1	406.8	0.05	41,209	610	0.0193	0.0100
74	Matamoros et al. 1999, C10-05S [60]	69.6	586.1	406.8	0.05	41,209	610	0.0193	0.0100
75	Matamoros et al. 1999, C10-20N [60]	65.5	572.3	513.7	0.21	41,209	610	0.0193	0.0100
76	Matamoros et al. 1999, C5-20N [60]	48.3	586.1	406.8	0.14	41,209	610	0.0193	0.0100
77	Matamoros et al. 1999, C5-40N [60]	38.1	572.3	513.7	0.36	41,209	610	0.0193	0.0100
78	Matamoros et al. 1999, C5-40S [60]	38.1	573.3	514.7	0.36	41,209	610	0.0193	0.0100
79	Aboutaha et al. 1999, ORC3 [61]	83	414	414	0.16	154,940	1829	0.0253	0.0070
80	Thomsen et al. 1994, A3 [62]	86.3	517.1	793	0.2	23,225.76	596.9	0.0245	0.0134
81	Thomsen et al. 1994, B2 [62]	83.4	455.1	793	0.1	23,225.76	596.9	0.0245	0.0152
82	Paultre et al. 2000, No. 1006025 [63]	93.3	430	391	0.28	93,025	2000	0.0215	0.0399
83	Paultre et al. 2000, No. 1006040 [63]	98.2	451	418	0.39	93,025	2000	0.0215	0.0399
84	Paultre et al. 2000, No. 10013015 [63]	94.8	451	391	0.14	93,025	2000	0.0215	0.0184
85	Paultre et al. 2000, No. 10013025 [63]	97.7	430	391	0.26	93,025	2000	0.0215	0.0184
86	Paultre et al. 2001, No. 1206040 [64]	109.2	446	438	0.41	93,025	2000	0.0215	0.0399
87	Paultre et al. 2001, No. 1008040 [64]	104.2	446	825	0.37	93,025	2000	0.0215	0.0298
88	Paultre et al. 2001, No. 1005552 [64]	104.5	446	744	0.53	93,025	2000	0.0215	0.0434
89	Paultre et al. 2001, No. 1006052 [64]	109.4	446	492	0.51	93,025	2000	0.0215	0.0399
90	Bechtoula et al. 2002, D1N30 [65]	37.6	461	485	0.3	62,500	625	0.0243	0.0109
91	Bechtoula et al. 2002, D1N60 [65]	37.6	461	485	0.6	62,500	625	0.0243	0.0109
92	Bechtoula et al. 2002, L1N60 [65]	39.2	388	524	0.57	360,000	1200	0.0169	0.0182
93	ZHANG. 2015, Z1 [33]	42.27	480	417.5	0.6	90,000	1750	0.0205	0.0180
94	ZHANG, 2015, Z2 [33]	42.27	480	417.5	0.3	90,000	1750	0.0205	0.0180
95	ZHANG. 2015, Z3 [33]	42.27	480	417.5	0.9	90,000	1750	0.0205	0.0180
96	ZHANG. 2015, Z4 [33]	42.27	480	417.5	0.9	90,000	1750	0.0205	0.0029
97	ZHANG. 2015, Z5 [33]	42.27	480	417.5	0.9	90,000	1750	0.0419	0.0180
98	ZHANG. 2015, Z6 [33]	42.27	480	417.5	0.6	90,000	1080	0.0205	0.0180

References

1. Priestley, M. Performance based seismic design. *Bull. N. Z. Soc. Earthq. Eng.* **2000**, *33*, 325–346. [CrossRef]
2. Sun, B.; Yan, P. Damage characteristics and seismic capacity of buildings during Nepal M s 8.1 earthquake. *Earthq. Eng. Eng. Vib.* **2015**, *14*, 571–578. [CrossRef]
3. Chopra, A.K.; Kan, C. Effects of stiffness degradation on ductility requirements for multistorey buildings. *Earthq. Eng. Struct. Dyn.* **1973**, *2*, 35–45. [CrossRef]
4. Ibarra, L.F.; Medina, R.A.; Krawinkler, H. Hysteretic models that incorporate strength and stiffness deterioration. *Earthq. Eng. Struct. Dyn.* **2005**, *34*, 1489–1511. [CrossRef]
5. Park, Y.J.; Reinhorn, A.M.; Kunnath, S.K. *IDARC: Inelastic Damage Analysis of Reinforced Concrete Frame–Shear-Wall Structures*; National Center for Earthquake Engineering Research: Buffalo, NY, USA, 1987.
6. Miramontes, D.; Merabet, O.; Reynouard, J. Beam global model for the seismic analysis of RC frames. *Earthq. Eng. Struct. Dyn.* **1996**, *25*, 671–688. [CrossRef]
7. Ozcebe, G.; Saatcioglu, M. Hysteretic shear model for reinforced concrete members. *J. Struct. Eng.* **1989**, *115*, 132–148. [CrossRef]
8. Sivaselvan, M.V.; Reinhorn, A.M. *Hysteretic Models for Cyclic Behavior of Deteriorating Inelastic Structures*; Multidisciplinary Center for Earthquake Engineering Research: Buffalo, NY, USA, 1999.

9. Karayannis, C.G.; Golias, E.; Naoum, M.C.; Chalioris, C.E. Efficacy and Damage Diagnosis of Reinforced Concrete Columns and Joints Strengthened with FRP Ropes Using Piezoelectric Transducers. *Sensors* **2022**, *22*, 8294. [CrossRef] [PubMed]
10. Kytinou, V.K.; Kosmidou, P.M.K.; Chalioris, C.E. Numerical analysis exterior RC beam-column joints with CFRP bars as beam's tensional reinforcement under cyclic reversal deformations. *Appl. Sci.* **2022**, *12*, 7419. [CrossRef]
11. Thomoglou, A.; Rousakis, T.; Karabinis, A. Numerical modeling of shear behavior of URM strengthened with FRCM or FRP subjected to seismic loading. In Proceedings of the 16th European Conference on Earthquake Engineering, Thessaloniki, Greece, 18–20 June 2018; pp. 18–21.
12. Roufaiel, M.S.; Meyer, C. Analytical modeling of hysteretic behavior of R/C frames. *J. Struct. Eng.* **1987**, *113*, 429–444. [CrossRef]
13. Chung, Y.S.; Meyer, C.; Shinozuka, M. Modeling of concrete damage. *Struct. J.* **1989**, *86*, 259–271.
14. Youssef, M.; Ghobarah, A. Strength deterioration due to bond slip and concrete crushing in modeling of reinforced concrete members. *Struct. J.* **1999**, *96*, 956–966.
15. Kunnath, S.K.; Reinhorn, A.M.; Park, Y.J. Analytical modeling of inelastic seismic response of R/C structures. *J. Struct. Eng.* **1990**, *116*, 996–1017. [CrossRef]
16. Rahnama, M.; Krawinkler, H. *Effects of Soft Soil and Hysteresis Model on Seismic Demands*; John A. Blume Earthquake Engineering Center: Standford, UK, 1993; Volume 108.
17. Sucuoglu, H.; Erberik, A. Energy-based hysteresis and damage models for deteriorating systems. *Earthq. Eng. Struct. Dyn.* **2004**, *33*, 69–88. [CrossRef]
18. Lee, C.S.; Jeon, J.S. Adaptive hysteretic model for reinforced concrete columns including variations in axial force and shear span length. *Earthq. Eng. Struct. Dyn.* **2021**, *50*, 4001–4031. [CrossRef]
19. Qu, Z.; Ye, L. Strength deterioration model based on effective hysteretic energy dissipation for RC-members under cyclic loading. In Proceedings of the Joint Conference Proceedings, 7th International Conference on Urban Earthquake Engineering (7CUEE) & 5th International Conference on Earthquake Engineering (5ICEE), Tokyo, Japan, 3–5 March 2010.
20. Huang, C.; Li, Y.; Gu, Q.; Liu, J. Machine learning–based hysteretic lateral force-displacement models of reinforced concrete columns. *J. Struct. Eng.* **2022**, *148*, 04021291. [CrossRef]
21. FEMA, P. Effects of Strength and Stiffness Degradation on Seismic Response. FEMA P440A 2009. Available online: https://www.nehrp.gov/pdf/femaP440A.pdf (accessed on 22 March 2023).
22. Sivaselvan, M.V.; Reinhorn, A.M. Hysteretic models for deteriorating inelastic structures. *J. Eng. Mech.* **2000**, *126*, 633–640. [CrossRef]
23. Elwood, K.J.; Eberhard, M.O. Effective Stiffness of Reinforced Concrete Columns. *ACI Struct. J.* **2009**, *106*, 476-484.
24. LeBorgne, M.; Ghannoum, W. Calibrated analytical element for lateral-strength degradation of reinforced concrete columns. *Eng. Struct.* **2014**, *81*, 35–48. [CrossRef]
25. Haselton, C.B.; Liel, A.B.; Taylor-Lange, S.C.; Deierlein, G.G. Calibration of model to simulate response of reinforced concrete beam-columns to collapse. *ACI Struct. J.* **2016**, *113*, 1141–1152. [CrossRef]
26. Lee, C.S.; Han, S.W. Computationally effective and accurate simulation of cyclic behaviour of old reinforced concrete columns. *Eng. Struct.* **2018**, *173*, 892–907. [CrossRef]
27. Dowell, O.; Seible, F.; Wilson, E.L. Pivot hysteresis model for reinforced concrete members. *ACI Struct. J.* **1998**, *95*, 607–617.
28. Clough, R.W. *Effect of Stiffness Degradation on Earthquake Ductility Requirements*; Report 66-16, Structural and Materials Research; Structural Engineering Laboratory, University of California: Berkeley, CA, USA, 1966.
29. Peng, F.; Han-Lin, Q.; Lie-Ping, Y. Discussion and definition on yield points of materials, members and structures. *Eng. Mech.* **2017**, *34*, 36–46.
30. Zhang, G.; Sun, B.; Bai, W.; Zhang, H. Prediction of the Yield Performance and Failure Mode of RC Columns under Cyclic-Load by PSO-BP Neural Network. *Buildings* **2022**, *12*, 507. [CrossRef]
31. Takemura, H.; Kawashima, K. Effect of Loading Hysteresis on Ductility Capacity of Reinforced Concrete Bridge Piers. *J. Struct. Eng.* **1997**, *43*, 849–858.
32. Liu, B.; Bai, S.; Lai, M. Experimental study of low-cycle behavior of concrete columns. *Earthq. Eng. Eng. Vib.* **1998**, *18*, 82–89.
33. Zhang, L.; Xie, X.; Zhang, H.; Mao, C. A new method for safety assessment of reinforced concrete frame structures after earthquakes by using damage index: Experimental research. *Earthq. Eng. Eng. Dyn.* **2015**, *35*, 15.
34. Brown, S.H. Multiple linear regression analysis: A matrix approach with MATLAB. *Ala. J. Math.* **2009**, *34*, 1–3.
35. Cameron, A.C.; Windmeijer, F.A. An R-squared measure of goodness of fit for some common nonlinear regression models. *J. Econom.* **1997**, *77*, 329–342. [CrossRef]
36. Monti, G.; Spacone, E. Reinforced concrete fiber beam element with bond-slip. *J. Struct. Eng.* **2000**, *126*, 654–661. [CrossRef]
37. Bayrak, O.; Sheikh, S. Confinement steel requirements for high strength concrete columns. In Proceedings of the 11th World Conference on Earthquake Engineering, Acapulco, Mexico, 23–28 June 1996.
38. Ang, B.; Priestley, M.; Park, R. *Ductility of Reinforced Bridge Piers under Seismic Loading*; Report 81-83; University of Canterbury: Christchurch, New Zealand, 1981. Available online: https://ir.canterbury.ac.nz/bitstream/handle/10092/3594/Thesis_fulltext.pdf;jsessionid=43A0460A599DE4FC4092416C1F3D7676?sequence=1 (accessed on 22 March 2023).
39. Tanaka, H. Effect of Lateral Confining Reinforcement on the Ductile Behaviour of Reinforced Concrete Columns. 1990. Available online: https://ir.canterbury.ac.nz/handle/10092/1241 (accessed on 22 March 2023).

40. Azizinamini, A.; Johal, L.S.; Hanson, N.W.; Musser, D.W.; Corley, W.G. Effects of Transverse Reinforcement on Seismic Performance of Columns—A Partial Parametric Investigation. Project No. CR 1988. Available online: https://nehrpsearch.nist.gov/static/files/NSF/PB89148068.pdf (accessed on 22 March 2023).
41. Zahn, F.A.; Park, R; Priestley, M.J.N. *Design of Reinforced Bridge Columns for Strength and Ductility*; Report 86-7; Department of Civil Engineering, University of Canterbury: Christchurch, New Zealand, 1986; 330p.
42. Watson, S. *Design of Reinforced Concrete Frames of Limited Ductility*; Report 89-4; Department of Civil Engineering, University of Canterbury: Christchurch, New Zealand, 1989; 232p.
43. Zhou, X.; Satoh, T.; Jiang, W.; Ono, A.; Shimizu, Y. Behavior of reinforced concrete short column under high axial load. *Trans. Jpn. Concr. Inst.* **1987**, *9*, 541–548.
44. Wight, J.K. *Shear Strength Decay in Reinforced Concrete Columns Subjected to Large Deflection Reversals*; University of Illinois at Urbana-Champaign: Champaign, IL, USA, 1973.
45. Gill, W.D.; Park, R.; Priestley, M.J.N. *Ductility of Rectangular Reinforced Concrete Columns With Axial Load*; Report 79-1; Department of Civil Engineering, University of Canterbury: Christchurch, New Zealand, 1979; 136p.
46. Soesianawati, M.T.; Park, R.; Priestley, M.J.N. *Limited Ductility Design of Reinforced Concrete Columns*; Report 86-10; Department of Civil Engineering, University of Canterbury: Christchurch, New Zealand, 1986; 208p.
47. Ohno, T.; Nishioka, T. An experimental study on energy absorption capacity of columns in reinforced concrete structures. *Doboku Gakkai Ronbunshu* **1984**, *1984*, 23–33. [CrossRef]
48. Kanda, M. Analytical study on elasto-plastic hysteretic behavior of reinforced concrete members. *Trans. Jpn. Concr. Inst.* **1988**, *10*, 257–264.
49. Arakawa, T.; Arai, Y.; Mizoguchi, M.; Yoshida, M. Shear resisting behavior of short reinforced concrete columns under biaxial bending-shear. *Trans. Jpn. Concr. Inst.* **1989**, *11*, 317–324.
50. Muguruma, H.; Watanabe, F.; Komuro, T. Applicability of high strength concrete to reinforced concrete ductile column. *Trans. Jpn. Concr. Inst.* **1989**, *11*, 309–316.
51. Sakai, Y. *Experimental Studies on Flexural Behavior of Reinforced Concrete Columns Using High-Strength Concrete*; Japan Concrete Institute: Tokyo, Japan, 1990.
52. Amitsu, S.; Shirai, N.; Adachi, H.; Ono, A. Deformation of reinforced concrete column with high or fluctuating axial force. *Trans. Jpn. Concr. Inst.* **1991**, *13*, 355–362.
53. Atalay, M.B.; Penzien, J. *The Seismic Behavior of Critical Regions of Reinforced Concrete Components as Influenced by Moment, Shear and Axial Force*; Earthquake Engineering Research Center, University of California Berkeley: Berkeley, CA, USA, 1975.
54. Umehara, H.; Jirsa, J.O. *Shear Strength and Deterioration of Short Reinforced Concrete Columns Under Cyclic Deformations*; PMFSEL Report No. 82-3; Department of Civil Engineering, University of Texas: Austin, TX, USA, 1982; 256p.
55. Saatcioglu, M.; Ozcebe, G. Response of reinforced concrete columns to simulated seismic loading. *Struct. J.* **1989**, *86*, 3–12.
56. Galeota, D.; Giammatteo, M.; Marino, R. Seismic resistance of high strength concrete columns. In Proceedings of the 11th World Conference on Earthquake Engineering, Acapulco, Mexico, 23–28 June 1996.
57. Xiao, Y.; Martirossyan, A. Seismic performance of high-strength concrete columns. *J. Struct. Eng.* **1998**, *124*, 241–251. [CrossRef]
58. Sugano, S. Seismic behavior of reinforced concrete columns which used ultra-high-strength concrete. In Proceedings of the Eleventh World Conference on Earthquake Engineering, Paper, Acapulco, Mexico, 23–28 June 1996; p. 1383.
59. Saatcioglu, M.; Grira, M. Confinement of reinforced concrete columns with welded reinforced grids. *Struct. J.* **1999**, *96*, 29–39.
60. Matamoros, A.B. *Study of Drift Limits for High-Strength Concrete Columns*; University of Illinois at Urbana-Champaign: Champaign, IL, USA, 1999.
61. Aboutaha, R.; Machado, R. Seismic resistance of steel-tubed high-strength reinforced-concrete columns. *J. Struct. Eng.* **1999**, *125*, 485–494. [CrossRef]
62. Thomson, J.H.; Wallace, J.W. Lateral load behavior of reinforced concrete columns constructed using high-strength materials. *Struct. J.* **1994**, *91*, 605–615.
63. Legeron, F.; Paultre, P. Behavior of high-strength concrete columns under cyclic flexure and constant axial load. *Struct. J.* **2000**, *97*, 591–601.
64. Paultre, P.; Légeron, F.; Mongeau, D. Influence of concrete strength and transverse reinforcement yield strength on behavior of high-strength concrete columns. *Struct. J.* **2001**, *98*, 490–501.
65. Kono, S.; Watanabe, F. Damage evaluation of reinforced concrete columns under multiaxial cyclic loadings. In Proceedings of the Second US-JAPAN Workshop on Performance-Based Earthquake Engineering Methodology for Reinforced Concrete Building Structures, Hokkaido, Japan, 11–13 September 2000; pp. 221–231.

buildings

MDPI

Article

Seismic Damage "Semaphore" Based on the Fundamental Period Variation: A Probabilistic Seismic Demand Assessment of Steel Moment-Resisting Frames

Sergio A. Díaz [1], Luis A. Pinzón [2,3,*], Yeudy F. Vargas-Alzate [4] and René S. Mora-Ortiz [1]

[1] División Académica de Ingeniería y Arquitectura, Universidad Juárez Autónoma de Tabasco, Villahermosa, Tabasco 86040, Mexico; alberto.diaz@ujat.mx (S.A.D.); rene.mora@ujat.mx (R.S.M.-O.)
[2] Scientific and Technological Research Center, Universidad Católica Santa María La Antigua, Panama City 0819, Panama
[3] Sistema Nacional de Investigación (SNI), Secretaría Nacional de Ciencia, Tecnología e Innovación (SENACYT), Panama City 0824, Panama
[4] Departament d'Enginyeria Civil y Ambiental, Universitat Politècnica de Catalunya Barcelona Tech (UPC), 08034 Barcelona, Spain; yeudy.felipe.vargas@upc.edu
* Correspondence: lpinzon@usma.ac.pa

Abstract: During strong earthquakes, structural damage usually occurs, resulting in a degradation of the overall stiffness of the affected structures. This degradation produces a modification in the dynamic properties of the structures, for instance, in the fundamental period of vibration (T_1). Hence, the variation of T_1 could be used as an indicator of seismic structural damage. In this article, a seismic damage assessment in four generic typologies of steel buildings was carried focused on verifying the variation of T_1. To do so, several seismic damage states were calculated using the maximum inter-story drift ratio, MIDR, and following the Risk-UE guidelines. Then, a series of probabilistic nonlinear static analyses was implemented using Monte Carlo simulations. The probabilistic approach allows one to vary the main mechanical properties of the buildings, thus analyzing in this research 4000 buildings (1000 building samples for each of the four generic typologies). The variation of T_1 was estimated using the capacity spectrum, and it was related to the MIDR for each damage state. As a main result of this study, the expected variation of T_1 for several damage states is provided. Finally, a proposal for a seismic damage preventive "semaphore" and fragility curves are presented. These results may be useful as parameters or criteria in the evaluation of on-site structural monitoring for steel buildings.

Keywords: fragility curves; fundamental period; maximum inter-story drift ratio; preventive "semaphore"; steel buildings

Citation: Díaz, S.A.; Pinzón, L.A.; Vargas-Alzate, Y.F.; Mora-Ortiz, R.S. Seismic Damage "Semaphore" Based on the Fundamental Period Variation: A Probabilistic Seismic Demand Assessment of Steel Moment-Resisting Frames. *Buildings* **2023**, *13*, 1009. https://doi.org/10.3390/buildings13041009

Academic Editors: Chao Li, Weiping Wen, Jian Zhong, Xiaowei Wang and Suiwen Wu

Received: 15 March 2023
Revised: 3 April 2023
Accepted: 7 April 2023
Published: 11 April 2023

1. Introduction

Earthquakes can cause significant consequences in exposed regions to high seismic hazards. Some of these regions have vulnerable structures due to the low quality of materials and construction practice. This can be related to socioeconomic conditions or the scant interest of governments in compliance with the standards required in construction codes. An alternative to prevent the possible effects of earthquakes in cities is to undertake studies related to the seismic–structural vulnerability of structures [1–3]. Proper knowledge about the vulnerability of buildings is fundamental to help engineers in assessing and strengthening existing structures [4]. In this respect, seismic vulnerability studies allow the performance of structures to be analyzed against expected actions. For a building, this mainly depends on its characteristics (structural system, number of stories, among others) and the level of seismic actions to which it will be subjected [5].

Seismic vulnerability in buildings can be reduced by complying with the performance level control criteria of current structural codes [6,7]. Assessing code compliance can be carried out from two perspectives: (1) evaluating and monitoring the structural health, and

(2) structural assessment through complex numerical models. The first approach is carried out with the instrumentation and monitoring of the building. Its objective is to verify the behavior of the structure accurately. In brief, the Structural Health Monitoring (SHM) process consists of measuring the evolution of representative structural parameters in a certain period of time. To do so, strategically located sensors are placed in the structure to be analyzed in order to collect temporary records of acceleration, velocity, or displacement, which serve to determine the fundamental vibration frequencies of the building. In this way, the information obtained helps plan maintenance activities, verify design hypotheses, reduce uncertainty about the structural elements, and guarantee structural safety on the damage control proposed by current regulations. These methodologies represent the most accurate way to study the structural vulnerability of a building. Still, they carry a high cost, which is why they are solutions for exceptional cases and not oriented to monitor an entire city [8–11]. The second approach is based on the determination of the structural response of buildings through numerical modelling using static or dynamic methods. Both types of analyses are powerful tools for understanding and quantifying the performance of structures for evaluating expected damage.

Current technological advances facilitate the processing and treatment of large amounts of data in a relative simplified manner. They allow complex probabilistic numerical models to be developed for civil structures and for nonlinear static (NLSA) and dynamic (NLDA) analyses to be computed affordably and in a reasonable amount of time.

In probabilistic approaches, both the uncertainties of seismic actions and building properties have been incorporated in previous studies [7,12] through computational algorithms, e.g., the Monte Carlo method [13–16]. As a result, a global view of the expected performance of buildings is obtained considering the main uncertainties in the implied variables.

The structural design codes regulate the buildings' structural safety, the materials' quality and the correct application of the design criteria. They also provide the design earthquake motion (design spectrum) and seismic parameters of the building (ductility Q, response modification Ro, redundancy ρ, and irregularity factors α). Furthermore, they suggest engineering demand parameters to control expected damages, e.g., the Maximum Inter-story Drift Ratio (MIDR) [17,18]. The MIDR is calculated as the maximum absolute difference in displacement between consecutive stories divided by the height at each level. This parameter has been related to the expected damage of buildings in several studies [17,19–21]. On the other hand, during strong earthquakes, structural damage usually occurs, resulting in a degradation of the overall stiffness of the affected structures. This degradation produces a modification in the dynamic properties of the structures, for example, in the fundamental period of vibration (T_1). In the last decade, variation of T_1 has been used as a parameter for seismic damage control, and several studies related to this topic have been developed. Regarding reinforced concrete buildings, research based on the evaluation of the fundamental period of undamaged and damaged structures [22–24], correlation of structural seismic damage with a fundamental period [23], prediction of the fundamental period of regular frame buildings [25], a fundamental-period-preserving retrofit procedure for low-rise buildings with supplemental inerters [26], and parametric studies on the variation of the fundamental period [27] have been carried out. As for steel buildings, researchers have proposed modifications to current formulations to approximate fundamental periods for seismic design of steel buildings assigned to high-risk categories that incorporate the change in system strength [28]. Furthermore, research related to the elongation of the period in buildings during seismic events has also been carried out [29]. These research studies show the importance of the T_1 parameter as a damage indicator in seismic evaluations of different types of buildings. From the above, establishing objective limit values in MIDR and T_1 parameters prevents damage, loss of functionality of the building, and human and economic losses due to high-intensity seismic action. These limit values range from moderate damage to the collapse of the building.

In line with the above, this research presents a proposal for a seismic damage preventive "semaphore" and fragility curves based on variations of T_1. To develop these elements, a seismic damage assessment considering four generic typologies of the steel buildings located in Mexico is performed with the purpose of verifying the variation of T_1. These buildings are located in two cities in Mexico, namely, Oaxaca City in Oaxaca state, and Tuxtla Gutiérrez City in Chiapas state. These are cities with very high and high seismic hazards, respectively [30]. Then, different seismic damage states are estimated using the MIDR and following the Risk-UE guidelines [31]. To do so, probabilistic nonlinear static analyses are implemented using Monte Carlo simulations. With the probabilistic approach, the uncertainty in the main mechanical properties of the four generic typologies of buildings is considered, basing this research on the study of 4000 building models (1000 building samples for each of the 4 typologies). The variation of T_1 is estimated using the capacity spectrum, and it is related to the MIDR for each damage state.

2. Buildings

Low-rise (3-story) and mid-rise (7-story) steel buildings located in Oaxaca (OA) and Tuxtla Gutiérrez (TG) cities in Mexico were used as case studies. These buildings were designed for office purposes [32] following the MDOC-CFE [30] regulations for OA and TG cities. The dead (DL) and live (LL) load criteria of the NTCDS-RCDF [33] and the design standards ANSI/AISC 360-16 [34] were employed in the projection of these structures. Table 1 shows the considered loads, and Tables 2 and 3 show the list of steel wide flange sections for beams and columns. Figure 1 shows a 2-D view of the main frames (Special Moment Frames, SMF). The SMFs satisfy the AISC criterion "strong-column–weak-beam", and the structural sections of the beams and columns meet the slenderness criterion of the AISC-341-16 [34]. The beams consider continuous lateral bracing for their compression flanges. The slabs of the buildings are considered rigid with a composite deck system (concrete slab–steel deck with shear connectors). Connections between elements are fully restrained (FR) [34]. The modal analysis and seismic response evaluation were performed in SeismoStruct [35]. Table 4 shows the foremost characteristics of the modal analysis.

Table 1. Dead and live loads of the 3- and 7-story buildings.

Load Types	Story	Load (kN/m^2)
Dead Load (DL)	Inter-story	6.5
	Roof	5.0
Live Load (LL)	Inter-story	2.5
(Office building)	Roof	1.0

Table 2. Steel W-type sections of the 3-story buildings.

City	Columns		Beams
	C1	C2	B1
Oaxaca (OA)	W14 × 74	W18 × 119	W12 × 72
Tuxtla Gutiérrez (TG)	W16 × 67	W18 × 97	W14 × 48

Table 3. Steel W-type sections of the 7-story buildings.

City	Columns			Beams	
	C1	C2	C3	B1	B2
Oaxaca (OA)	W16 × 100	W18 × 192	W21 × 201	W12 × 53	W14 × 61
Tuxtla Gutiérrez (TG)	W16 × 89	W18 × 119	W21 × 147	W14 × 61	W16 × 57

The modal characteristics show that the structural response of these buildings is dominated by their fundamental period of vibration. Therefore, it was expected that their seismic responses would be consistent using both NLSA and NLDA [7,15,36]. Based on the above and to simplify the probabilistic approach of this study, the analyses were carried out using the NLSA in SeismoStruct [35] and the capacity spectrum method of the ATC-40 [37]. The NLSA adopted are not novel like the NLDA, but that does not detract from the validity and simplicity, especially for evaluations of regular buildings with structural responses dominated by their first mode of vibration or fundamental period, as the buildings analyzed here [7,15,36]. The NLDA has advantages but entails greater complexity and high computational cost with probabilistic analysis if one does not use a well-established computational tool. The implementation of simple tools has been sought to carry out seismic evaluations in a practical way in the field of applied structural engineering [15,36,38].

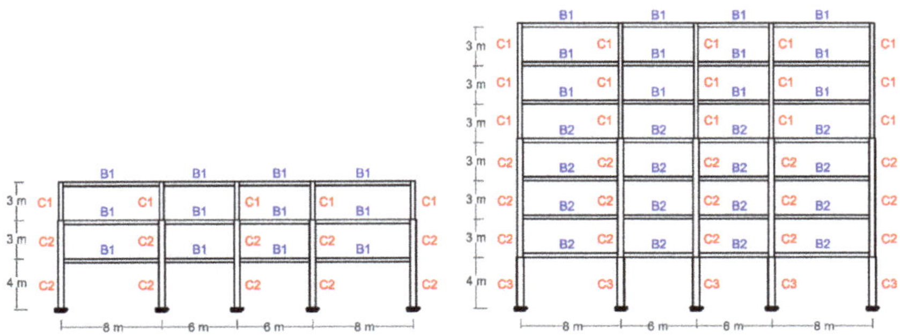

Figure 1. A 2-D view of the main frame evaluated of each building.

Table 4. The main characteristics of the modal analysis of the 3- and 7-story buildings.

City	Stories	T_1 (s) *	PF_1 *	α_1 *	W (kN) *
Oaxaca (OA)	3	0.51	1.30	0.89	327.38
	7	0.89	1.39	0.82	831.04
Tuxtla Gutiérrez (TG)	3	0.56	1.27	0.91	325.89
	7	1.04	1.33	0.84	823.95

* T_1: fundamental period; PF_1: modal participation factor for T_1; α_1: modal mass coefficient; W: total weight of the building.

The linear and non-linear behaviors of the beams and columns were modelled following the fiber approach, where each fiber is associated with uniaxial stress (σ)–strain (ε) relationships. Thus, the cross-section behavior is defined by a uniaxial bilinear σ-ε model with kinematic strain hardening, which is commonly used in the modelling of structural steel elements [35]. The deterministic five model-calibrating parameters used were as follows: Modulus of elasticity, $E_s = 2.00 \times 10^8$ kN/m^2; Yield strength, $f_y = 396{,}448.54$ kN/m^2; Strain hardening parameter, $\mu = 0.01$; Fracture/buckling strain = 0.10, and Specific weight, $\gamma = 78.00$ kN/m^3. Finally, two performance criteria of the sections were defined: (1) yielding of steel (ε_y), steel strains larger than the ratio between yield strength and modulus of elasticity ($\varepsilon_y = f_y/E_s$); and (2) fracture of steel (ε_u), steel strains larger than the fracture strain, which in this study was $\varepsilon_u = 0.06$.

3. Probabilistic Variables

The randomness in the mechanical properties of the cross-sections and the seismic actions represent the variables that provide more significant uncertainty in the structural response of buildings [16,39]. Thus, a set of probabilistic numerical models is generated to represent the random nature of the expected behavior of the buildings. These models consider the uncertainties in the mechanical properties of the steel W-type sections that are relevant to the seismic response. In summary, the three variables influencing the linear and non-linear response are (1) yield strength, f_y; (2) modulus of elasticity, E_s; and (3) fracture strain, ε_u. Table 5 shows the mean values ($\overline{\mu}$), the coefficients of variation (CVs), and standard deviations ($\overline{\sigma}$) for each variable. The Monte Carlo method was used [40,41] to generate the random sets. In addition, mean values of the three variables were used to perform and conclude about deterministic approaches.

Table 5. Mean values and coefficients of variation of the variables used for probabilistic sampling with the Monte Carlo method.

Variable	Mean ($\overline{\mu}$) *	Coefficients of Variation (CV) *	Standard Deviations ($\overline{\sigma}$) *
Yield strength, f_y (kN/m^2)	396,448.54	0.066	26,165.60
Modulus of elasticity, E_s (kN/m^2)	200,000,000	0.039	7,800,000
Fracture strain, ε_u	0.06	0.155	0.0093

* Based on reports by Schmidt and Bartlett [42] and Bartlett et al. [43] for statistics of steel mechanical properties.

Variables shown in Table 5 follow a Normal Probability Distribution (NPD), and the sampling was limited to a range of $\overline{\mu} \pm 2\overline{\sigma}$. Thus, overestimated, or underestimated values of the variables were excluded. Likewise, the steel sections (beams and columns) of the same story of the buildings were considered with a correlation of 0.65, since they could be from the same batch of steel production [14]. Sections of different stories were considered with null correlation. Therefore, a set of 1000 random samples for each of the three variables was generated. Figure 2 shows an example of the NPD with a 0.65 correlation and NPD with a null correlation of the variable f_y in the beams and columns. Additionally, Figure 2 shows the assumed NPD and truncated NPD with the histogram of the samples obtained through the Monte Carlo method. Good agreement between the histogram of the samples and the target NPDs can be seen. As pointed out above, 1000 structural random samples were used. This number was determined as follows: Several random samples were generated according to the truncated NPD. For every 100 new samples, the mean value and the standard deviation of the overall samples were obtained. Once 1000 samples were reached, no significant variations were obtained in their mean value and standard deviation by adding more samples. Therefore, 1000 was considered an adequate number of samples representing the predefined truncated NPD. This was attributed to the fact that the Monte Carlo method is based on the Latin Hypercube Sampling (LHS) technique [44], and this LHS technique avoids duplicating case combinations so that fewer samples adequately represent the target NPD.

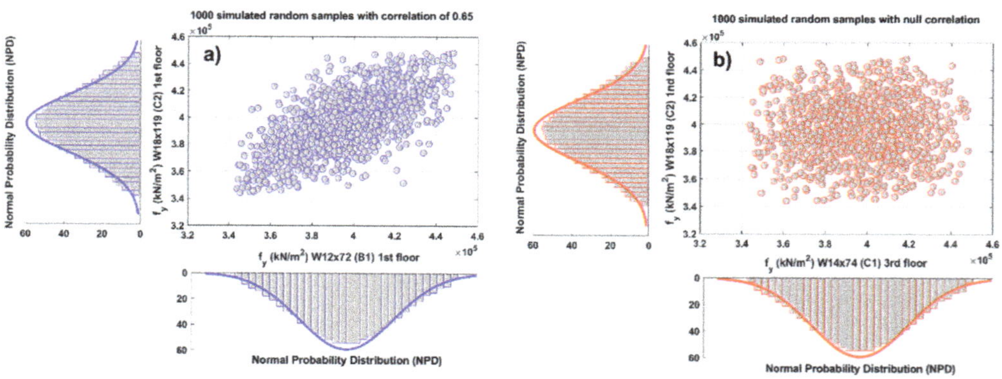

Figure 2. An example of the (**a**) NPD with a 0.65 correlation and (**b**) NPD with a null correlation of the variable f_y in the beams and columns in the low-rise (3-story) steel buildings in Oaxaca (OA).

Finally, the probabilistic models were generated assigning the 1000 random variables of f_y, E_s, and ε_u to each of the beams and columns of the models. This process was carried out through a special function for creating multiple files (SPF Creator) in SeismoStruct [35]. In this research, 4000 steel buildings were analyzed.

4. Nonlinear Static Analysis

The generation of probabilistic models, the automatic execution, and the NLSA sequential analysis were implemented with the SeismoBatch function introduced in SeismoStruct [35]. Then, the output files of the analyses were extracted, and the probabilistic capacity curves of the four buildings studied were obtained. Figures 3 and 4 show the capacity curves of the deterministic and probabilistic cases corresponding to the 3-story and 7-story buildings in OA and TG cities, respectively. The capacity curves were presented in the base shear (V)—roof displacement δ) and base shear (V)—maximum inter-story drift ratio (MIDR) formats.

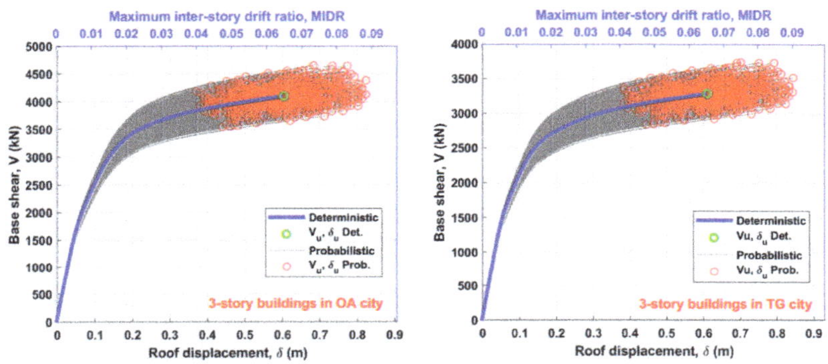

Figure 3. Deterministic and probabilistic capacity curves of the 3-story buildings in OA and TG cities.

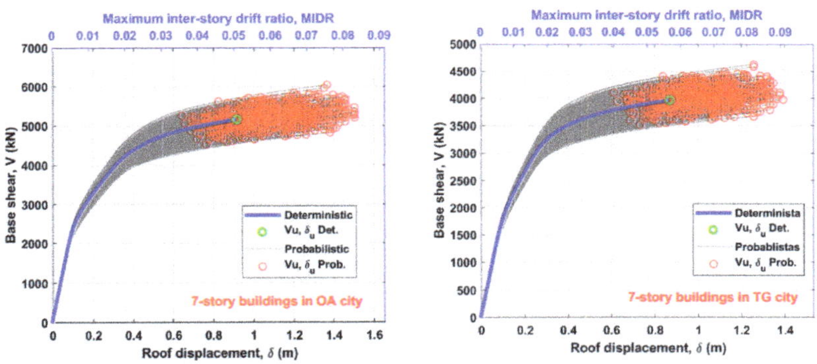

Figure 4. Deterministic and probabilistic capacity curves of the 7-story buildings in OA and TG cities.

5. Capacity Spectra and Damage States

Based on the ATC-40 [37] and by considering the values from the modal analysis (Table 4), each capacity curve in V-δ format was transformed into the Capacity Spectrum (CS) in Spectral Acceleration (S_a)–Spectral Displacement (S_d), where $S_d = \delta/PF_1$, and $S_a = V/(W^*\alpha_1)$. Then, the four non-null damage states ($DS_{non\text{-}null}$) of the RISK UE guidelines [31] were obtained. These were determined based on the yield (S_{dy}) and ultimate (S_{du}) spectral displacements as follows: Slight = $0.7S_{dy}$, Moderate = S_{dy}, Extensive = S_{dy} + 0.25 (S_{du}–S_{dy}), and Complete = S_{du}. The following equation proposed by Diaz et al. [15] was used to calculate the yield point (S_{dy} and S_{ay}) of the capacity spectrum:

$$S_{dy} = \frac{[2A_{sc} - (S_{du} \cdot S_{du})]}{[(K_i \cdot S_{du}) - S_{au}]} \text{ and } S_{ay} = K_i \cdot S_{dy} \quad (1)$$

where the variables are characteristics of the capability spectrum: K_i is the initial slope, A_{sc} is the area under the curve, and S_{du} and S_{au} are the ultimate capacity points.

Finally, for the S_{ds} of each non-null damage state, the respective MIDR value was obtained. Figures 5 and 6 show the deterministic and probabilistic capacity spectrum of the buildings in OA and TG cities, in S_a–S_d and S_a–MIDR formats, together with the respective $DS_{non\text{-}null.}$ Table 6 shows the minimum, mean, and maximum MIDR values from the probabilistic and the deterministic cases for the non-null damage states following the Risk-UE guidelines. The colors in the Tables 6–11 indicate the $DS_{non\text{-}nulls}$ [31]: slight damage (green color); moderate damage (yellow color); extensive damage (orange color) and complete damage (red color) and, are used in the conceptualization the Preventive "Semaphore" for Seismic Damage (PSSD) and fragility curves in the next sections.

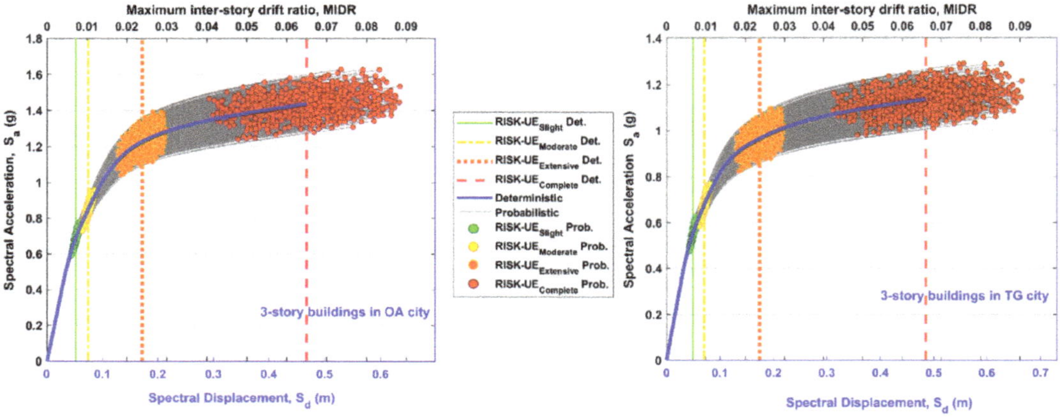

Figure 5. Deterministic and probabilistic capacity spectra of the 3-story buildings in OA and TG cities and their respective $DS_{non\text{-}null}$.

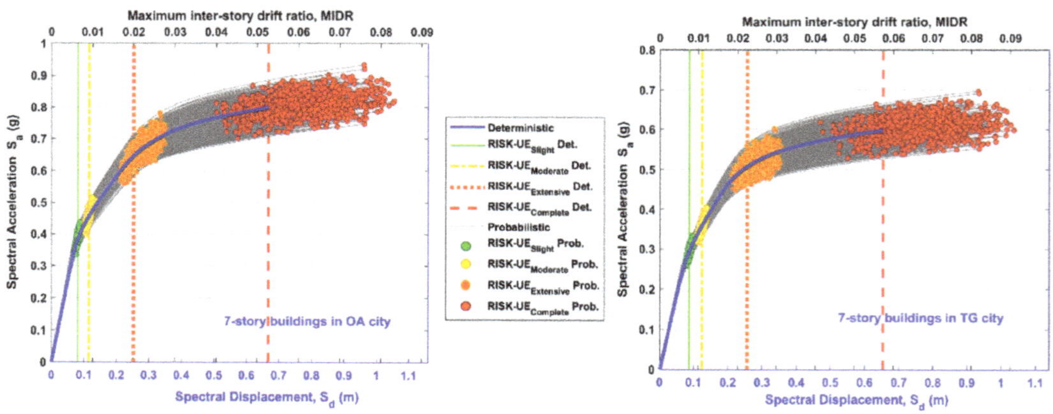

Figure 6. Deterministic and probabilistic capacity spectra of the 7-story buildings in OA and TG cities and their respective $DS_{non\text{-}null}$.

Table 6. Minimum, mean, maximum probabilistic, and deterministic values of the MIDR for each $DS_{non\text{-}null}$ from the Risk-UE guideline in the buildings.

City	Stories	MIDR (DS_{Slight})				MIDR ($DS_{Moderate}$)			
		Min	Max	Mean	Det.	Min	Max	Mean	Det.
Oaxaca (OA)	3	0.0060	0.0083	0.0071	0.0070	0.0085	0.0118	0.0101	0.0102
	7	0.0054	0.0074	0.0063	0.0063	0.0078	0.0106	0.0089	0.0090
Tuxtla Gutiérrez (TG)	3	0.0056	0.0078	0.0067	0.0067	0.0080	0.0112	0.0095	0.0095
	7	0.0066	0.0089	0.0076	0.0076	0.0094	0.0128	0.0108	0.0108
City	**Stories**	**MIDR ($DS_{Extensive}$)**				**MIDR ($DS_{Complete}$)**			
		Min	Max	Mean	Det.	Min	Max	Mean	Det.
Oaxaca (OA)	3	0.0176	0.0298	0.0237	0.0239	0.0406	0.0884	0.0647	0.0650
	7	0.0168	0.0276	0.0219	0.0199	0.0399	0.0832	0.0610	0.0527
Tuxtla Gutiérrez (TG)	3	0.0175	0.0298	0.0237	0.0236	0.0424	0.0905	0.0662	0.0659
	7	0.0184	0.0311	0.0224	0.0247	0.0412	0.0909	0.0664	0.0572

6. T_1 Variation

The initial slope of the capacity spectrum in S_a–S_d format is directly related to the fundamental natural period of vibration, T_1, of the building, through the following equation [37]:

$$S_{dy} = \frac{{T_1}^2}{4\pi^2}(S_{ay} \times g) \quad (2)$$

where g is the gravity of acceleration. Based on the above equation, the new fundamental period T_{1i} for all points (S_{ai} and S_{di}) in the capacity spectrum can be obtained as follows:

$$T_{1i} = \sqrt{\frac{4\pi^2 * S_{di}}{S_{ai} * g}} = 2\pi\sqrt{\frac{S_{di}}{S_{ai} * g}} \quad (3)$$

The representation of the different structural periods in S_a–S_d format can be plotted using diagonal lines. Each line that agrees with the initial slope of the capacity spectrum represents the fundamental period T_1 of the building. In this way, the variation of the T_1 period can be observed from the non-linear zone of the capacity spectrum, which is related to the structural damage of the building. Figure 7 shows the capacity spectrum in S_a–S_d format for the deterministic cases with the different values of period represented with dotted lines.

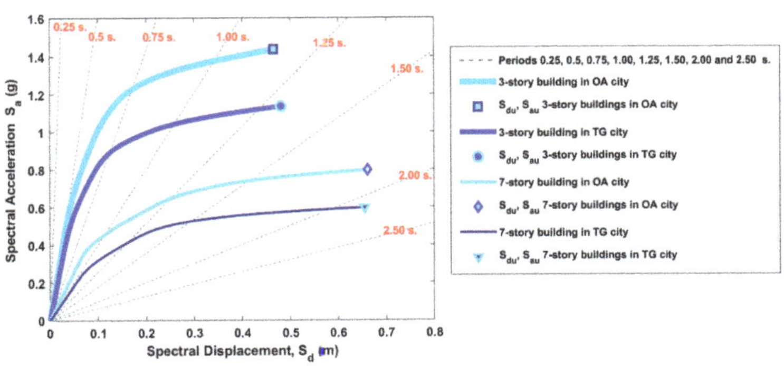

Figure 7. Capacity spectrum of the deterministic cases in OA and TG cities. The dotted lines represent different periods.

Equation (3) was used in both the deterministic and probabilistic capacity spectra for calculating the fundamental period T_{1i} in the buildings. Figures 8 and 9 show the T_{1i}–MIDR relationship curves of the four buildings studied. The deterministic case (vertical lines) and probabilistic points of the four $DS_{non\text{-}null}$ are also shown. The percentage increase in the T_{1i} for the four $DS_{non\text{-}null}$ of the deterministic case is also displayed. Additionally, vertical lines are plotted to indicate the MIDR of the service state (S_{state}) and collapse prevention state (CP_{state}) defined by the Mexican Seismic Design Guide [30]. It was observed that the S_{state} was lower than the deterministic and probabilistic cases of the DS_{Slight}, while the CP_{state} was in the range of the $DS_{Extensive}$. Thus, for the buildings studied here, S_{state} limited the damage correctly, and CP_{state} agreed with the damage expected before collapse.

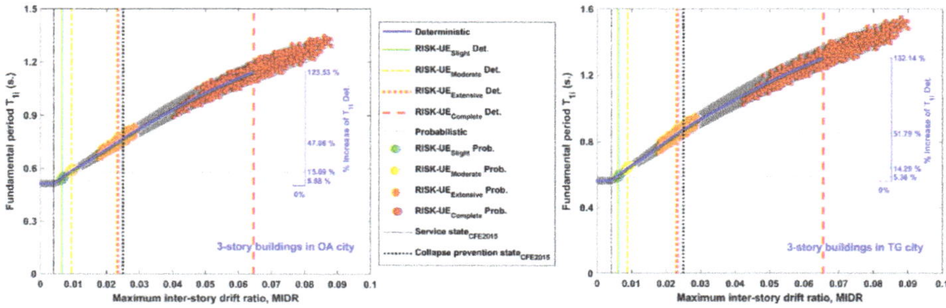

Figure 8. Deterministic and probabilistic T_{1i}–MIDR relationship curves of the 3-story buildings in OA and TG cities and their respective $DS_{non\text{-}null}$.

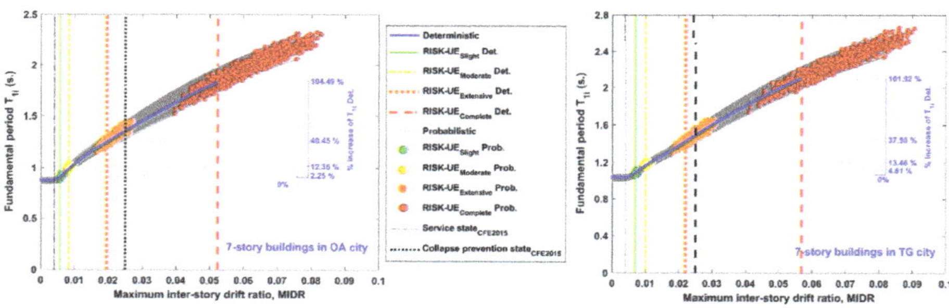

Figure 9. Deterministic and probabilistic T_{1i}–MIDR relationship curves of the 7-story buildings in OA and TG cities and their respective $DS_{non\text{-}null}$.

Table 7 shows the minimum, mean, and maximum values of the fundamental periods T_{1i} from the probabilistic and the deterministic cases for each $DS_{non\text{-}null}$ of the Risk-UE guidelines [31]. It was observed that the deterministic case agreed with the probabilistic mean.

Table 7. Minimum, mean, maximum probabilistic, and deterministic values of the T_{1i} for each $DS_{non\text{-}null}$ from the Risk-UE guideline in the buildings.

City	Stories	T_{1i} (DS_{Slight})				T_{1i} ($DS_{Moderate}$)			
		Min	Max	Mean	Det.	Min	Max	Mean	Det.
Oaxaca (OA)	3	0.52	0.57	0.54	0.54	0.56	0.61	0.59	0.59
	7	0.90	0.96	0.91	0.91	0.95	1.05	1.00	1.00
Tuxtla Gutiérrez (TG)	3	0.57	0.62	0.59	0.59	0.61	0.67	0.64	0.64
	7	1.05	1.14	1.09	1.09	1.13	1.23	1.18	1.18
City	**Stories**	**T_{1i} ($DS_{Extensive}$)**				**T_{1i} ($DS_{Complete}$)**			
		Min	**Max**	**Mean**	**Det.**	**Min**	**Max**	**Mean**	**Det.**
Oaxaca (OA)	3	0.65	0.85	0.75	0.75	0.89	1.35	1.14	1.14
	7	1.14	1.46	1.30	1.25	1.52	2.33	1.96	1.82
Tuxtla Gutiérrez (TG)	3	0.73	0.96	0.85	0.85	1.03	1.55	1.30	1.30
	7	1.30	1.66	1.48	1.43	1.74	2.66	2.24	2.10

7. Preventive "Semaphore" for Seismic Damage

Considering the fundamental period of the building, T_1, as the starting point of the period variation in the capacity spectrum, the percentage increase of the period (% $T_{1i}(DS_{non\text{-}null})$) could be determined for each non-null damage state ($DS_{non\text{-}null}$ = Slight; Moderate; Extensive and Complete) as follows:

$$\%\ T_{1i}\left(DS_{non-null}\right) = \frac{\left[T_{1i}\left(DS_{non-null}\right) - T_1\right]}{T_1} \times 100 \tag{4}$$

The $\%T_{1i}$ ($DS_{non\text{-}null}$) of the deterministic case in the buildings is shown in Figures 8 and 9. Furthermore, the $\%T_{1i}$ ($DS_{non\text{-}null}$) was determined for the probabilistic cases of the buildings. Table 8 shows the obtained values.

Table 8. Minimum, mean, maximum probabilistic, and deterministic value of the $\%T_{1i}$ for each $DS_{non\text{-}null}$ from the Risk-UE guideline in the buildings.

City	Stories	$\%T_{1i}$ (DS_{Slight})				$\%T_{1i}$ ($DS_{Moderate}$)			
		Min	Max	Mean	Det.	Min	Max	Mean	Det.
Oaxaca (OA)	3	3.62	8.10	6.11	5.88	11.14	16.70	14.44	15.69
	7	3.98	7.33	4.52	2.25	9.28	18.49	15.01	12.36
Tuxtla Gutiérrez (TG)	3	3.29	7.24	5.50	5.36	11.12	16.38	14.03	14.29
	7	3.02	8.64	5.96	4.81	10.70	17.41	14.62	13.46
Average		3.48	7.83	5.52	4.57	10.56	17.25	14.53	13.95
City	**Stories**	$\%T_{1i}$ ($DS_{Extensive}$)				$\%T_{1i}$ ($DS_{Complete}$)			
		Min	Max	Mean	Det.	Min	Max	Mean	Det.
Oaxaca (OA)	3	25.94	64.59	45.79	47.06	70.79	162.16	121.65	123.53
	7	29.55	69.48	48.77	40.45	72.00	172.17	124.97	104.49
Tuxtla Gutiérrez (TG)	3	28.65	70.37	50.66	51.79	79.60	174.49	131.76	132.14
	7	24.87	63.89	43.82	37.50	66.55	162.61	118.09	101.92
Average		27.25	67.08	47.26	44.20	72.24	167.86	124.12	115.52

Considering the average values of the $\%T_{1i}$ ($DS_{non\text{-}null}$), from the probabilistic analysis in the four buildings studied, a Preventive "Semaphore" for Seismic Damage (PSSD) was proposed. Table 9 shows the proposal for low-rise and mid-rise steel buildings with a structural system of "Special Moment Frames, SMF". In the PSSD presented, the following analogy between the $DS_{non\text{-}null}$ from the Risk-UE guidelines [31] and the performance levels defined in the Vision 2000 report [45] was proposed:

- Null damage ≈ Operational Limit (OL)
- Slight damage (green color) ≈ Immediate Occupancy (IO)
- Moderate damage (yellow color) ≈ Life Safety (LS)
- Extensive damage (orange color) ≈ Collapse Prevention (CP)
- Complete damage (red color) ≈ Complete Collapse (CC)

Likewise, a criterion of the expected damage or expected operating condition in the buildings is established when an increase in the $\%T_{1i}$ is detected in an SHM assessment.

Considering the relationship between T1i and MIDR (Figures 8 and 9) as a validation parameter of PSSD, it was observed that the green color was consistent with the MIDR of the service state (S_{state}), whereas the orange color was consistent with the MIDR of the collapse prevention state (CP_{state}), both defined by the Mexican Seismic Design Guide [31]. In addition, the range of values presented by the PSSD was in accordance with the Structural Warning System (SWS) proposed in [46,47]. The SWS is a Structural Health Monitoring (SHM) system that has been developed in Mexico to evaluate instrumented buildings of less than 25 stories whose dynamic response is dominated by fundamental modes of vibration [48]. It should be noted that there is a need to perform tests in future studies to verify the accuracy and good engineering practicability of the PSSD in comparison with SHM in steel buildings with the characteristics of those studied here.

Table 9. Preventive "Semaphore" for Seismic Damage (PSSD) based on the average $\%T_{1i}$ ($DS_{non\text{-}null}$) probabilistic of the four buildings studied.

PSSD (Risk-UE Guideline)		**Null Damage**	**Slight Damage**	**Moderate Damage**	**Extensive Damage**	**Complete Damage**
PSSD (Vision 2000 Report)		**Operational Limit (OL)**	**Immediate Occupancy (IO)**	**Life Safety (LS)**	**Collapse Prevention (CP)**	**Complete Collapse (CC)**
Minimum values	%T1i	<3.48%	≥3.48%	≥10.56%	≥27.25%	≥72.24%
Mean values		<5.52%	≥5.52%	≥14.53%	≥47.26%	≥124.12%
Maximum values		<7.83%	≥7.83%	≥17.25%	≥67.08%	≥167.86%

8. Fragility Curves

MIDR and $\%T_{1i}$ clouds from the probabilistic analysis allowed for the development of Fragility Curves (FC) for each of the four damage states. In brief, the fragility curve represents the probability of the $DS_{non\text{-}null}$ being exceeded as a function of the MIDR or $\%T_{1i}$ ($P[DS_{non\text{-}null}/\text{MIDR or } \%T_{1i}]$). The FCs are obtained as follows: (i) each cloud is sorted in ascending order, and the number of points in the cloud is normalized from 0 to 1; and (ii) a Lognormal Cumulative Distribution function (Logncdf) is fitted using the Mean Squared Error (MSE). Then, the Logncdf function with lower MSE is used for fitting. Each Logncdf function corresponds to a fragility curve of each $DS_{non\text{-}null}$ and is completely defined by the μ and σ parameters; μ is the mean value of the MIDR or $\%T_{1i}$ thresholds, and σ represents its standard deviation. Tables 10 and 11 present the m and s parameters obtained for each fragility curve of the studied buildings. Figures 10 and 11 show the fragility curves as a function of the MIDR and $\%T_{1i}$ for each $DS_{non\text{-}null}$ following the Risk-UE guidelines [31].

Table 10. The μ and σ of the fragility curves (FC) as a function of the $\%T_{1i}$ for the studied buildings.

City	**Stories**	**FC_{Slight}**		**$FC_{Moderate}$**		**$FC_{Extensive}$**		**$FC_{Complete}$**	
		μ ($\%T_{1i}$)	σ	μ ($\%T_{1i}$)	σ	μ ($\%T_{1i}$)	σ	μ ($\%T_{1i}$)	σ
Oaxaca (OA)	3	6.13	0.12	14.50	0.06	45.53	0.18	121.58	0.15
	7	4.50	0.25	15.05	0.10	48.31	0.16	124.34	0.16
Tuxtla Gutiérrez (TG)	3	5.53	0.12	14.08	0.06	50.51	0.17	131.16	0.15
	7	5.94	0.16	14.67	0.07	43.42	0.17	117.61	0.15
Average		5.53	0.16	14.58	0.07	46.94	0.17	123.67	0.15

Table 11. The μ and σ of the fragility curves (FC) as a function of the MIDR for the studied buildings.

City	**Stories**	**FC_{Slight}**		**$FC_{Moderate}$**		**$FC_{Extensive}$**		**$FC_{Complete}$**	
		μ (MIDR)	σ	μ (MIDR)	σ	μ (MIDR)	σ	μ (MIDR)	σ
Oaxaca (OA)	3	0.007	0.17	0.010	0.05	0.024	0.12	0.064	0.17
	7	0.006	0.10	0.009	0.07	0.022	0.12	0.061	0.16
Tuxtla Gutiérrez (TG)	3	0.006	0.14	0.010	0.06	0.024	0.12	0.066	0.17
	7	0.008	0.18	0.011	0.06	0.025	0.12	0.066	0.16
Average		0.0068	0.15	0.010	0.06	0.024	0.12	0.064	0.165

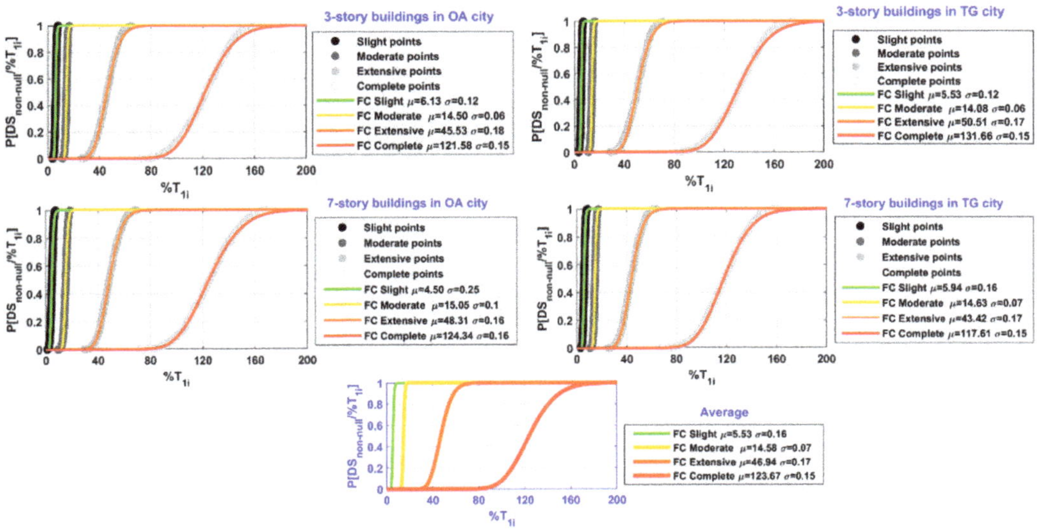

Figure 10. Probabilistic fragility curves of the $DS_{non\text{-}null}$ for the buildings in OA and TG cities as functions of the $\%T_{1i}$.

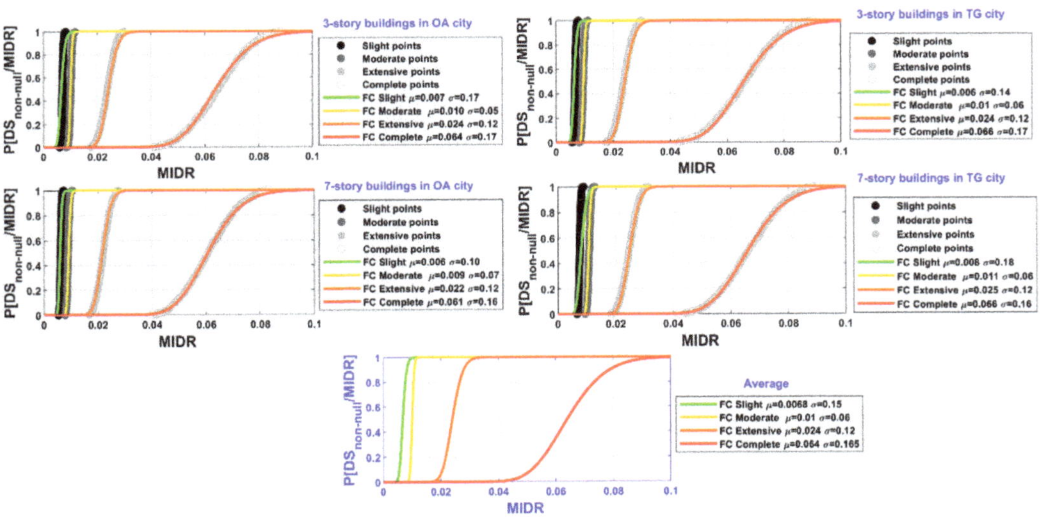

Figure 11. Probabilistic fragility curves of the $DS_{non\text{-}null}$ for the buildings in OA and TG cities as functions of the MIDR.

9. Discussion and Conclusions

This article presents a probabilistic study of the fundamental period (T_1) variation of steel buildings based on seismic damage. One low-rise (3-story) and one mid-rise (7-story) steel building located in two cities in México were studied. The seismic performance of the buildings was obtained through probabilistic nonlinear static analyses. Uncertainties in the yield strength, f_y, modulus of elasticity, E_s, and ultimate strain, ε_u, of the structural sections were considered via Monte Carlo simulation. The T_1 variation was estimated using the capacity spectrum [37], and the seismic damage was defined by the maximum inter-story drift ratio, MIDR, and damage states of the Risk-UE guidelines [31].

In the nonlinear static analysis of the buildings, the seismic actions are not considered as in nonlinear dynamic analysis. However, due to the number of stories and symmetry of the buildings analyzed, the structural response is dominated by their fundamental period of vibration. Thus, the compatibility of results between the static and dynamic approaches is assumed to be adequate. Moreover, the NLSA was performed by implementing a probabilistic approach, which provides a complete perspective of expected seismic performance considering the randomness in the main mechanical properties of the beams and columns of the buildings. In addition, the probabilistic clouds allow the trends and relationships between the variables of interest in the linear and non-linear performance of buildings to be analyzed.

Based on the classic ATC-40 equations, the T_1 variation in the probabilistic capacity spectra of the buildings can be easily obtained and related to its respective MIDR for each $DS_{non\text{-}null}$ of the Risk UE guidelines. As a result of this study, practical tools used for seismic assessment in low-rise and mid-rise steel buildings were proposed: (1) a Preventive "Semaphore" of Seismic Damage (PSSD), and (2) Fragility Curves (FCs). The PSSD proposes a percentage increase of the fundamental period ($\%T_{1i}$) for the four $DS_{non\text{-}null}$ (slight, moderate, extensive, and complete) buildings analyzed here. The PSSD can be helpful as a reference or criteria to determine the health of the building through structural monitoring. Finally, the FC developed are an interesting contribution to determining the probabilities of exceedance of the $DS_{non\text{-}null}$ thresholds. These FC have a novel approach based on MIDR or $\%T_{1i}$ for seismic action in low-rise and mid-rise steel buildings with SMF structural systems. It should be noted that in order to spread this methodology, it would be necessary to carry out analyses with different structural typologies and number of stories and compare the results with on-site measurements.

Author Contributions: Conceptualization, methodology, software, validation, and formal analysis, S.A.D., L.A.P. and Y.F.V.-A. Investigation, resources, and data curation, R.S.M.-O. Writing—original draft preparation, S.A.D. Writing—review and editing, L.A.P. and Y.F.V.-A. Visualization, supervision, and project administration, S.A.D., L.A.P. and R.S.M.-O. Funding acquisition, L.A.P. All authors have read and agreed to the published version of the manuscript.

Funding: This study was funded by the Secretaría Nacional de Ciencia, Tecnología e Innovación (SENACYT) of Panama.

Data Availability Statement: Data are contained within the article.

Acknowledgments: L.A.P. would like to thank SENACYT and SNI for their support. S.A.D. and R.S.M.-O. thank the Academic Group: Risk Assessment and Sustainability of Civil Works (UJAT-CA-287) in the Universidad Juárez Autónoma de Tabasco (UJAT) and SNI-CONACYT in México for their support, and they thank Erick Garcia and Rosa Estrada, civil engineering students from the professional practices program at UJAT.

Conflicts of Interest: The authors declare no conflict of interest.

References

1. Aguilar-Meléndez, A.; Pujades, L.G.; Barbat, A.H.; Ordaz, M.G.; De la Puente, J.; Lantada, N.; Rodríguez-Lozoya, H.E. A probabilistic approach for seismic risk assessment based on vulnerability functions. Application to Barcelona. *Bull. Earthq. Eng.* **2019**, *17*, 1863–1890. [CrossRef]
2. Romero, D.Z.; Akbas, B.; Budiman, J.; Shen, J. Consideration of economic vulnerability in seismic performance evaluation of structures. *Bull. Earthq. Eng.* **2020**, *18*, 3351–3381. [CrossRef]
3. Ramírez-Eudave, R.; Ferreira, T.M.; Vicente, R. Parameter-based seismic vulnerability assessment of Mexican historical buildings: Insights, suitability, and uncertainty treatment. *Int. J. Disaster Risk Reduct.* **2022**, *74*, 102909. [CrossRef]
4. Ferreira, T.; Rodrigues, H. *Seismic Vulnerability Assessment of Civil Engineering Structures at Multiple Scales. From Single Buildings to Large-Scale Assessment*; A volume in Woodhead Publishing Series in Civil and Structural Engineering; Woodhead Publishing: Sawston, UK, 2022; pp. 1–383. [CrossRef]
5. Li, S.Q.; Liu, H.B. Vulnerability prediction model of typical structures considering empirical seismic damage observation data. *Bull. Earthq. Eng.* **2022**, *20*, 5161–5203. [CrossRef]

6. Barbat, A.H.; Carreño, M.L.; Pujades, L.G.; Lantada, N.; Cardona, O.D.; Marulanda, M.C. Seismic vulnerability and risk evaluation methods for urban areas. A review with application to a pilot area. *Struct. Infrastruct. Eng.* **2010**, *6*, 17–38. [CrossRef]
7. Díaz, S.A.; Pujades, L.G.; Barbat, A.H.; Hidalgo-Leiva, D.A.; Vargas, Y.F. Capacity, damage and fragility models for steel buildings. A probabilistic approach. *Bull. Earthq. Eng.* **2018**, *16*, 1209–1243. [CrossRef]
8. Roghaei, M.; Zabihollah, A. An Efficient and Reliable Structural Health Monitoring System for Buildings after Earthquake. *APCBEE Procedia* **2014**, *9*, 309–316. [CrossRef]
9. Fujino, Y.; Siringoringo, D.M.; Ikeda, Y.; Nagayama, T.; Mizutani, T. Research and Imple-mentations of Structural Monitoring for Bridges and Buildings in Japan. *Engineering* **2019**, *5*, 1093–1119. [CrossRef]
10. Alva, R.E.; Pujades, L.G.; González-Drigo, R.; Luzi, G.; Caselles, O.; Pinzón, L.A. Dynamic Monitoring of a Mid-Rise Building by Real-Aperture Radar Interferometer: Advantages and Limitations. *Remote Sens.* **2020**, *12*, 1025. [CrossRef]
11. Gopinath, V.K.; Ramadoss, R. Review on structural health monitoring for restoration of heritage buildings. *Mater. Today Proc.* **2021**, *43*, 1534–1538. [CrossRef]
12. Vargas-Alzate, Y.F.; Pujades, L.G.; Barbat, A.H.; Hurtado, J.E. An efficient methodology to estimate probabilistic seismic damage curves. *J. Struct. Eng. ASCE* **2019**, *145*, 04019010. [CrossRef]
13. Vamvatsikos, D.; Fragiadakis, M. Incremental dynamic analysis for estimating seismic performance sensitivity and uncertainty. *Earthq. Eng. Struct. Dyn.* **2010**, *39*, 141–163. [CrossRef]
14. Kazantzi, A.K.; Vamvatsikos, D.; Lignos, D.G. Seismic performance of a steel moment-resisting frame subject to strength and ductility uncertainty. *Eng. Struct.* **2014**, *78*, 69–77. [CrossRef]
15. Díaz, S.A.; Pujades, L.G.; Barbat, A.H.; Vargas, Y.F.; Hidalgo-Leiva, D.A. Energy damage index based on capacity and response spectra. *Eng. Struct.* **2017**, *152*, 424–436. [CrossRef]
16. Vargas-Alzate, Y.F.; Lantada, N.; González-Drigo, R.; Pujades, L.G. Seismic Risk Assessment Using Stochastic Nonlinear Models. *Sustainability* **2020**, *12*, 1308. [CrossRef]
17. Pinzón, L.A.; Vargas-Alzate, Y.F.; Pujades, L.G.; Diaz, S.A. A drift-correlated ground motion intensity measure: Application to steel frame buildings. *Soil Dyn. Earthq. Eng.* **2020**, *132*, 106096. [CrossRef]
18. Wang, F.; Shi, Q.X.; Wang, P. Research on the Physical Inter-story Drift Ratio and the Damage Evaluation of RC Shear Wall Structures. *KSCE J. Civ. Eng.* **2021**, *25*, 2121–2133. [CrossRef]
19. Kenari, M.S.; Celikag, M. Correlation of Ground Motion Intensity Measures and Seismic Damage Indices of Masonry-Infilled Steel Frames. *Arab. J. Sci. Eng.* **2019**, *44*, 5131–5150. [CrossRef]
20. Aljawhari, K.; Gentile, R.; Freddi, F.; Galasso, C. Effects of ground-motion sequences on fragility and vulnerability of case-study reinforced concrete frames. *Bull. Earthq. Eng.* **2021**, *1*, 6329–6359. [CrossRef]
21. Vargas-Alzate, Y.F.; Hurtado, J.E.; Pujades, L.G. New insights into the relationship between seismic intensity measures and nonlinear structural response. *Bull. Earthq. Eng.* **2022**, *20*, 2329–2365. [CrossRef]
22. Masi, A.; Vona, M. Experimental and numerical evaluation of the fundamental period of undamaged and damaged RC framed buildings. *Bull. Earthq. Eng.* **2010**, *8*, 643–656. [CrossRef]
23. Anastasia, K.E.; Athanasios, I.K. Correlation of Structural Seismic Damage with Fundamental Period of RC Buildings. *Open J. Civ. Eng.* **2013**, *3*, 5–67. [CrossRef]
24. Ditommaso, R.; Vona, M.; Gallipoli, M.R.; Mucciarelli, M. Evaluation and considerations about fundamental periods of damaged reinforced concrete buildings. *Nat. Hazards Earth Syst. Sci.* **2013**, *13*, 1903–1912. [CrossRef]
25. Aninthaneni, P.V.; Dhakal, R.P. Prediction of Fundamental Period of Regular Frame Buildings. *Bull. N. Z. Soc. Earthq. Eng.* **2016**, *49*, 175–189. [CrossRef]
26. Lu, W.-T.; Phillips, B.M. A fundamental-period-preserving seismic retrofit methodology for low-rise buildings with supplemental inerters. *Eng. Struct.* **2022**, *266*, 114583. [CrossRef]
27. Sarma, S.; Sundar Das, T.; Bora, A.; Bharadwaj, K. Parametric Study on the Variation of Time Period of RC MRF Buildings. In *Recent Advances in Earthquake Engineering*; Lecture Notes in Civil Engineering; Kolathayar, S., Chian, S.C., Eds.; Springer: Berlin/Heidelberg, Germany, 2022; Volume 175, pp. 415–426. [CrossRef]
28. Harris, J.L.; Michel, J.J.L. Approximate Fundamental Period for Seismic Design of Steel Buildings Assigned to High-Risk Categories. *Pract. Period. Struct. Des. Constr.* **2019**, *24*, 04019023. [CrossRef]
29. Gallipoli, M.R.; Stabile, T.A.; Guéguen, P.; Mucciarelli, M.; Comelli, P.; Bertoni, M. Fundamental period elongation of a RC building during the Pollino seismic swarm sequence. *Case Stud. Struct. Eng.* **2016**, *6*, 45–52. [CrossRef]
30. Comisión Nacional de Electricidad (CFE). MDOC-CFE. Manual de Diseño de Obras Civiles. Diseño por Sismos. 2015, pp. 1–745. Available online: https://www.gob.mx/ineel/articulos/manual-de-diseno-de-obras-civiles-diseno-por-sismo-logro-de-la-ingenieria-de-mexico (accessed on 3 April 2023).
31. Milutinovic, Z.V.; Trendafiloski, G.S. Risk-UE An Advanced Approach to Earthquake Risk Scenarios with Applications to Different European Towns. 2003. Available online: https://cordis.europa.eu/project/id/EVK4-CT-2000-00014 (accessed on 3 April 2023).
32. Arcos-Díaz, D.; Díaz, S.A.; Pinzón, L.A.; Jesús, H.; Mora-Ortiz, R.S. Seismic performance assessment based on the interstory drift of steel buildings. *Lat. Am. J. Solids Struct.* **2022**, *19*, e431. [CrossRef]

33. Gaceta Oficial del Gobierno de México. NTCDS-RCDF. Norma Técnica Complementaria para la Revisión de la Seguridad Estructural de las Edificaciones de la Ciudad de México. 2017, pp. 1–712. Available online: https://smie.com.mx/smie-2022/informacion-tecnica/normas-tecnicas-complementarias.php (accessed on 3 April 2023).
34. American Institute of Steel Construction. ANSI/AISC 360-16. Specification for Structural Steel Buildings. 2016. Available online: https://www.aisc.org/Specification-for-Structural-Steel-Buildings-ANSIAISC-360-16-Download (accessed on 3 April 2023).
35. Seismosoft. SeismoStruct. Civil Engineering Software for Structural Assessment and Structural Retrofitting. 2021. Available online: https://seismosoft.com/products/seismostruct/ (accessed on 3 April 2023).
36. Fernández, R.; Yamin, L.; D'Ayala, D.; Adhikari, R.; Reyes, J.C.; Juan Echeverry, J.; Fuentes, G. A simplified component-based methodology for the seismic vulnerability assessment of school buildings using nonlinear static procedures: Application to RC school buildings. *Bull. Earthq. Eng.* **2022**, *20*, 6555–6585. [CrossRef]
37. Applied Technology Council. ATC-40. Seismic Evaluation and Retrofit of Concrete Buildings. 1996. Available online: https://www.atcouncil.org/pdfs/atc40toc.pdf (accessed on 3 April 2023).
38. Pujades, L.G.; Vargas-Alzate, Y.F.; Barbat, A.H.; González-Drigo, J.R. Parametric model for capacity curves. *Bull. Earthq Eng.* **2015**, *13*, 1347–1376. [CrossRef]
39. Jalayer, F.; De Risi, R.; Manfredi, G. Bayesian Cloud Analysis: Efficient structural fragility assessment using linear regression. *Bull. Earthq. Eng.* **2015**, *13*, 1183–1203. [CrossRef]
40. Hurtado, J.E.; Barbat, A.H. Monte Carlo techniques in computational stochastic mechanics. *Arch. Comput. Methods Eng.* **1998**, *5*, 3–29. [CrossRef]
41. Rubinstein, R.Y.; Kroese, D.P. *Simulation and the Monte Carlo Method*, 3rd ed.; John Wiley & Sons, Inc.: New York, NY, USA, 2016; pp. 1–414. Available online: https://onlinelibrary.wiley.com/doi/book/10.1002/9781118631980 (accessed on 3 April 2023).
42. Schmidt, B.J.; Bartlett, F.M. Review of resistance factor for steel: Data collection. *Can. J. Civ. Eng.* **2002**, *29*, 98–108. [CrossRef]
43. Bartlett, F.M.; Dexter, R.J.; Graeser, M.D.; Jelinek, J.J.; Schmidt, B.J.; Galambos, T.V. Updating standard shape material properties database for design and reliability. *ASCI Eng. J.* **2003**, *40*, 1–14. Available online: https://www.aisc.org/Updating-Standard-Shape-Material-Properties-Database-for-Design-and-Reliability (accessed on 3 April 2023).
44. Iman, R.L. Appendix A: Latin hypercube sampling. In *Encyclopedia of Statistical Sciences*; Wiley: New York, NY, USA, 1999; Volume 3, pp. 408–411.
45. Structural Engineers Association of California. SEAOC. Vision 2000: Performance Based Seismic Engineering of Buildings. 1995. Available online: https://www.seaoc.org/store/ViewProduct.aspx?id=11238558 (accessed on 3 April 2023).
46. Aldama, B.D. Proceso Automatizado Para Determinar el Estado Estructural en Edificios Instrumentados. Master's Thesis, Universidad Nacional Autónoma de México (UNAM), Ciudad de México, México, 2009. (In Spanish).
47. Murià-Vila, D.; Aldama, B.D.; Loera, S. Structural warning for instrumented buildings. In Proceedings of the 14th European Conference on Earthquake Engineering, Ohrid, North Macedonia, 30 August–3 September 2010; Volume 2. Available online: https://www.tib.eu/en/search/id/TIBKAT:667434461/14th-European-Conference-on-Earthquake-Engineering?cHash=1727db18b7fdcc71af94e1729c032324 (accessed on 3 April 2023).
48. Murià-Vila, D.; Aldama-Sánchez, B.D.; García-Illescas, M.Á.; Rodríguez Gutiérrez, G. Monitoring of a rehabilitated building in soft soil in Mexico and structural response to the September 2017 earthquakes: Part 1: Structural health monitoring system. *Earthq. Spectra* **2021**, *37*, 2737–2766. [CrossRef]

buildings

MDPI

Article

Experimental Investigation of the Tensile Properties with Bending of CFRP Tendons in Suspension Bridges

Lijun Jia [1], Wenchao Zhang [1], Jiawei Xu [1] and Yang Jiang [2,*]

1 Department of Bridge Engineering, College of Civil Engineering, Tongji University, Shanghai 200092, China
2 Shanghai Municipal Engineering Design Institute (Group) Co., Ltd., Shanghai 200092, China
* Correspondence: 2011510@tongji.edu.cn; Tel.: +86-18019293757

Abstract: Carbon-fiber-reinforced polymer (CFRP) has gradually become a new material to replace traditional steel due to its outstanding advantages. Because of its poor transverse stress performance, there is a reduction effect on the tensile strength in the bending state. To study the mechanical properties of CFRP tendons subjected to combined tension and bending at the saddle of a suspension bridge, a series of bond-type anchorages were made. Specimens with different diameters of CFRP tendons were tensioned on the device with different bending radius saddles. The test results revealed that the tensile properties were significantly affected by the severity of the bending of the CFRP tendons, including the failure mode, fracture force, and stress distribution. The highest reduction in fracture force was found at the bending radius of 3 m, of up to 38.05%. Furthermore, the tensile properties were also found to be influenced by the diameter of CFRP tendons. It was found that increasing the bending radius was more conducive to improving the performance of CFRP tendons with a smaller diameter. When the bending radius increased from 3 to 12 m, the efficiency coefficient (the ratio of the fracture force to the ultimate force) of D8, D10, and D14 increased by 11.21%, 7.74%, and 2.26%, respectively. Decreasing the bending radius leads to unevenness of the stress distribution and increasing the diameter of the CFRP tendon leads to brittleness and difficulties in anchoring, thus resulting in the decrease in the efficiency coefficient. In addition, the ratio of the bending radius to the tendon diameter was less than 2.4, the efficiency coefficient of the specimen was less than 80%, and the specimen mostly suffered shear failure. Furthermore, the finite element (FE) models validated by the test results were used to reveal the stress state and study the effect of contact friction on the properties of CFRP tendons. The FE results show that the CFRP tendons with a smaller bending radius presented higher shear stress concentrations. As the contact friction increased, the load-bearing capacity of CFRP tendons decreased significantly.

Keywords: carbon-fiber-reinforced polymer (CFRP) tendons; tensile properties with bending; efficiency coefficient; contact friction; finite element (FE) analysis; suspension bridges

Citation: Jia, L.; Zhang, W.; Xu, J.; Jiang, Y. Experimental Investigation of the Tensile Properties with Bending of CFRP Tendons in Suspension Bridges. *Buildings* **2023**, *13*, 988. https://doi.org/10.3390/buildings13040988

Academic Editors: Junjie Zeng and Gianfranco De Matteis

Received: 13 January 2023
Revised: 24 February 2023
Accepted: 6 April 2023
Published: 8 April 2023

1. Introduction

Carbon-fiber-reinforced polymer (CFRP) composites have gained popularity in bridge engineering [1–3]. As the span of suspension bridges increases continuously, the ratio of the self-weight stress of the main cable to the allowable stress increases, limiting the load-bearing efficiency and economic benefits [4,5]. With outstanding advantages of high strength, high fatigue resistance, corrosion resistance, and light weight, CFRP composites are considered to be substitutes for steel cable in cable structures [6,7]. In addition, the cable-anchored structure in suspension bridges provides a service in water environments and a reciprocating load in actual engineering, and scholars have conducted related research, which shows that the composite material has excellent characteristics in a complex environment [8,9]. Many studies have explored the feasibility of applying CFRP cables to extremely long-span bridges [10,11]. When CFRP tendons are employed as the main cables in suspension bridges, the tendons in service may be used in the bending area. In suspension bridges, local structures, such as the main cable saddle and cable distribution saddle, cause main cable tension, bending, lateral extrusion, and friction with the saddle. Thus, the stress distribution of the main cable in the bending region is different from that at other locations. Furthermore, considering the anisotropy of CFRP composites, the shear strength of CFRP composites is far below the tensile strength [12]. It is critical to conduct corresponding studies on the mechanical properties of CFRP tendons under combined tension and bending.

The flexural strength of CFRP composites has been previously studied, and it was considered a significant concern [13]. Some scholars have experimentally studied the performance of the reinforced members with composite materials and have also conducted parametric studies of the reinforced members by means of FEM [14,15]. In addition, bonding of CFRP composites to the tension face of a beam has also become a significant flexural strengthening method [16–18]. However, these studies were conducted on CFRP straps and CFRP laminates, and they mainly focused on CFRP composites applied in prestressed and reinforced concrete structures. Limited studies have been carried out on the performance of CFRP tendons subjected to combined tension and bending applied in cable structures. In addition, these studies of CFRP composites focused on composite bending, which is different from the study of CFRP tendons under combined tension and bending at the saddle of suspension bridges.

Some research programs have been conducted to investigate the flexural behavior of the CFRP tendon and CFRP cables. Studies showed that CFRP cables exhibited excellent tensile properties, but bending over a small radius could not achieve the performance of wire rope [19,20]. Han et al. [21] studied the effect of diameter on transverse mechanical properties of the transverse enhanced CFRP tendon. The results showed that increasing the diameter of the CFRP tendon reduced the transverse mechanical properties. Arczewska et al. [22] investigated the correlation between tensile strength due to bending and direct tensile strength. Fang et al. [23] studied the mechanical behavior of CFRP wires subjected to combined tension and bending through transverse load tests. The test results showed that the failure of the wires was caused by a fiber tensile fracture at the loading position. The same conclusion was drawn by Cai et al., namely, that interfacial failure led to a reduction in the ultimate strength of unidirectional CFRP composites under transverse tensile loading [24]. Additionally, CFRP tendons in service inevitably contact the saddle, which may affect the ultimate strength of CFRP tendons [25–27]. However, limited studies have been conducted on the performance of CFRP tendons under combined tension and bending considering friction. Furthermore, shear forces from the hanger ropes may threaten the main cable fabricated by CFRP tendons, adversely affecting the main cable. Liu et al. [28] carried out tests to study the performance of the main cable fabricated by CFRP wires at the cable clamp, and the test results showed that the bending efficiency of the CFRP wire was high, with an average of 96.0%. However, the test results did not consider the effect of transverse loads. To date, some research programs have been conducted to investigate the flexural behavior of FRP composites through transverse load tests [29,30], which can provide references for the research on the effects of shear loads on main cables.

The mechanical performance of the main cable fabricated by CFRP tendons is complicated at saddle locations. The literature above mainly focused on the load effect on the mechanical performance of CFRP cables. The information about CFRP tendons at saddle locations subjected to combined tension and bending is relatively limited, especially with friction. Moreover, the fundamental research on the CFRP tendon will also provide references for the tensile and bending properties of the main cable fabricated by CFRP tendons. Additionally, Xiang et al. [31] performed drop-weight tests on eight CFRP cable specimens. The results indicated that the complete strand and a single CFRP wire presented similar longitudinal and transverse ultimate loads and deformation capacity. Therefore, it is essential to conduct corresponding experimental research on an individual CFRP tendon to further study the mechanical performance of CFRP tendons at saddle locations.

Based on the preceding statements, this study was conducted to investigate the tensile and bending mechanical properties of CFRP tendons at saddle locations. Static load tests were performed to investigate the fracture force, failure mode, and load–strain relationship of CFRP tendons with different diameters and bending radii. A corresponding finite element (FE) model was developed based on ANSYS software and validated by the test results. The stress distributions of the CFRP tendon were then obtained, and parametric FE analyses were conducted to study the effect of the friction at the CFRP and saddle interface on the CFRP tendon behavior. Based on the tests and FE results, the failure mechanisms and mechanical properties of the CFRP tendon subjected to combined tension and bending could be analyzed. This research can help promote the widespread application of the main cable fabricated by CFRP tendons in suspension bridges.

2. Experimental Programs

2.1. Materials

The steel pipe material examined was No. 45 steel based on GB/T 699-2015, and its mechanical properties are shown in Table 1. The embossed CFRP tendons were used to obtain superior anchorage performance due to the significant friction and mechanical engagement with the bonding medium [32]. The mechanical properties of CFRP tendons provided by the manufacturer are typical values, as shown in Table 2. The hardness and mechanical properties of epoxy resin are similar to those of the CFRP tendon and resin matrix, respectively [33]. Therefore, epoxy resin, which exhibited superior anchoring performance, was used for CFRP tendon anchoring in the tests. According to the manufacturer, the elastic modulus, the average bond strength, and the compressive strength of the used epoxy resin are 2.6 GPa, 20.9 MPa, and 106.6 MPa, respectively. Both the curing temperature and working temperature of epoxy resin are room temperature, and epoxy resin was cured for three days.

Table 1. Mechanical properties of steel pipe.

Material	Elastic Modulus (GPa)	Tensile Strength (MPa)	Yield Strength (MPa)	Elongation (%)
No.45 steel	210	600	355	16

Table 2. Material properties of CFRP tendon.

X_τ (MPa)	X_C (MPa)	Y_τ (MPa)	Y_C (MPa)	S_L (MPa)	S_τ (MPa)	E_X (GPa)	E_y (GPa)	G_{Xy} (GPa)	G_{XZ} (GPa)	G_{yZ} (GPa)	v_{XY}	v_{YZ}
2300	1440	57	228	71	12	150	10.5	6.2	6.2	7.1	0.27	0.02

Note: The fiber longitudinal direction was defined as material orientation X, and the other two directions perpendicular to the fiber direction were defined as material orientations Y and Z; X_τ = tensile strength along the fiber direction; X_C = compressive strength along the fiber direction; Y_τ = tensile strength perpendicular to the fiber direction; Y_C = compressive strength perpendicular to the fiber direction; S_L = shear strength along the fiber direction; S_τ = shear strength perpendicular to the fiber direction; E_X = elastic modulus along the fiber direction; E_y = elastic modulus perpendicular to the fiber direction; G_{Xy}, G_{XZ}, G_{yZ} = shear modulus; and $v_{XY} = v_{XZ}$, v_{YZ} = Poisson's ratio.

2.2. Specimens and Preparation

A total of 48 bond-type anchorage specimens with different diameters of CFRP tendons were prepared. The bond-type anchorage composed of steel pipe, CFRP tendon, and epoxy resin was used, as shown in Figure 1. The nominal diameters of CFRP tendons were 5, 8, 10, and 14 mm, as shown in Figure 2. In the preparation process of specimens, the anchoring position corresponding to the steel pipe and the tendon was firstly determined and marked. After the tendon was positioned, one end of the steel pipe was sealed, and the other end was poured with epoxy resin. The steel pipe oscillated continuously during pouring to ensure the bonding and tightness between the CFRP tendon and epoxy resin. Then the tendon was inserted from one end and stopped at the mark. The plug at the other end was sealed when the CFRP tendon was installed at the marked position. Finally, the strain gauges were attached at the designated position (relative to the external radius of the tendon). Figure 3 shows more details of strain gauges. Figure 4 shows the flow chart of the whole preparation process.

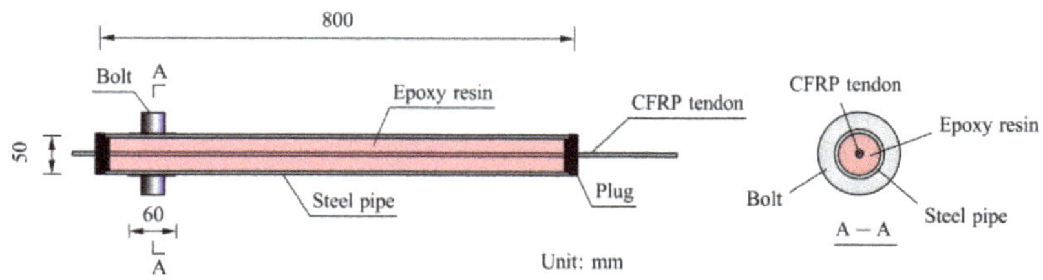

Figure 1. Bond-type anchorage (mm).

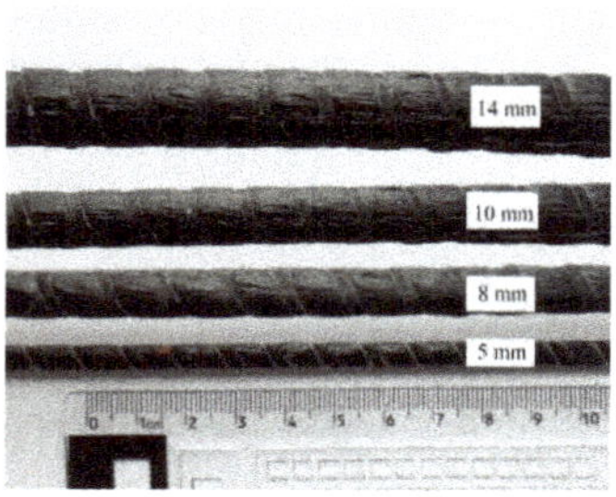

Figure 2. CFRP tendon (mm).

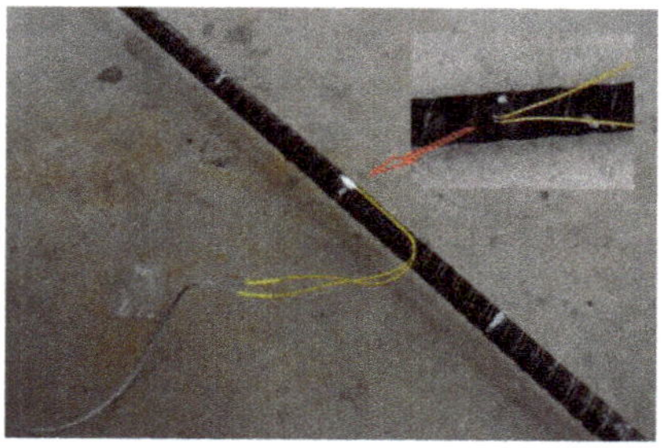

Figure 3. Details of strain gauges.

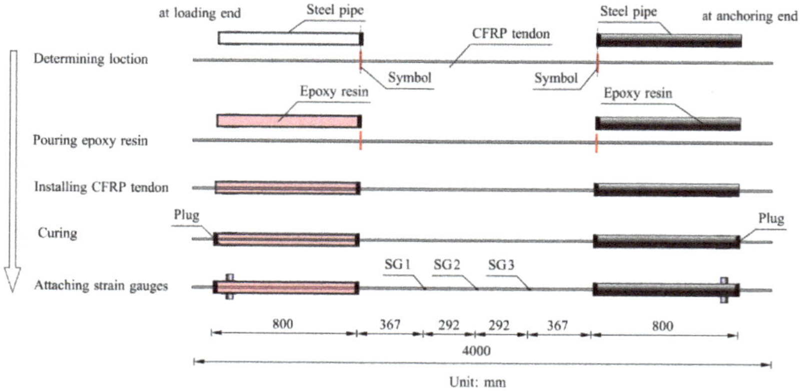

Figure 4. The preparation process of specimens (mm).

2.3. Test Setup and Procedure

The specimens were used in the static load tests conducted on the concrete device, as shown in Figure 5. The concrete loading device was divided into the outer protective frame and curved saddle, and its material was C40 concrete reinforcement. The curved saddle was placed in the middle of the outer protective frame. There were three saddles with bending radii of 3, 8, and 12 m. In addition, a set of unbent specimens were set up for comparison. The parameters of the saddles were selected based on a practical suspension bridge. Three slotted holes were opened on both sides of the outer protective frame for CFRP tendon anchoring channels. The corners of the three saddles were chamfered to prevent the CFRP tendon from being affected by local stress concentrations during loading.

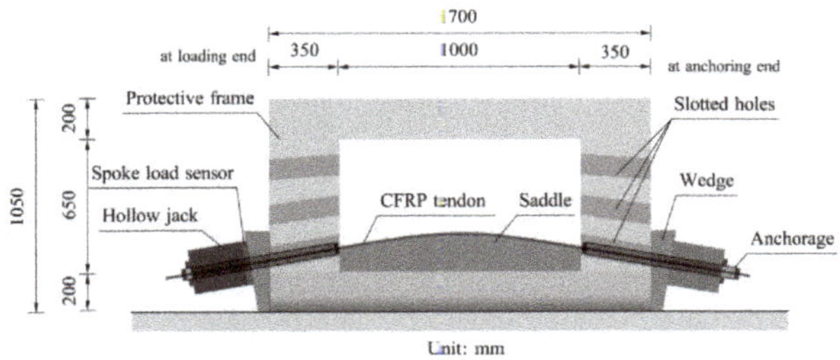

(a)

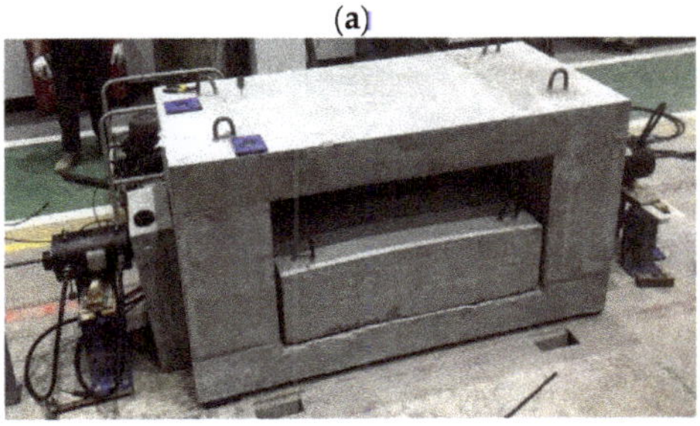

(b)

Figure 5. Test loading device: (**a**) schematic; (**b**) photograph (mm).

The test used one hollow jack with a range of 600 kN, a maximum tension stroke of 150 mm, and one spoke load sensor with a measuring range of 1600 kN. Before the tensile tests, a preloading process with 5% of the ultimate tensile force was conducted. The loading speed was based on JC/T 2404-2017 [34], and each level was loading for 30 s. The fracture force, slippage of CFRP tendons, and strain in the free lengths of CFRP tendons were measured. Meanwhile, different failure modes of specimens were observed.

3. Results and Discussions

3.1. General

Loading applied to all specimens was continuous until specimens failed. These specimens showed no apparent changes at the early loading stage. With the increase in load, the intermittent sound of fibers was heard. Then, the specimens failed with a thud sound, and the anchorage and jack were ejected from the test device together. In order to analyze the causes of the failure modes of the specimens, the typical failure figure of the specimens was compared with the finite elements and shown in the finite element part. An efficiency coefficient η was introduced to evaluate the tensile properties of CFRP tendons in the bending state. The efficiency coefficient [Equation (1)] was used to evaluate the reduction effect of bending on the ultimate tensile force of the CFRP tendon, where F is average fracture force and F_{cu} is ultimate tensile force. Table 3 lists the results for each group of specimens in the static load tests.

$$\eta = \frac{F}{F_{cu}} \tag{1}$$

Table 3. Results of testing.

R (m)	D (mm)	Group	F_1 (kN)	F_{cu} (kN)	η_1 (%)
3	5	R3D5	35.12	45.16	77.78
	8	R3D8	69.76	115.61	60.34
	10	R3D10	104.79	180.64	58.01
	14	R3D14	184.03	354.06	51.98
8	5	R8D5	33.12	45.16	73.33
	8	R8D8	75.74	115.61	65.52
	10	R8D10	111.78	180.64	61.88
	14	R8D14	187.03	354.06	52.82
12	5	R12D5	36.13	45.16	80.00
	8	R12D8	82.72	115.61	71.55
	10	R12D10	118.76	180.64	65.75
	14	R12D14	192.03	354.06	54.24
0	5	R0D5	46.16	45.16	102.22
	8	R0D8	112.62	115.61	97.41
	10	R0D10	154.69	180.64	85.64
	14	R0D14	227.04	354.06	64.12

Note: R = bending radius of saddle; D = diameter of CFRP tendon; F_1 = average fracture force of CFRP tendon by testing; F_{cu} = ultimate tensile force of CFRP tendon; η_1 = efficiency coefficient by testing = F_1/F_{cu}.

3.2. Fracture Force of the Specimens

Figure 6 shows the average fracture force histogram of these specimens. As seen in the figures, when compared with the unbent specimens R0, the fracture force of the specimens with the bending radius of 3 m decreased by 23.91%, 38.05%, 32.26%, and 18.94%, respectively. In addition, the fracture force of CFRP tendons with the same diameter increased as the bending radius increased. The test results suggest that the fracture force of CFRP tendons was significantly reduced due to bending. In addition, the fracture force of R3D5 (35 kN) exceeded that of R8D5 (33 kN) and R12D5 (36 kN). The abnormal data can be interpreted as follows: the failure position was related to material defect, loading position, stress concentrations, and steel pipe port effect; the ultimate tensile force of the CFRP tendon with a diameter of 5 mm was low, and it had large discreteness.

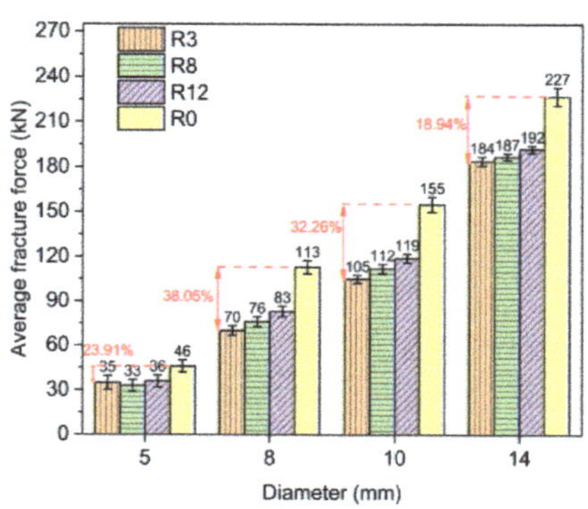

Figure 6. Average fracture force histogram of specimens.

3.3. The Ratio of the Bending Radius to the Tendon Diameter

Figure 7 shows the relationship between the efficiency coefficient and the ratio of the bending radius to the tendon diameter. It can be seen that the efficiency coefficient is linearly related to the ratio of bending radius to the tendon diameter, except the slipping failure of the specimens with a 14 mm CFRP tendon. Moreover, combined with the failure modes of the specimens, it was observed that the ratio of the bending radius to the tendon diameter was below 2.4, the efficiency coefficient was less than 80%, and the failure modes of the specimens were mostly shear failure. Therefore, it is suggested that the ratio of the bending radius to the tendon diameter should be more than 2.4 in engineering practice.

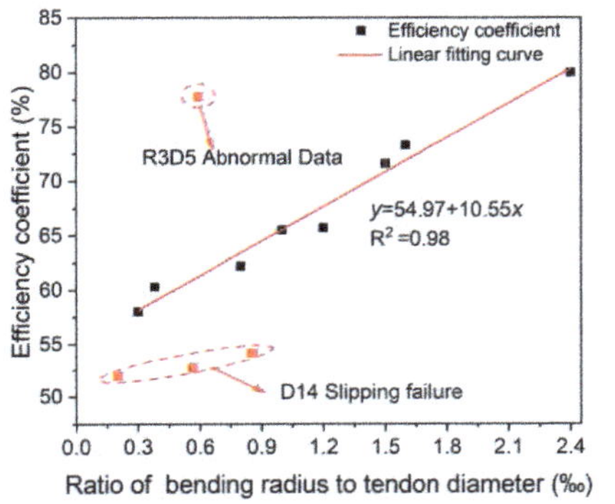

Figure 7. Relationship between the efficiency coefficient and the ratio of the bending radius to the tendon diameter.

4. Numerical Simulation

4.1. General

A finite element (FE) model was established using ANSYS version 18.2 mechanical software to validate the test results and extend the research. Figure 8 presents the outline of the simulation model. The critical dimensions and parameters were determined from the experimental tests. The failure criterion for CFRP tendons uses the Tsai–Wu failure criterion. According to the structural and mechanical characteristics of the test device, the contact interface between the CFRP tendon and saddle should be addressed in numerical simulation. Three features of the interface can be applied: (1) impenetrability constraints are set between surfaces; (2) the contact interface can transmit normal pressure and tangential friction; (3) normal tension can hardly be transmitted. The stress distributions and the effects of the contact interface on the mechanical properties of CFRP tendons were studied by FE analysis.

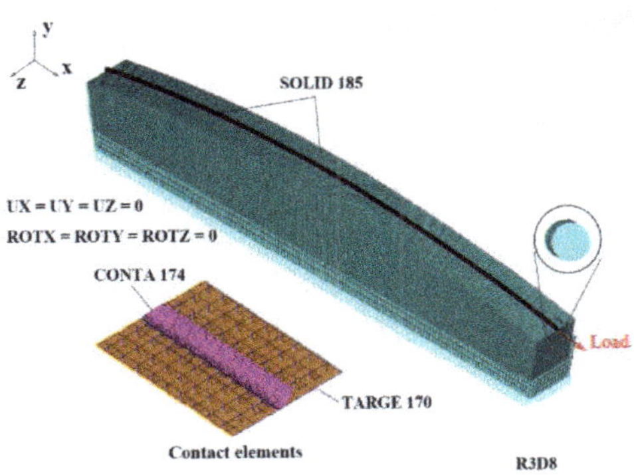

Figure 8. FE model of CFRP tendon and concrete saddle.

4.2. Materials

Two parts of the FE model were specified as having different materials, and the CFRP tendon was assigned with given values for each material parameter based on Table 2. The axial and radial mechanical properties of the CFRP tendon are different. Therefore, the anisotropic elastic material model was chosen, and the primary fiber direction was defined as material orientation X. In addition, CFRP material is different from a traditional steel material, because it is an anisotropic material; therefore, the maximum stress and maximum strain criteria applied to traditional material are no longer applicable. At present, the Tsai–Wu criterion is often used for the simulation of composite materials. The Tsai–Wu criterion can reflect the interconnection between the damage strength X, Y, and Z, and can more realistically reflect the damage pattern of composite materials [35,36].

4.3. Elements

Advanced three-dimensional element SOLID185 was adopted to simulate the CFRP tendon and concrete saddle based on previous studies [37]. The CFRP tendon was discretized with axial and radial mesh sizes of 4 mm and around 1 mm, respectively, as shown in Figure 8. A mesh sensitivity analysis was conducted before selecting the mesh sizes. The interface (between CFRP tendon and saddle) in Figure 8 was simulated in ANSYS as contact elements. Surface-to-surface contact pairs (CONTA 174 and TARGE 170) simulated the interactions. The contact parameters were normal contact stiffness factor (FKN = 0.1) and initial closure factor (ICONT = 0.1) [32]. The friction coefficient (FC) between the CFRP tendon and concrete saddle was 0.2 based on the test results [38], preferably reflecting the real contact state. Default settings set the other contact parameters. The contact nonlinearity was calculated by an augmented Lagrangian contact algorithm. Therefore, it is accurate to simulate the variable radial compression stress and tangential friction force between the contact interface.

4.4. Boundary Conditions and Loading

The boundary conditions and loading positions are displayed in Figure 8. On the bottom of the concrete saddle, all rotations and displacements (UX = UY = UZ = 0, ROTX = ROTY = ROTZ = 0) were constrained. The displacement of the CFRP tendon cross-section nodes at the anchoring end was constrained based on the actual constraint of the anchorage. A uniform load was applied over the cross-section of the other end of the CFRP tendon.

4.5. Model Validation

The efficiency coefficient of the FE models was compared with the corresponding test results to validate the reliability of the FE models. Table 4 lists the results for each group of specimens in the static load tests and FE analysis. The results suggest that the simulated values of the efficiency coefficient for four bending radii agreed well with the corresponding test results; in addition to the abnormal data, the maximum value of deviation (δ) was less than 6.0%. Therefore, it can be concluded that the established FE models are reliable.

4.6. Load–Strain Curves of CFRP Tendons

Figure 9 shows the load–strain curves of typical specimens investigated experimentally and numerically, drawn only before the fracture load. CFRP tendons with a diameter of 8, 10, and 14 m were selected as the research objects, and the rest of the specimens had the same trend of change. CFRP tendons did not reach their total capacity due to bending. From the strain values, the loading end had the maximum deformability. At the same time, most load–strain curves showed linear behavior. The same trend was also observed from the numerical simulation results, and the values were very close. The strain dispersion for the CFRP tendons was not significant during the lower loading stage (before 20% F_{cu}), and it increased with the load after that. As the load increased, significant nonlinear segments appeared at the end part of the load–strain curves of some specimens. This is because as the load increased, cracks appeared on the surface of the CFRP tendon, which may lead to the deviation in the strain gauge direction from the longitudinal direction of the tendon. Han et al. [30] pointed out that this kind of nonlinear segment was the typical characteristic of the splitting failure for the unidirectional-fiber-reinforced composites. In addition, a spiral curve appeared at the lower loading stage of specimen R8D8 due to the long continuous loading time in the experiment; thus, it may lead to anchorage loosening or jack oil leakage in a short time. For specimen R12D10, SG1 changed significantly when the load exceeded 90 kN. This is because many cracks appeared, and the screwed wires lost their mutual constraint, leading the CFRP tendon to burst into multiple strands.

Table 4. Results of FE models.

R (m)	D (mm)	Group	F_2 (kN)	η_2 (%)	δ (%)
3	5	R3D5	32.11	71.11	8.57
	8	R3D8	72.75	62.93	4.29
	10	R3D10	103.79	57.46	0.95
	14	R3D14	179.03	50.56	2.72
8	5	R8D5	34.12	75.56	3.03
	8	R8D8	77.74	67.24	2.63
	10	R8D10	114.77	63.54	2.68
	14	R8D14	190.03	53.67	1.60
12	5	R12D5	37.13	82.22	2.78
	8	R12D8	83.72	72.41	1.20
	10	R12D10	120.76	66.85	1.68
	14	R12D14	200.03	56.50	4.17
0	5	R0D5	44.16	97.78	4.35
	8	R0D8	114.61	99.14	1.77
	10	R0D10	158.68	87.85	2.58
	14	R0D14	240.04	67.80	5.73

Note: F_2 = average fracture force of CFRP tendon by FE analysis; η_1 = efficiency coefficient by test = F_1/F_{cu}; η_2 = efficiency coefficient by FE analysis = F_1/F_{cu}; and δ = deviation = $|(\eta_2 - \eta_1)/\ \eta_1|$.

4.7. Failure Mode

Determining the failure modes of the specimens was one of the most critical outcomes in the research, and offered more understanding of load transfer. Figure 10 shows the comparative analysis of the finite element and test failure modes of the specimens. It can be seen from the figure that the failure mode is related to the stress distribution of the CFRP tendons. For the R3 saddle, specimens R3D5 and R3D8 presented an uneven fracture section, while for the R8 saddle, specimens R8D8 and R8D10 presented typical shear failure. The results show that a larger bending radius led to a more uniform stress distribution on the CFRP tendon cross-section, thus resulting in synchronous fracture of the fibers. For the R12 saddle, specimen R12D8 presented shear fractures, while multiple parallel longitudinal cracks were observed on the surface of R12D10 and R12D14, with lengths ranging from 2 to 7 cm. From the stress distribution of the CFRP tendons after failure of the finite element model, it can be seen that the surface stress distribution of the CFRP tendons where splitting damage occurs is more uniform, and the larger the diameter of the CFRP tendons, the more likely it is that splitting damage occurs. In addition, for the specimens of the 14 mm CFRP tendon, the CFRP tendons finally had varying degrees of slip. This is attributed to the fact that the CFRP tendon with a larger diameter causingmore epoxy resin out of the steel pipe during the installation, resulting in insufficient adhesion between the CFRP tendon and residual bonding medium. Based on the above analysis, it can be seen that increasing the bending radius contributed to the uniformity of the stress distribution. Additionally, increasing the diameter of the CFRP tendon made it brittle and difficult to anchor, so the splitting failure and slipping failure were observed.

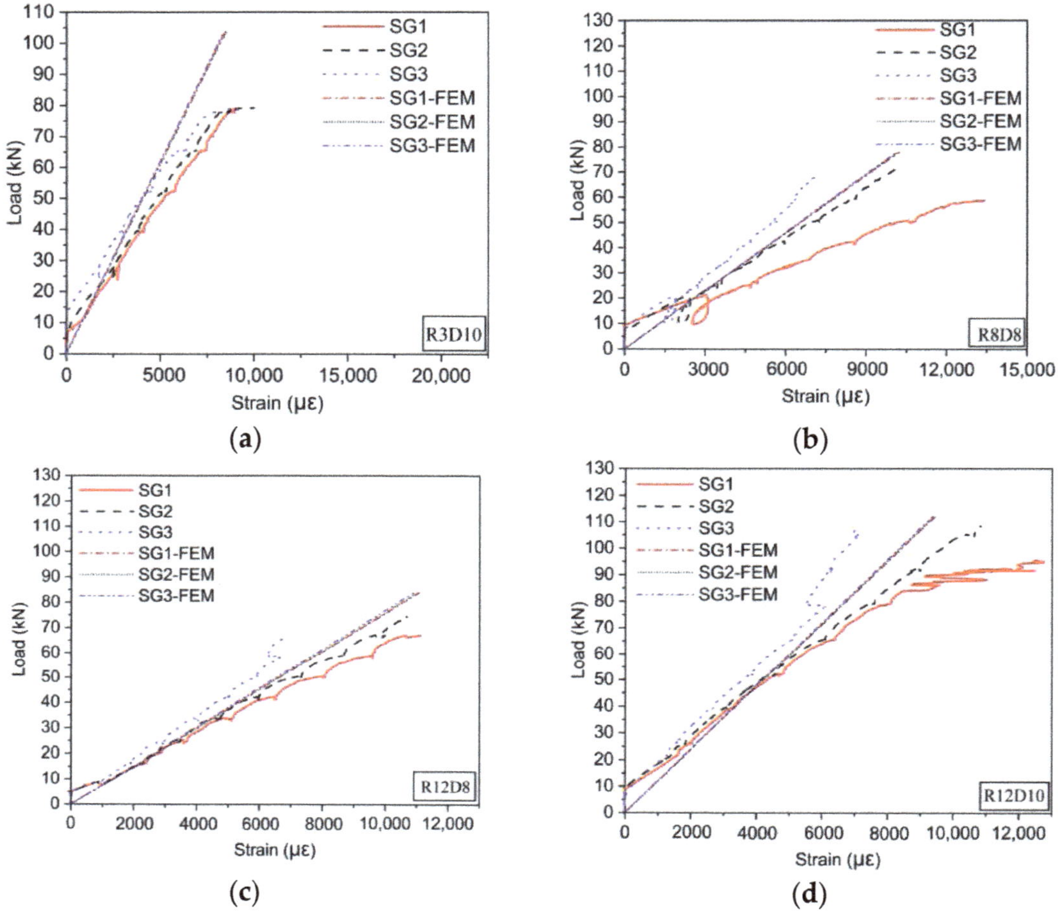

Figure 9. Load–strain curves of CFRP tendons: (**a**) R3D10; (**b**) R8D8; (**c**) R12D8; (**d**) R12D10.

4.8. Parameter Evaluation

4.8.1. Effect of CFRP Tendon Diameter and Bending Radius on Stress Distribution

The stress distributions of CFRP tendons were analyzed using the FE models, as shown in Figure 11. The von Mises cloud diagrams show that the stress was maximized at the anchoring end (R3) and loading end (R8 and R12). With increasing bending radius, the stress gradient along the length of the CFRP tendon decreased, and the stress distribution was more uniform. An increase in diameter led to a reduction in peak stress, indicating that CFRP tendons with a smaller diameter were more likely to provide the advantages of CFRP composites.

Figure 12 shows the friction stress and contact compressive stress distribution of CFRP tendons by numerical simulation, where SFRI is friction stress and PRES is compressive stress. The friction and contact compressive stress distributions of CFRP tendons were uniform except at the loading end. Although the load-bearing capacity of CFRP tendons decreased with the decrease in bending radius, the friction stress and compressive stress increased significantly, as shown in Figure 12a. The results show that reducing the bending radius had significant adverse effects on the properties of CFRP tendons. The friction stress and contact compressive stress increased with the increase in CFRP tendon diameter, as shown in Figure 12b. However, the effect of friction stress caused by increasing CFRP tendon diameter on the properties of CFRP tendons is unclear. This is because the load-bearing capacity of CFRP tendons increased accordingly, increasing the contact compressive stress.

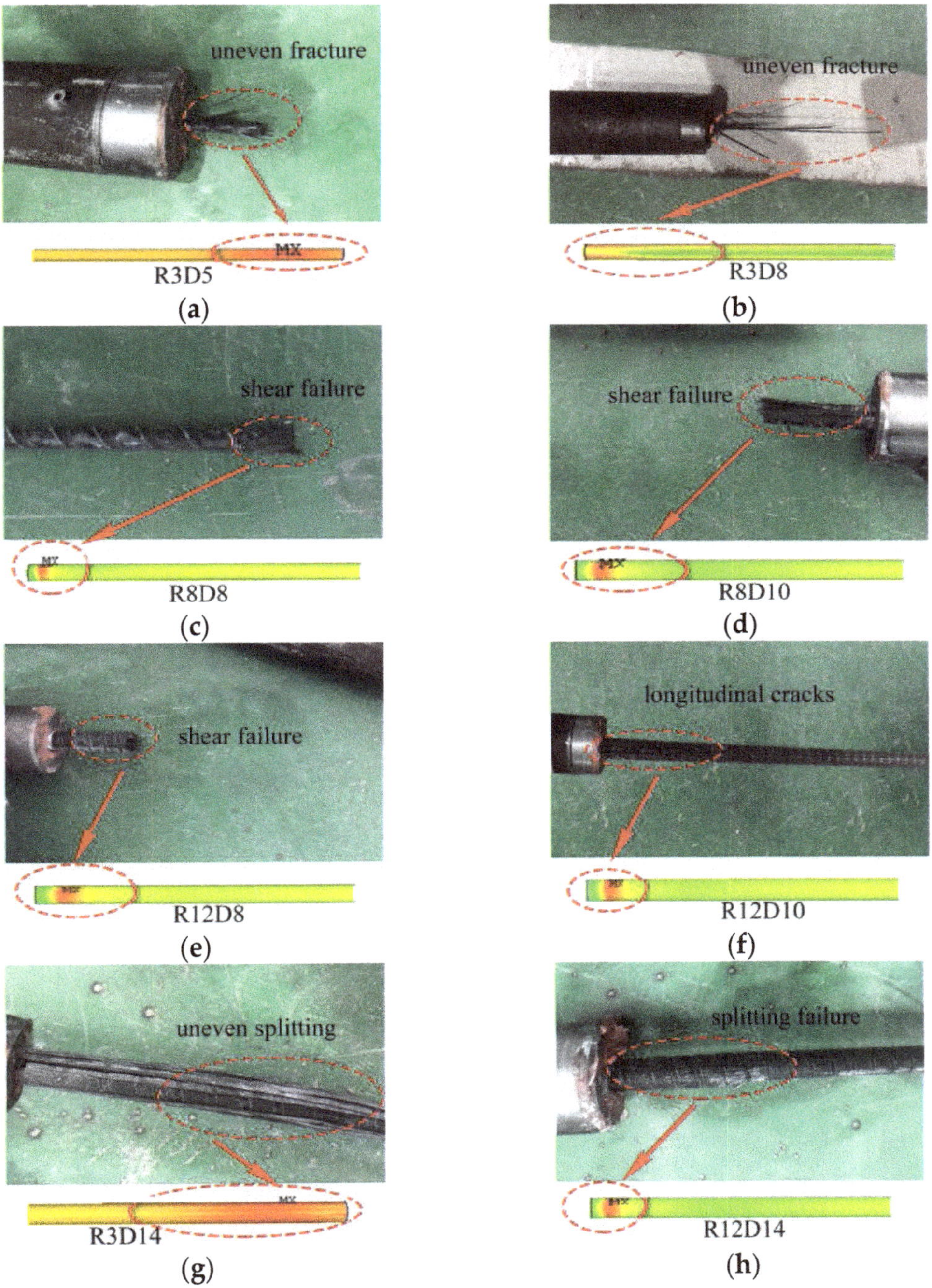

Figure 10. Failure mode of typical specimens: (**a**) R3D5; (**b**) R3D8; (**c**) R8D8; (**d**) R8D10; (**e**) R12D8; (**f**) R12D10; (**g**) R3D14; (**h**) R12D14.

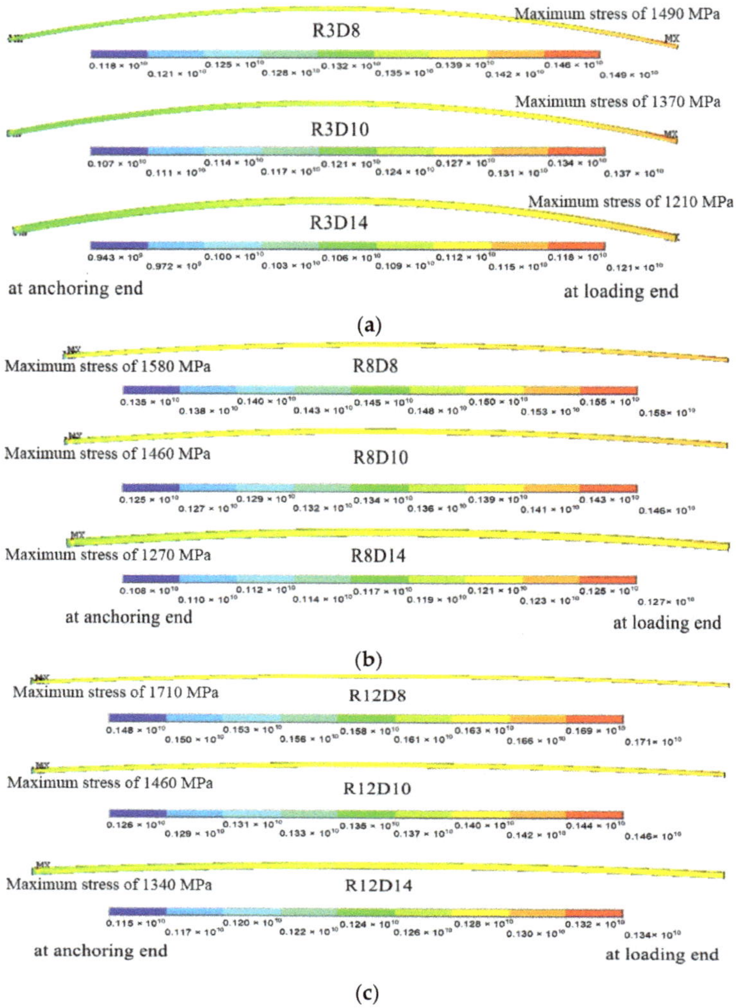

Figure 11. Stress distribution of CFRP tendons: (**a**) R3; (**b**) R8; (**c**) R12.

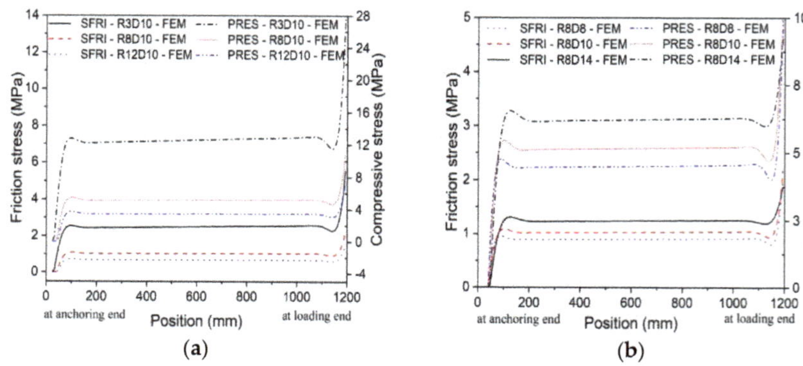

Figure 12. Friction stress and contact compressive stress distributions of CFRP tendons: (**a**) D10; (**b**) R8.

4.8.2. Effect of Bending Radius of the Saddle on Efficiency Coefficient

Figure 13 compares the efficiency coefficient of CFRP tendons simulated by FE analysis with the test results. In addition to the abnormal data, the efficiency coefficient increased nonlinearly with increasing bending radius. When the bending radius increased from 3 to 12 m, the efficiency coefficient of D8 increased from 62.93% to 72.41%, an increase of close to 10%. However, when the bending radius increased from 3 to 12 m, the efficiency coefficient of D14 increased by only 5.96%. The results show that increasing the bending radius was more conducive to improving the performance of CFRP tendons with a smaller diameter.

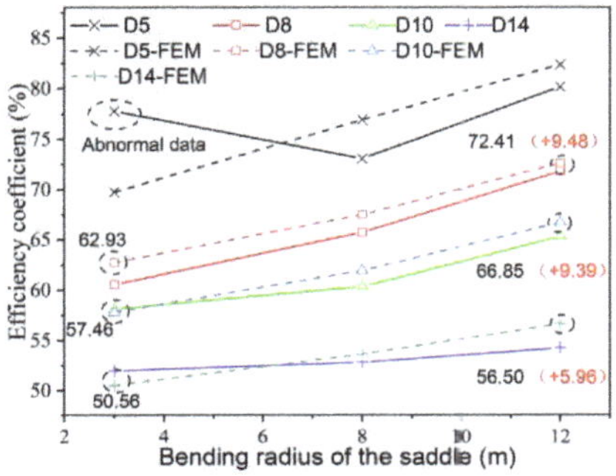

Figure 13. Curves of bending radius and efficiency coefficient.

Figure 14 shows the effect of saddle bending radius on the axial stress and shear stress distributions of CFRP tendons in FE analysis. The axial stress of R3D8 reached a minimum because the CFRP tendon was subjected to significant shear stress, which caused the ultimate tensile force to be significantly reduced. In addition, as the bending radius decreased, the shear stress increased sharply. It can be observed that CFRP tendons with a smaller bending radius presented higher shear stress concentrations at both ends.

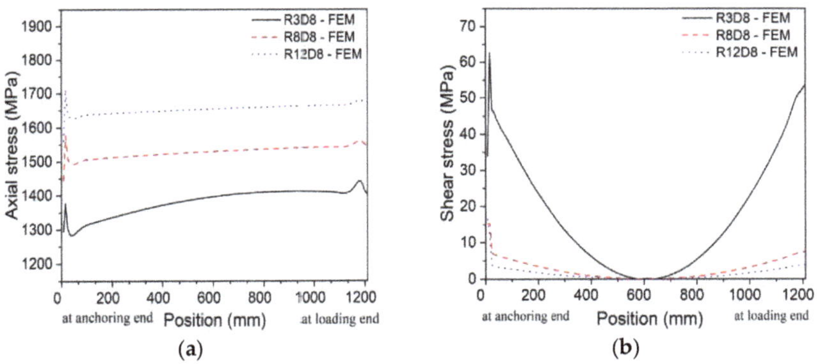

Figure 14. Effect of bending radius on the stress distributions of CFRP tendons: (**a**) axial stress; (**b**) shear stress.

The test and FE results show that decreasing bending radius increased the force perpendicular to the fiber direction, which had a significant adverse effect on the properties of CFRP tendons. This is because the shear strength of CFRP tendons is only 7% of the tensile strength. Combined with the failure modes in the test, CFRP tendons were extruded by epoxy resin in the steel pipe. When the CFRP tendon extended out of the steel pipe, the radial extrusion constraint suddenly disappeared, resulting in transverse expansion of CFRP wires. Interface failure caused the CFRP tendon to lose load-bearing capacity due to shear stress, and the tensile fracture of the fibers led to the final fracture. In engineering practice, it is suggested that bending radius of CFRP tendons should be more than or equal to 8 m, and the shear force concentrations should be concerned.

4.8.3. Effect of Diameter of the CFRP Tendon on Efficiency Coefficient

Figure 15 compares the efficiency coefficient of CFRP tendons with different diameters simulated by FE analysis with the test results. In addition to the abnormal data, as the diameter of CFRP tendons increased, the efficiency coefficient of all specimens decreased significantly. When the diameter of CFRP tendons reached 14 mm, the corresponding efficiency coefficients of R3 and R12 were 50.56% and 56.50%, respectively, which were very low. The results show that increasing the diameter of CFRP tendons had a significant adverse effect on performance.

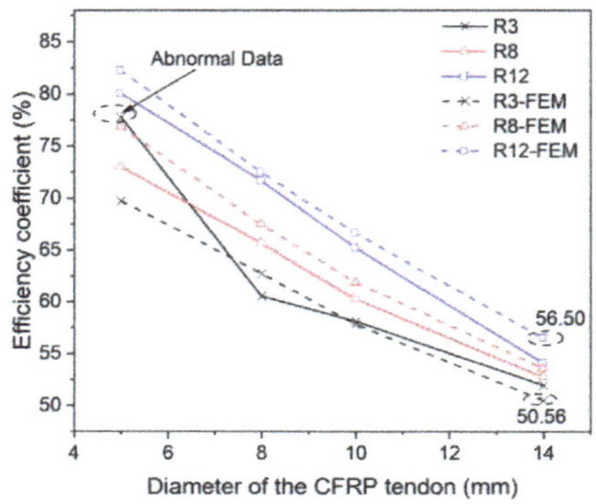

Figure 15. Curves of diameter and efficiency coefficient.

Figure 16 shows the effect of diameter on the axial stress and shear stress distributions of CFRP tendons in FE analysis. As the diameter decreased, the axial stress increased significantly, indicating that CFRP tendons with a smaller diameter were more likely to provide the advantages of CFRP composites. The trend of the three curves in Figure 16a was consistent due to the uniform distribution of axial stress. Although the change in diameter had little effect on the shear stress, CFRP tendons still presented shear stress concentrations at both ends.

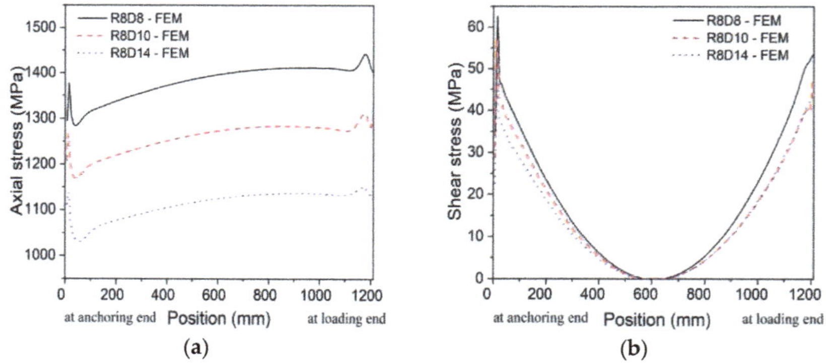

Figure 16. Effect of diameter on the stress distributions of CFRP tendons: (**a**) axial stress; (**b**) shear stress.

The test and FE results show that CFRP tendons with a smaller diameter had superior properties. Combined with the previous conclusion, CFRP tendons with a larger diameter had worse flexibility, thus resulting in fracture by tension and local shear stress concentrations. Therefore, CFRP tendons with a diameter smaller than or equal to 8 mm are preferred in main cables for holding a high efficiency coefficient.

4.8.4. Effect of Friction Coefficient (FC) on Efficiency Coefficient

Considering the limitation of the experiment, the effect of contact friction between the CFRP tendon and saddle on the properties of CFRP tendons was studied by changing the friction coefficient (FC) in FE analysis. The efficiency coefficient–friction coefficient relationship of CFRP tendons is shown in Figure 17. When increasing the friction coefficient, the efficiency coefficient decreased significantly, revealing the adverse effect of friction on the properties of CFRP tendons. As the friction coefficient increased from 0.1 to 0.5, the corresponding efficiency coefficient of D14 decreased from 59.84% to 25.04%, a decrease of 34.8%. Therefore, increasing the friction coefficient, especially for the CFRP tendon with a larger diameter, degraded its mechanical properties under combined loads.

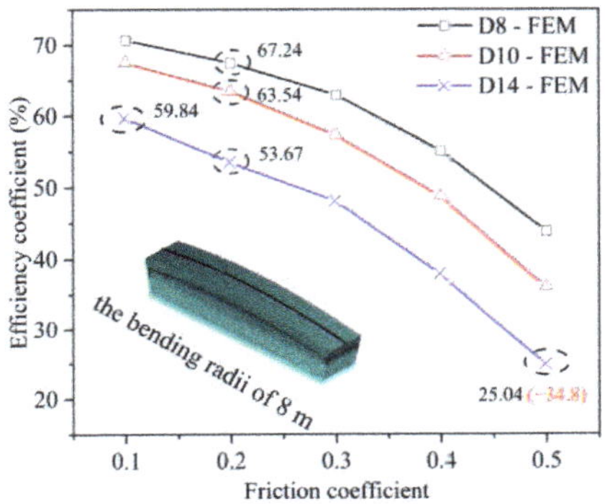

Figure 17. Curves of friction coefficient and efficiency coefficient.

Figure 18 shows the effect of friction on the axial stress and shear stress distributions of CFRP tendons in FE analysis. The change in friction coefficient had little effect on the shear stress but significantly affected the axial stress. As the friction coefficient increased, the load-bearing capacity of CFRP tendons was significantly reduced; thus, the axial stress with the friction coefficient of 0.5 reached a minimum. The results suggest that CFRP tendons with a smaller friction coefficient had superior properties. Therefore, it is suggested that some low friction materials should be coated between the CFRP tendon and saddle contact surface to reduce the friction coefficient.

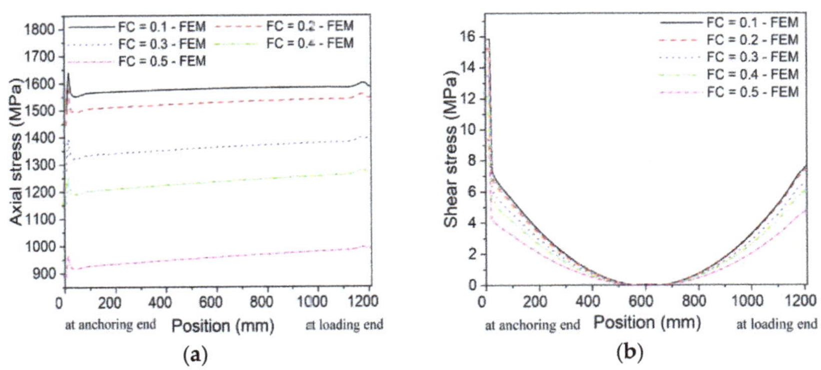

Figure 18. Effect of friction coefficient on the stress distribution of CFRP tendons: (**a**) axial stress; (**b**) shear stress.

5. Conclusions

In this research, a series of static load tests were carried out to study the mechanical properties of CFRP tendons subjected to combined tension and bending, and the research was extended through FE simulations. Based on the results, the conclusions are summarized as follows:

1. The fracture force and deformability of the specimens were observed to be significantly decreased with the decrease in the bending radius. Meanwhile, the CFRP tendon diameter also was found to have major influencing effects on the ultimate capacity of the specimens.
2. The efficiency coefficients (the ratio of the fracture force to the ultimate force) were found to be significantly affected by the increases in the bending radius. In the test, the efficiency coefficients of D8, D10, and D14 increased by 11.21%, 7.74%, and 2.26%, respectively, when the bending radius was increased from 3 to 12 m. Meanwhile, the efficiency coefficient of D14 increased by only 2.26%. Therefore, increasing the bending radius was more conducive to improving the performance of CFRP tendons with a smaller diameter.
3. The failure modes were found to be influenced by the bending radius and the CFRP tendon diameter. The test results show that increasing the bending radius contributed to the uniformity of the stress distribution. The interface failure caused the CFRP tendon to lose load-bearing capacity due to shear stress, and the tensile fracture of the fibers led to the final fracture. Additionally, increasing the diameter of the CFRP tendon made it brittle and difficult to anchor, so the splitting failure and slipping failure were observed. The ratio of the bending radius to the tendon diameter was below 2.4, the efficiency coefficient was less than 80%, and the failure modes of the specimens were mostly shear failure.
4. Combined with the FE model analysis, it was shown that the CFRP tendons with a smaller bending radius presented higher shear stress concentrations at both ends and had a significant adverse effect on the contact friction stress. The change in friction coefficient had little effect on the shear stress but had a significant effect on the axial stress of CFRP tendons. As the friction coefficient increased, the load-bearing capacity of CFRP tendons was significantly reduced, and the axial stress with the friction coefficient of 0.5 reached a minimum. In addition, the results of the FE models agreed well with the test results, and provided a reliable basis for CFRP composites used in practical engineering.

Author Contributions: Conceptualization, L.J.; methodology L.J.; formal analysis, W.Z.; investigation, W.Z.; writing—original draft preparation, W.Z.; writing—review and editing, J.X.; supervision, Y.J.; project administration, Y.J.; Data curation, J.X.; Visualization, J.X. All authors have read and agreed to the published version of the manuscript.

Funding: This work was supported by the National Natural Science Foundation of China (51878488).

Data Availability Statement: Some or all data, models, or code that support the findings of this study are available from the corresponding author upon reasonable request (load-strain data and FE data).

Acknowledgments: The authors are greatly indebted to the anonymous reviewers for their valuable comments and suggestions, which greatly helped in improving the overall quality of this manuscript.

Conflicts of Interest: The authors declare no conflict of interest.

References

1. Rohleder, W.J., Jr.; Tang, B.; Doe, T.A.; Grace, N.F.; Burgess, C.J. Carbon Fiber-Reinforced Polymer Strand Application on Cable-Stayed Bridge, Penobscot Narrows, Maine. *Transp. Res. Rec.* **2008**, *2050*, 169–176. [CrossRef]
2. Xiong, W.; Cai, C.S.; Zhang, Y.; Xiao, R. Study of super long span cable-stayed bridges with CFRP components. *Eng. Struct.* **2011**, *33*, 330–343. [CrossRef]
3. Yang, Y.Q.; Fahmy, M.F.M.; Guan, S.J.; Pan, Z.H.; Zhan, Y.; Zhao, T.D. Properties and applications of FRP cable on long-span cable-supported bridges: A review. *Compos. Part B-Eng.* **2020**, *190*, 107934. [CrossRef]
4. Jiang, Y.; Jia, L. Study on Applicable Spans and Static Performance of the Suspension Bridges using CFRP Cables. In Proceedings of the Advances in Civil Engineering, PTS 1-6, Uttar Pradesh, India, 14–16 October 2011; pp. 1115–1119.

5. Jiang, Y.; Jia, L. Contrastive Analysis on Limitation Span between Suspension Bridge using steel and CFRP Cable. In Proceedings of the Advanced Research on Civil Engineering and Material Engineering, Wuhan, China, 25–26 August 2012; pp. 57–60.
6. Meiarashi, S.; Nishizaki, I.; Kishima, T.I. Life-cycle cost of all-composite suspension bridge. *J. Compos. Constr.* **2002**, *6*, 206–214. [CrossRef]
7. Liu, Y.; Zwingmann, B.; Schlaich, M. Carbon Fiber Reinforced Polymer for Cable Structures-A Review. *Polym.-Basel* **2015**, *7*, 2078–2099. [CrossRef]
8. Xian, G.; Guo, R.; Li, C.; Wang, Y. Mechanical performance evolution and life prediction of prestressed CFRP plate exposed to hygrothermal and freeze-thaw environments. *Compos. Struct.* **2022**, *293*, 115719. [CrossRef]
9. Lal, H.M.; Uthaman, A.; Li, C.; Xian, G.; Thomas, S. Combined effects of cyclic/sustained bending loading and water immersion on the interface shear strength of carbon/glass fiber reinforced polymer hybrid rods for bridge cable. *Constr. Build. Mater.* **2022**, *314*, 125587. [CrossRef]
10. Wang, X.; Wu, Z. Integrated high-performance thousand-metre scale cable-stayed bridge with hybrid FRP cables. *Compos. Part B-Eng.* **2010**, *41*, 166–175. [CrossRef]
11. Feng, B.; Wang, X.; Wu, Z.S. Static and Fatigue Behavior of Multitendon CFRP Cables with Integrated Anchorages. *J. Compos. Constr.* **2019**, *23*, 4019051. [CrossRef]
12. Al-Mayah, A.; Soudki, K.; Plumtree, A. Mechanical behavior of CFRP rod anchors under tensile loading. *J. Compos. Constr.* **2001**, *5*, 128–135. [CrossRef]
13. Sudarisman; Davies, I.J. The effect of processing parameters on the flexural properties of unidirectional carbon fibre-reinforced polymer (CFRP) composites. *Mater. Sci. Eng. A-Struct. Mater. Prop. Microstruct. Process.* **2008**, *498*, 65–68. [CrossRef]
14. Rozylo, P.; Falkowicz, K.; Wysmulski, P.; Debski, H.; Pasnik, J.; Kral, J. Experimental-Numerical Failure Analysis of Thin-Walled Composite Columns Using Advanced Damage Models. *Materials* **2021**, *14*, 1506. [CrossRef] [PubMed]
15. Debski, H.; Samborski, S.; Rozylo, P.; Wysmulski, P. Stability and Load-Carrying Capacity of Thin-Walled FRP Composite Z-Profiles under Eccentric Compression. *Materials* **2020**, *13*, 2956. [CrossRef]
16. Kadhim, M.M.A.; Jawdhari, A.; Nadir, W.; Cunningham, L.S. Behaviour of RC beams strengthened in flexure with hybrid CFRP-reinforced UHPC overlays. *Eng. Struct.* **2022**, *262*, 114356. [CrossRef]
17. Peng, K.-D.; Huang, B.-T.; Xu, L.-Y.; Hu, R.-L.; Dai, J.-G. Flexural strengthening of reinforced concrete beams using geopolymer-bonded small-diameter CFRP bars. *Eng. Struct.* **2022**, *256*, 113992. [CrossRef]
18. Vo-Le, D.; Tran, D.T.; Pham, T.M.; Ho-Huu, C.; Nguyen-Minh, L. Re-evaluation of shear contribution of CFRP and GFRP sheets in concrete beams post-tensioned with unbonded tendons. *Eng. Struct.* **2022**, *259*, 114173. [CrossRef]
19. de Menezes, E.A.W.; da Silva, L.V.; Cimini, C.A.; Luz, F.F.; Amico, S.C. Numerical and Experimental Analysis of the Tensile and Bending Behaviour of CFRP Cables. *Polym. Polym. Compos.* **2017**, *25*, 643–649. [CrossRef]
20. Luz, F.F.; Menezes, E.A.W.; Silva, L.V.; Cimini, C.A., Jr.; Marczak, R.J.; Amico, S.C. Bending behavior of CFRP cables in the nonlinear displacement range. *J. Braz. Soc. Mech. Sci. Eng.* **2019**, *42*, 1–7. [CrossRef]
21. Han, Q.; Wang, L.; Xu, J. Experimental research on mechanical properties of transverse enhanced and high-temperature-resistant CFRP tendons for prestressed structure. *Constr. Build. Mater.* **2015**, *98*, 864–874. [CrossRef]
22. Arczewska, P.; Polak, M.A.; Penlidis, A. Relation between Tensile Strength and Modulus of Rupture for GFRP Reinforcing Bars. *J. Mater. Civ. Eng.* **2019**, *31*, 4018362. [CrossRef]
23. Fang, Y.W.; Fang, Z.; Jiang, Z.W.; Jiang, R.N.A.; Zhou, X.H. Investigation on failure behavior of carbon fiber reinforced polymer wire subjected to combined tension and bending. *Compos. Struct.* **2021**, *267*, 113927. [CrossRef]
24. Cai, D.A.; Wang, X.P.; Shi, Y.H.; Hao, X.F.; Qian, Y.; Zhou, G.M. A New Interfacial Model for Transverse Mechanical Properties of Unidirectional Fiber Reinforced Composites. *Fibers Polym.* **2021**, *22*, 430–441. [CrossRef]
25. Hwash, M.; Knippers, J. Load-Bearing Capacity of Deviated CFRP Strips. *J. Compos. Constr.* **2014**, *18*, 4013055. [CrossRef]
26. Suwei, H.O.U.; Ping, Z.; Shizhong, Q.; Cuijuan, L.I. Experimental Investigation of Friction Properties between CFRP Main Cable and Saddle of Suspension Bridge. *J. Southwest Jiaotong Univ.* **2011**, *46*, 391–397.
27. Fan, H.F.; Vassilopoulos, A.P.; Keller, T. Experimental and numerical investigation of tensile behavior of non-laminated CFRP straps. *Compos. Part B-Eng.* **2016**, *91*, 327–336. [CrossRef]
28. Liu, M.H.; Qiang, S.Z.; Xu, G.P.; Ren, W.P. Study and Prototype Design of a Suspension Bridge with Ultra-Long Span and CFRP Main Cables. *J. Highw. Transp. Res. Dev.* **2014**, *8*, 47–56.
29. Wang, X.; Wang, Z.H.; Wu, Z.S.; Cheng, F. Shear behavior of basalt fiber reinforced polymer (FRP) and hybrid FRP rods as shear resistance members. *Constr. Build. Mater.* **2014**, *73*, 781–789. [CrossRef]
30. Han, Q.H.; Wang, L.C.; Xu, J. Experimental research on fracture behaviors of damaged CFRP tendons: Fracture mode and failure analysis. *Constr. Build. Mater.* **2016**, *112*, 1013–1024. [CrossRef]
31. Xiang, Y.; Fang, Z.; Wang, C.L.; Zhang, Y.; Fang, Y.W. Experimental Investigations on Impact Behavior of CFRP Cables under Pretension. *J. Compos. Constr.* **2017**, *21*, 4016087. [CrossRef]
32. Rizkalla, S.H.; Busel, J.P. Prestressing Concrete Structures with FRP Tendons (ACI 440.4R-04). In Proceedings of the Structures Congress 2005, New York, NY, USA, 20–24 April 2005.
33. Fang, Z.; Zhang, K.; Tu, B. Experimental investigation of a bond-type anchorage system for multiple FRP tendons. *Eng. Struct.* **2013**, *57*, 364–373. [CrossRef]

34. *JC/T 2404-2017*; Chinese Standards. Test Method for Tensile Behavior of Continuous Fiber-Reinforced Ceramic Composites at Room Temperature. China Standards Press: Beijing, China, 2017.
35. Tsai, S.W.; Wu, E.M. A General Theory of Strength for Anisotropic Materials. *J. Compos. Mater.* **1971**, *5*, 58–80. [CrossRef]
36. Hashin, Z. Failure criteria for unidirectional fiber composites. *J. Appl. Mech.-Trans. Asme* **1980**, *47*, 329–334. [CrossRef]
37. Cai, D.S.; Xu, Z.H.; Yin, J.; Liu, R.G.; Liang, G. A numerical investigation on the performance of composite anchors for CFRP tendons. *Constr. Build. Mater.* **2016**, *112*, 848–855. [CrossRef]
38. Yu, T.L.; Zhang, L.Y. Friction Loss of Externally Prestressed Concrete Beams with Carbon Fiber-Reinforced Polymer Tendons. In Proceedings of the Advances in Structures, PTS 1-5, Hamburg, Germany, 28–30 March 2011; pp. 3701–3706.

buildings

MDPI

Article

Collapse Assessment of Mid-Rise RC Dual Wall-Frame Buildings Subjected to Subduction Earthquakes

Marco F. Gallegos [1], Gerardo Araya-Letelier [2,3,*], Diego Lopez-Garcia [1,4,*] and Pablo F. Parra [5]

1 Department of Structural & Geotechnical Engineering, Pontificia Universidad Catolica de Chile, Santiago 7820436, RM, Chile
2 School of Civil Construction, Faculty of Engineering, Pontificia Universidad Catolica de Chile, Santiago 7820436, RM, Chile
3 Concrete Innovation Hub UC (CIHUC), Pontificia Universidad Catolica de Chile, Santiago 7820436, RM, Chile
4 Research Center for Integrated Disaster Risk Management (CIGIDEN) ANID FONDAP 1522A0005, Santiago 7820436, RM, Chile
5 Facultad de Ingenieria y Ciencias, Universidad Adolfo Ibanez, Santiago 7941169, RM, Chile
* Correspondence: gerardo.araya@uc.cl (G.A.-L.); dlg@ing.puc.cl (D.L.-G.)

Abstract: In Chile, office buildings are typically reinforced concrete (RC) structures whose lateral load-resisting system comprises core structural walls and perimeter moment frames (i.e., dual wall-frame system). In the last 20 years, nearly 800 new dual wall-frame buildings have been built in the country and roughly 70% of them have less than ten stories. Although the seismic performance of these structures was deemed satisfactory in previous earthquakes, their actual collapse potential is indeed unknown. In this study, the collapse performance of Chilean code-conforming mid-rise RC buildings is assessed considering different hazard levels (i.e., high and moderate seismic activity) and different soil types (i.e., stiff and moderately stiff). Following the FEMA P-58 methodology, 3D nonlinear models of four representative structural archetypes were subjected to sets of Chilean subduction ground motions. Incremental dynamic analysis was used to develop collapse fragilities. The results indicate that the archetypes comply with the 'life safety' risk level defined in ASCE 7, which is consistent with the observed seismic behavior in recent mega-earthquakes in Chile. However, the collapse risk is not uniform. Differences in collapse probabilities are significant, which might indicate that revisions to the current Chilean seismic design code might be necessary.

Keywords: performance-based earthquake engineering; collapse assessment; mid-rise building; Chilean RC dual wall-frame system; subduction seismicity

Citation: Gallegos, M.F.; Araya-Letelier, G.; Lopez-Garcia, D.; Parra, P.F. Collapse Assessment of Mid-Rise RC Dual Wall-Frame Buildings Subjected to Subduction Earthquakes. *Buildings* **2023**, *13*, 880. https://doi.org/10.3390/buildings13040880

Academic Editors: Chao Li, Weiping Wen, Jian Zhong, Xiaowei Wang and Suiwen Wu

Received: 22 February 2023
Revised: 16 March 2023
Accepted: 19 March 2023
Published: 28 March 2023

1. Introduction

In recent decades, Chilean reinforced concrete (RC) mid- and high-rise buildings were subjected to strong subduction earthquakes (e.g., M_w 8.0 1985 Valparaiso, M_w 8.8 2010 Maule, M_w 8.2 2014 Iquique, M_w 8.3 2015 Illapel, and M_w 7.6 2016 Chiloe [1]), and the seismic behavior of these buildings was deemed satisfactory. During the 2010 earthquake, only 2.0% of RC buildings with nine or more stories and only 0.4% of buildings with three or more stories suffered severe damage due to strong shaking [2]. In particular, this event caused only a few partial collapses (e.g., the O'Higgins Tower) and one total collapse (i.e., the Alto Rio building) [3], as shown in Figure 1.

The main objective of modern seismic design codes is to prevent collapse. Structural collapse, either partial or total, is the leading cause of casualties, injuries, and economic losses, as well as downtime and environmental impacts [4]. For these reasons, quantification of the collapse probability of code-conforming structures (e.g., buildings) is a very important issue in earthquake engineering, particularly in countries where the seismic activity is high and the seismic behavior of their buildings has not been adequately characterized [5].

(a) (b)

Figure 1. Buildings damaged by the M_W 8.8 2010 Chilean earthquake: (**a**) O'Higgins office building; (**b**) Alto Rio residential building.

In Chile, observations of the effects of recent earthquakes [3,6,7] indicate that the 'collapse prevention' target performance level of the current seismic design regulations for buildings [8,9] was achieved. However, in these regulations the 'collapse prevention' limit state is defined in qualitative terms rather than in quantitative metrics. The seismic design code NCh 433 [8] states that "although presenting damage, buildings should avoid collapse when subjected to exceptionally intense ground motions", but "exceptionally intense ground motions" are not defined, and neither are the target collapse metrics. In other words, quantitative acceptance criteria such as those defined in ASCE 7-22 [10] (i.e., a probability of collapse in 50 years of less than 1% and a probability of collapse conditional to the Maximum Considered Earthquake (MCE) of less than 10%) are not specified in the Chilean regulations.

Chilean RC mid- and high-rise buildings can be characterized by two types of structural systems. On the one hand, the structural system of residential buildings consists of a large number of shear walls, typically in a 'fish-bone' plan layout. This structural system is rarely found outside Chile, which is why it is sometimes referred to as the 'Chilean building'. On the other hand, the structural system of office buildings consists of core shear walls and perimeter moment frames. This structural system is commonly known as a dual wall-frame system and is frequently used in other earthquake-prone regions. The floor system typically consists of flat post-tensioned slabs (17–20 cm thickness) and the span lengths are relatively large (8–10 m). The perimeter moment frames are mainly intermediate moment frames (IMFs) designed for less than 25% of the base shear and the core walls are special structural walls (SWs) [11]. The core walls in office buildings are shorter and thicker than those in residential buildings but the wall density (i.e., the ratio of the area of the walls to the total floor area) is similar in both types of buildings. As a consequence of a commonly adopted criterion in Chilean engineering practice, the wall density is always greater than 0.1% [12].

The level of seismic hazard is high in Chile, mainly due to subduction-type earthquakes, but research to quantitatively assess the actual level of seismic collapse protection of buildings is still limited. To the best of the authors' knowledge, the collapse probability of Chilean RC buildings, assessed per the Performance-Based Earthquake Engineering (PBEE) framework proposed by the Pacific Earthquake Engineering Research Center (PEER) to implement the FEMA P-58 [13] methodology, has been analyzed in just two studies [14,15]. Within the FEMA P-58 framework, the current state-of-the-art procedure for collapse assessment is based on the FEMA P695 [16] methodology. Araya-Letelier et al. [14] analyzed the collapse risk of a code-conforming 16-story dual wall-frame building located in Santiago.

The mean annual frequency of collapse, λ_c, was found to be 2.61×10^{-5} and the collapse probability in 50 years, $P_c(50)$, was found to be 0.13%. Cando et al. [15] analyzed the seismic performance of a suite of four code-conforming 20-story shear-wall buildings located in Santiago. The λ_c values were found to range from 2.7×10^{-4} to 6.9×10^{-4} and the values of $P_c(50)$ were found to range from 1.3% to 3.4%. The latter probabilities were not only higher than those calculated by Araya-Letelier et al. [14] but also exceeded the 1% in 50 years target collapse probability indicated in ASCE 7-22 [10]. According to Cando et al. [15], these relatively high $P_c(50)$ values may be related to the assumed soil type (soil type B in [14], soil type C in [15], as defined by the Chilean seismic design code [9]) and small differences in the seismic hazard model. It must be noted that in each of these studies, only a single building archetype located at a particular location on a specific soil type was analyzed. In other words, the potential influence of different seismicity levels and soil types on the collapse probability when buildings are subjected to subduction earthquakes was not accounted for. Moreover, since these studies focused on the performance of tall buildings, the quantitative collapse performance of mid-rise dual wall-frame RC office buildings in Chile remains unclear.

The potential impact of these and other relevant variables (e.g., structural system, building height, modeling parameters, seismicity source, etc.) on the building's seismic performance has been evaluated to a limited extent. Some studies have analyzed buildings of different heights [17–20]. For instance, Dabaghi et al. [19] found that the collapse capacity decreases as the number of stories increases. Terzic et al. [20] examined low- and mid-rise office buildings and found that, although the seismic response (e.g., floor accelerations, story drifts, and residual drifts) is smaller in shorter buildings than in taller structures, shorter buildings suffer significantly higher levels of structural and nonstructural damage, which of course leads to higher repair costs per floor. In addition, Marafi et al. [21] pointed out that modeling assumptions (e.g., deformation capacity) and the level of axial load on the walls have a significant influence on the collapse risk of dual wall-frame buildings. Few studies have focused on subduction ground motions [22–24]. For example, Medalla et al. [24] revealed that the collapse probability of mid- and high-rise steel moment-frame (MF) buildings in subduction-prone locations could be up to two times higher than that in crustal-prone locations. The seismic response of dual wall-frame buildings on soft soils that are subjected to subduction ground motions has received relatively little attention. In one of these studies, Marafi et al. [25] analyzed the seismic performance of buildings considering the Seattle basin effects and Cascadia subduction earthquakes. High collapse probabilities were found due to the simultaneous effects of spectral acceleration, spectral shape and duration, and basin effects. This research review shows that some variables (e.g., type of seismic demand, height of the building, soil type, etc.) have a direct influence on the seismic performance of buildings, especially on the collapse potential, which needs to be further evaluated.

It is clear then that despite acceptable behavior during recent earthquakes, the collapse capacity of RC buildings subjected to Chilean subduction ground motions is still largely unknown. This is the main motivation for the present study, which aims at quantitatively characterizing the collapse capacity of Chilean RC dual wall-frame buildings. Emphasis is placed on mid-rise buildings, which represent a relevant proportion of current office buildings. The objective of this paper is to provide more insights into the collapse capacity of RC mid-rise dual wall-frame buildings by taking into account the influence of the level of seismic hazard and soil type. In particular, this paper focuses on post-2010 Chilean buildings subjected to Chilean subduction ground motions. To the best of the authors' knowledge, the collapse capacity of the class of buildings analyzed in this paper has not yet been quantitatively characterized. Even though such collapse capacity is expected to be acceptable (the empirical evidence indicates that pre-2010 buildings performed well when subjected to a huge earthquake and post-2010 buildings are expected to perform even better), quantitative characterizations are nevertheless necessary. The methodology used to evaluate the collapse capacity is described in Section 2. The specific research objectives are (1) to select a structural layout representative of Chilean RC dual wall-frame buildings (Section 3); (2) to design and model a suite of four building archetypes located in different seismic zones (i.e., high and moderate seismicity) and on different soil types (i.e., very stiff and moderately stiff soil), following post-2010 seismic provisions (Section 3); (3) to select and scale hazard-consistent ground motions (Section 4); and (4) to develop collapse fragilities and combine them with the respective hazard in order to assess the seismic collapse capacities (Section 5). The seismic behavior was evaluated using the well-established FEMA P-58 methodology [13] by performing Incremental Dynamic Analysis (IDA) [26] of 3D nonlinear structural models. The record-to-record (RTR) variability was accounted for by sets of carefully selected Chilean subduction ground motions. Values of several collapse-related metrics (i.e., probability of collapse conditional to the MCE, Collapse Margin Ratio (CMR), λ_c, $P_c(50)$) were calculated. Finally, all relevant findings are discussed in Section 6.

2. Collapse Assessment Methodology

Despite significant advances in the structural design and analysis of buildings, deterministic predictions of building collapse under earthquake shaking are still not possible. Rather, the evaluation of the seismic collapse performance should be addressed through probabilistic criteria [27]. For this research, probabilistic seismic collapse assessments were developed using the FEMA P-58 methodology [13] within the PBEE-PEER framework. The principal result was the earthquake-induced building-specific collapse fragility function, which is the probability of triggering a structural collapse based on a specific ground-motion (GM) intensity measure (IM). The collapse fragility functions were developed based on the IDA [26] and the IM adopted was the pseudo-spectral acceleration ordinate at the fundamental period of the structure, $S_a(T_1)$. Based on recent studies, $S_a(T_1)$ is still an adequate predictor of collapse for stiff RC structures whose seismic response is dominated by the fundamental vibration mode [28,29] (e.g., mid-rise RC dual wall-frame buildings). Consequently, for each archetype building model and GM record, nonlinear response history (NLRH) analyses at increasing $S_a(T_1)$ levels were performed until collapse. Further details of this methodology are provided in the following subsections.

2.1. Estimation of $S_a(T_1)$ Intensities That Trigger Collapse

Initially, all possible (global or local) building collapse types were identified. Current state-of-the-art studies describe alternatives that can be used to identify earthquake-induced structural collapse [13,16,26]. In this study, both non-simulated and simulated collapses were assessed (further details can be found in Sections 3.3 and 3.4). Subsequently, mathematical 3D models, which capture all failure modes of the structures, were developed. These models took into account the effects of material deterioration on strength and stiffness, as well as geometric nonlinearities (P-delta effects).

In the next step, structural 3D models were subjected to NLRH analyses using sets of subduction GM records (from Chile) that were scaled to increasing $S_a(T_1)$ values at 0.05 g steps until collapse. The sets of subduction GM records represented the aleatory uncertainty of the seismic hazard (called RTR variability), which is the dominant source of uncertainty compared to epistemic uncertainty (i.e., modeling uncertainty) [13]. Still, important epistemic uncertainty, such as damage and material decay due to, for example, climate factors, should be incorporated in future studies. To reduce the significant computational costs, the IDAs were performed only along the shorter horizontal plan direction, i.e., the direction along which the collapse fragility was greater. The lowest $S_a(T_1)$ value at which either the global or local structural collapse criteria were met was recorded for each structural model and each GM. Since earthquake-induced collapse should be evaluated in a probabilistic way, the dispersion of the $S_a(T_1)$ collapse intensity values was calculated to develop the respective building-specific collapse fragility functions.

2.2. Estimation of Collapse Fragility Functions

Earthquake-induced building-specific collapse fragility functions provide the probability of structural collapse based on the GM intensity $S_a(T_1)$, which is represented by $P(C \mid S_a(T_1)$, and assume a lognormal probability function (PF) [30–32]. These PFs are defined by two parameters: (1) the median (i.e., median collapse capacity, $\hat{\theta}$); and (2) the logarithmic standard deviation (i.e., dispersion, $\hat{\beta}$). Both parameters were estimated using the Maximum Likelihood Method (MLM) [33]. Then, Kolmogorov–Smirnov (K–S) and Lilliefors goodness-of-fit tests, both at a 5% significance level, were used to evaluate the quality of the lognormal PFs. The K–S test is a distribution-free (non-parametric) test used for continuous distributions [34], whereas the Lilliefors test, recommended by FEMA P-58 [13], is a more severe (compared to the K–S test) test and is recommended when the parameters are not specified but estimated from the sample [35].

2.3. Collapse Performance Metrics: $P(C \mid S_a(T_1)_{MCE})$, CMR, λ_c, and $P_c(50)$

The probability of collapse based on the MCE hazard level, $P(C \mid S_a(T_1)_{MCE})$, is obtained from the collapse fragility curves. This collapse performance metric is used to assess safety considerations, for instance, US code-conforming buildings must not have values greater than 10% [10]. Another collapse performance metric is the Collapse Margin Ratio (CMR), which is defined by FEMA P695 [16] as the median collapse capacity $\hat{\theta}$ divided by $S_a(T_1)_{MCE}$, as shown in Equation (1).

$$\mathrm{CMR} = \frac{\hat{\theta}}{S_a(T_1)_{MCE}} \tag{1}$$

A complementary collapse performance metric is the mean annual frequency of collapse, λ_c, which represents the average number of earthquake-induced structural collapses of a given structure per year. This λ_c value is a combination of the product of the collapse fragility and the seismic hazard (Equation (2), where $d\lambda_{Sa}/dS_a$ is the derivative of the seismic hazard curve).

$$\lambda_c = \int_0^{\infty} P(C|S_a(T_1)) \cdot \left| \frac{d\lambda_{S_a}}{dS_a} \right| \cdot dS_a \tag{2}$$

Finally, the probability of one earthquake-induced collapse in 50 years, $P_c(50)$, is obtained using Equation (3) and assumes a Poisson process. This performance metric is the collapse potential in 50 years and is also used to analyze safety considerations. For instance, US code-conforming buildings must not have $P_c(50)$ values greater than 1% [10].

$$P_c(50) = 1 - e^{-\lambda_c \cdot 50} \tag{3}$$

2.4. Deaggregation of λc

Lastly, the contribution of each $S_a(T_1)$ value to λ_c is obtained through the deaggregation of λ_c. Deaggregated λ_c values based on $S_a(T_1)$ are obtained using Equation (4), whose terms have been previously defined.

$$\text{Deaggregated } \lambda_c(S_a) = P(C|S_a(T_1)) \cdot \left| \frac{d\lambda_{S_a}}{dS_a} \right| \cdot dS_a \tag{4}$$

3. Methodology to Define Code-Conforming Archetype Buildings

3.1. Statistical Evaluation of Chilean RC Buildings

The archetype buildings were selected based on a statistical analysis of the available data on RC buildings in the Chilean national database of buildings [36]. The database contains disaggregated information on buildings with at least 3 stories that were built between 2002 and 2020 (both years included) such as the location, construction material used, number of stories, equivalent floor area, etc. In more detail, there are 8078 RC buildings of which roughly 10% are office buildings and roughly 90% are residential buildings. Of these, 470 are office buildings located in the Metropolitan Region (MR-Region) and 62 are office buildings located in the Valparaiso Region (V-Region), which are the most densely populated regions in the country. Figure 2a presents the classification of these office buildings as a function of the number of stories. The International Building Code (IBC) [37] defines a high-rise building as a building that is more than 22.86 m (75 ft) above the lowest level (Figure 2b). Considering an average story height of 2.5 m, this may correspond to a building of 10 stories or more. Therefore, a mid-rise building can be defined as a building that has more than 3 stories but no more than 9 stories. In Figure 2a, it is noted that 359 office buildings are classified as mid-rise buildings, whereas 173 office buildings are classified as high-rise buildings (i.e., ~70% and ~30%, respectively).

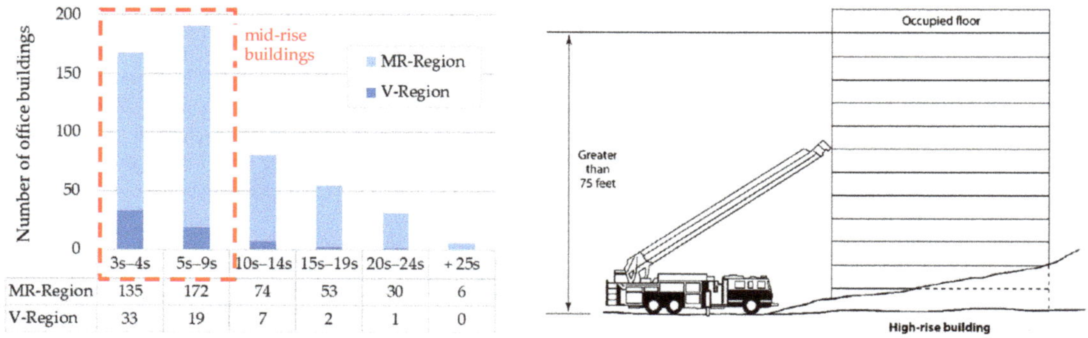

(a) (b)

Figure 2. (**a**) Chilean RC buildings 2002–2020: number of office buildings and number of stories. (**b**) Definition of high-rise building from IBC-2021 [37]. MR-Region = Metropolitan Region; V-Region = Valparaiso Region.

This research focused on mid-rise buildings for two reasons. First, the seismic response of mid-rise buildings can be expected to be dominated by the fundamental mode of vibration. Such a response generates a somewhat uniform distribution of story drift ratios, which may affect the collapse performance and economic losses [38]. Second, most of the existing RC office buildings are mid-rise buildings (roughly 70%), which makes them of great interest on a regional level in terms of their collapse performance. Furthermore, recent research [39] has also highlighted the need to study the actual seismic performance of mid-rise RC shear-wall buildings, which are a typical structure in Chile. From the database [36], it is also noted that the equivalent floor area is independent of the number of stories for RC office buildings, with an average value of 1000 m^2.

The influences of the hazard level and soil type were also assessed due to their impact on seismic design requirements, seismic hazard, and, possibly, collapse performance. Hence, two different cities, Santiago and Vina del Mar, were considered. Based on the Chilean seismic design provisions for buildings [8], Santiago and Vina del Mar are located in seismic zones 2 (moderate seismicity zone) and 3 (high seismicity zone), respectively. Furthermore, two different soil types were considered based on the Chilean seismic soil classification code [9], i.e., soil types B and D. In more detail, soil type B is fractured rock or very dense/firm soil, whereas soil type D is moderately dense/firm soil (soil type D is softer than soil type B). For the archetypes the naming convention of Bld-07-zX-sY was used, where 07 designates the number of stories, zX denotes the seismic zone (i.e., z2 or z3), and sY represents the soil type (i.e., sB or sD). For example, Bld-07-z3-sD denotes a 7-story archetype building located in seismic zone 3 on soil type D.

3.2. Structural Layout and Code-Conforming Design Procedure of the Archetype Buildings

The structural layout adopted was a minor simplification of the actual layout of a representative mid-rise office RC building. This building was selected because its geometric characteristics (i.e., height and equivalent floor area) and dimensions, as well as the arrangement of its structural elements (i.e., basement walls, shear walls, beams, and columns), were consistent with the results of the statistical survey and Chilean practices. In this sense, the archetype buildings have seven stories and three basement levels. The typical floor dimensions are 24 m × 40 m (960 m^2), whereas the beam span lengths are 8.0 m. The underground levels are surrounded by 20 cm thick perimeter walls. Figure 3a shows a typical floor plan view and Figure 3b shows a typical underground floor plan view. Likewise, Figure 3c depicts the lateral transverse view and a 3D view is provided in Figure 3d.

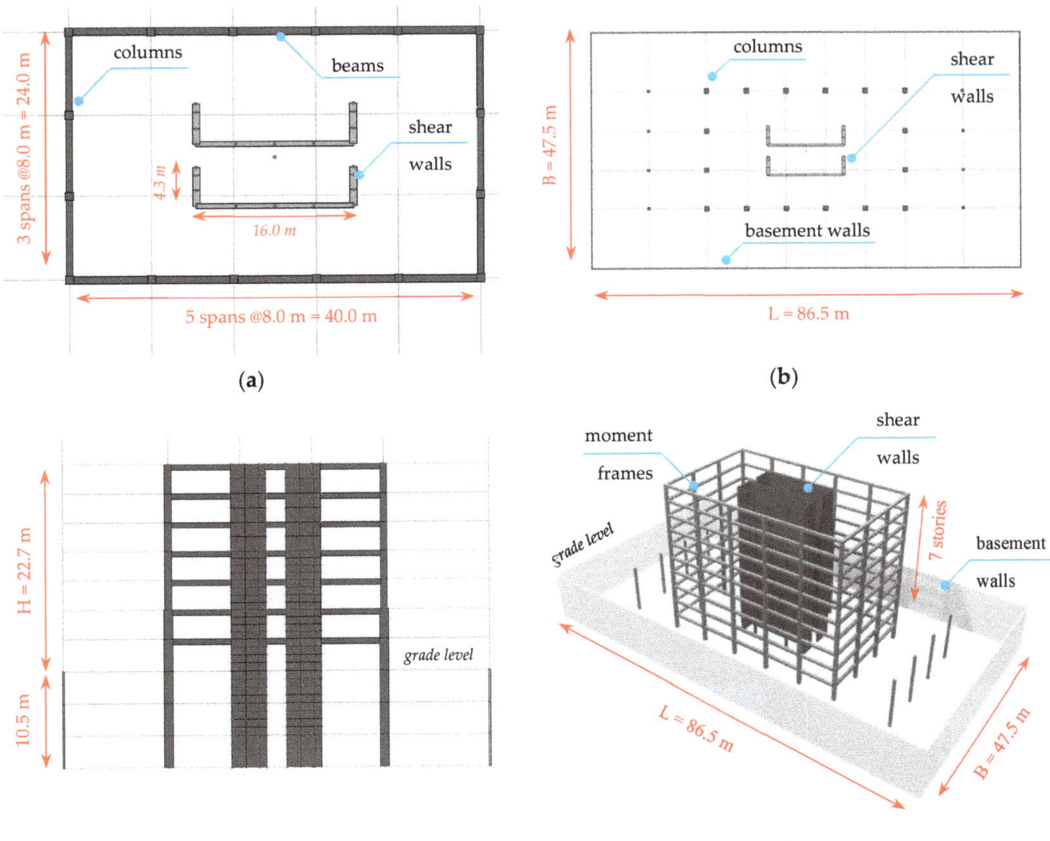

Figure 3. Archetype buildings. (**a**) Typical floor plan view. (**b**) Underground floor plan view. (**c**) Lateral transverse view. (**d**) 3D view.

The slabs are post-tensioned slabs with thicknesses of 22 cm and 20 cm in the underground and other levels, respectively. The story height of the underground stories and the first story is 3.50 m, whereas the story height of the remaining stories is 3.20 m. The structural system comprises two core C-shaped SWs and IMFs at the perimeter. Core walls are continuous walls from the foundation to the top. Chilean professional engineering practices and prescriptive code-based procedures were considered in the design of the archetype buildings [12]. The design was developed using the software ETABS [40], a commercial software that is used for structural design [41]. Concrete and steel bars with a nominal strength (f'_c) of 35 MPa and nominal yield strength (f_y) of 420 MPa, respectively, were assumed. A dead load of 2 kPa and a live load of 3 kPa were used for the underground levels, whereas 5 kPa and 2 kPa were considered for the typical stories and the roof, respectively.

The design process was based on the current Chilean seismic code NCh 433 [8], which was updated after the 2010 earthquake [9]. Therefore, the archetype buildings considered in this study are, in a strict sense, representative of post-2010 buildings. However, the statistical survey described in Section 3.1 provides data on characteristics (e.g., number of stories and floor area) that were not influenced by the 2010 earthquake. Hence, these data are representative of both pre- and post-2010 buildings and, consequently, so is the architectural layout of the archetypes.

The code-confirming design process was based on a modal response spectrum analysis. Seismic forces are provided by an elastic spectrum divided by an effective response modification factor (R^{eff}). The initial estimate of the response modification factor (R^*) depends exclusively on the building's fundamental period, the structural system, and the soil type. If the base-shear demand, V_b, divided by the seismic weight of the building, W, (normalized base-shear demand, C) satisfies both the upper (C_{max}) and lower (C_{min}) limits (i.e., $C_{min} \leq C \leq C_{max}$), then $R^{eff} = R^*$. Or else, R^* is modified so that $C_{min} \leq C \leq C_{max}$, and in this case, R^{eff} is set equal to the modified value of R^*. The relevant design factors are summarized in Table 1.

Table 1. Design parameters.

Archetype	Transverse Direction				Longitudinal Direction				C_{min}
	T (s)	R^*	R^{eff}	C (%)	T (s)	R^*	R^{eff}	C (%)	(%)
Bld-07-z2-sB	0.923	8.6	6.5	5.0	0.465	6.2	6.2	6.0	5.0
Bld-07-z2-sD	0.835	6.1	6.1	7.0	0.426	3.7	3.7	11.2	6.0
Bld-07-z3-sB	0.913	8.6	6.5	6.7	0.464	6.2	6.2	8.1	6.7
Bld-07-z3-sD	0.673	5.9	5.9	9.8	0.357	3.6	3.6	15.7	8.0

T = fundamental period; R^* = response modification factor; R^{eff} = effective response modification factor; C = normalized base-shear demand.

It must be noted that in the transverse direction, the seismic design of the archetypes on soil type B (i.e., Bld-07-z2-sB and Bld-07-z3-sB) was controlled by the minimum base-shear requirement (C_{min}), which depends on the importance factor (equal to 1.0 due to occupation category II), seismic zone, and soil type. As explained later, the design of the remaining archetypes in the transverse direction (i.e., Bld-07-z2-sD and Bld-07-z3-sD) was controlled by the axial deformation demand on the RC core walls. Figure 4 shows the elastic and reduced code-conforming design spectra (NCh 433) for each archetype building. The vertical dashed lines indicate the fundamental period of vibration (T_1) in the transverse direction (i.e., the direction of analysis). The reduced values of $S_a(T_1)$ used in the design of the structural members are indicated in each figure.

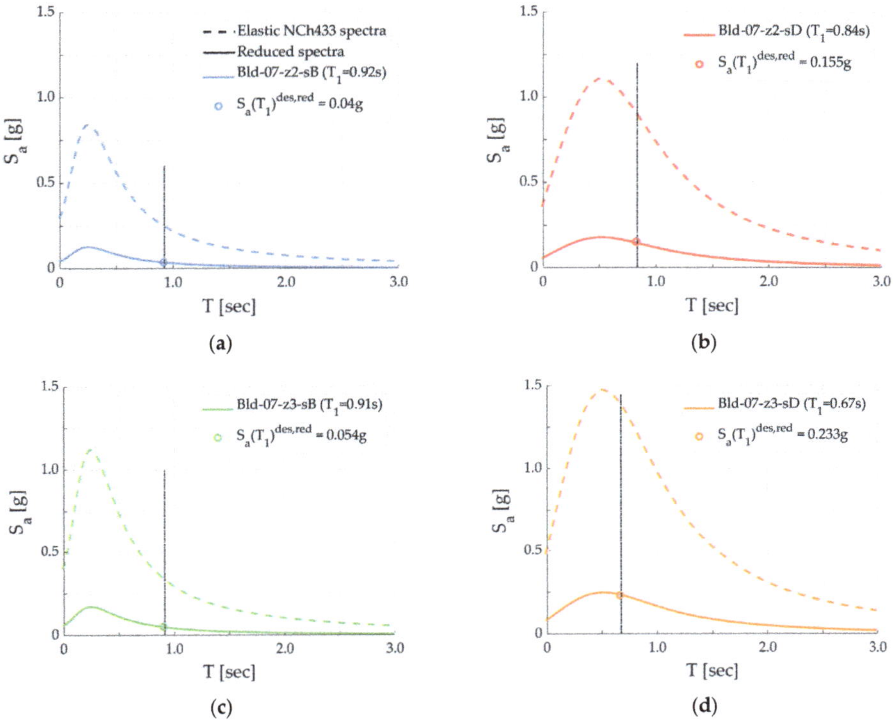

Figure 4. Comparison of elastic NCh 433 and reduced design spectra (transverse direction). (**a**) Bld-07-z2-sB; (**b**) Bld-07-z2-sD; (**c**) Bld-07-z3-sB; (**d**) Bld-07-z3-sD. Note: in all subfigures, dashed lines and continuous lines represent the elastic NCh433 spectrum and reduced spectrum, respectively.

Moreover, the Design-Basis Earthquake (DBE) uniform hazard spectra (UHS) (i.e., UHS with a 10% probability of exceedance in 50 years) and elastic code (NCh 433) spectra are shown in Figure 5 for each archetype for comparison purposes. As a reference, the DBE-UHS $S_a(T_1)$ values are indicated in each figure. The UHS and elastic NCh 433 spectra of the archetype buildings on soil type B (Figure 5a,c) have similar spectral shapes but differences in the values of the spectral ordinates are evident. On the other hand, the UHS and elastic code spectra of the archetype buildings on soil type D (Figure 5b,d) have dissimilar spectral shapes but the corresponding spectral ordinates are very close to each other at the fundamental period. It is worth mentioning that the elastic spectra specified in the current Chilean seismic code [8,9] are not probabilistic in nature but calibrated for the structural demands observed in recent earthquakes (1985 and 2010 earthquakes [12]). Therefore, differences between the DBE-UHS and elastic code spectra were expected. The process used to obtain the UHS spectra is explained in Section 4.

The member dimensions are summarized in Table 2. The length of the web and flanges of the core walls are the same in all archetype buildings (see Figure 3). The corresponding thicknesses were defined so that wall densities were greater than 0.1% in all stories (common Chilean engineering practice [12]). Beams and columns were designed per the Chilean standard DS 60 [11], which refers to ACI 318-08 [42]. Perimeter frames are IMFs because they were designed for less than 25% of the story shear at all stories. As required by the current Chilean codes, boundary elements (special, SBE, and ordinary, OBE) were provided at the ends of the flanges of the core walls of all archetypes. SBEs were provided only at the first two stories below ground level and at the first three (archetypes on soil type B) and two (archetypes on soil type D) stories above ground level. As commonly implemented in Chilean practice, the thickness of the shear walls was set to be constant along the entire height. Regarding the IMFs, all beams at all stories had the same cross-section. All columns from the third underground story to the second story had the same cross-section and all columns from the third story to the top story also had the same cross-section (Table 2). It is noted that archetypes Bld-07-z2-sB, Bld-07-z2-sD, and Bld-07-z3-sB had similar member sections and, hence, similar fundamental periods (T_1 values of 0.92 s, 0.84 s, and 0.91 s, respectively). On the other hand, archetype Bld-07-z3-sD had larger member sections (walls, beams, and columns) and was a significantly stiffer structure (T_1 = 0.67 s).

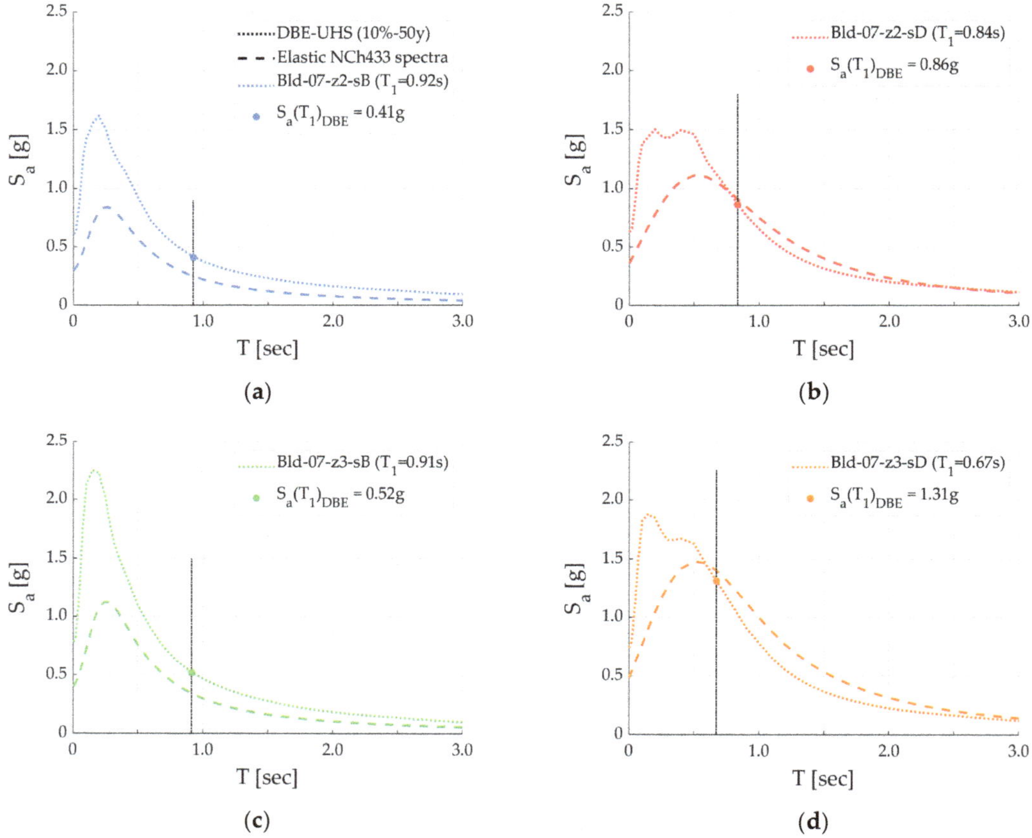

Figure 5. Comparison of DBE-UHS and elastic NCh 433 spectra. (**a**) Bld-07-z2-sB; (**b**) Bld-07-z2-sD; (**c**) Bld-07-z3-sB; (**d**) Bld-07-z3-sD. Note: in all subfigures, dot lines and dashed lines represent the DBE-UHS spectrum and the elastic NCh433 spectrum, respectively.

Table 2. Member cross-sections.

Archetype	Core Walls				Beams	Columns	
	Flanges		Webs		(b × h)	(b × h)	
	l	t	l	l		us3–s2°	s3°–s7°
Bld-07-z2-sB	4.2	0.35	16.0	0.25	0.6 × 0.5	0.7 × 0.7	0.6 × 0.6
Bld-07-z2-sD	4.2	0.45	16.0	0.30	0.6 × 0.5	0.7 × 0.7	0.6 × 0.6
Bld-07-z3-sB	4.2	0.35	16.0	0.25	0.6 × 0.5	0.7 × 0.7	0.6 × 0.6
Bld-07-z3-sD	4.2	0.55	16.0	0.40	0.7 × 0.6	1.0 × 1.0	0.8 × 0.8

l = length; t = thickness; b = width; h = height; us3 = third underground story; s2° = second story; s3° = third story; s7° = top story. Note: all dimensions are in meters.

3.3. Modeling Details

The software Perform-3D [43] was used to create the 3D mathematical models, taking into account the latest nonlinear (NL) modeling guidelines [44–50]. Perform-3D was chosen because it provides an adequate balance between accuracy and computational costs. Very recent studies (e.g., [51,52]) used Perform-3D to analyze RC shear-wall buildings, indicating that this software is adequate for the seismic analysis of the archetype buildings considered in this investigation. Figure 6a shows a typical Perform-3D model of the archetype buildings. The expected values of the material strength and member stiffness were considered in the structural models. The 'shear wall' element (Figure 6b) was used to model the shear walls because the walls were slender and had no openings. This macro-element (4 nodes and 24 degrees of freedom) is typically used in engineering practice and research studies [53,54] to model the NL response of flexure-controlled RC walls subjected to both lateral and vertical loading.

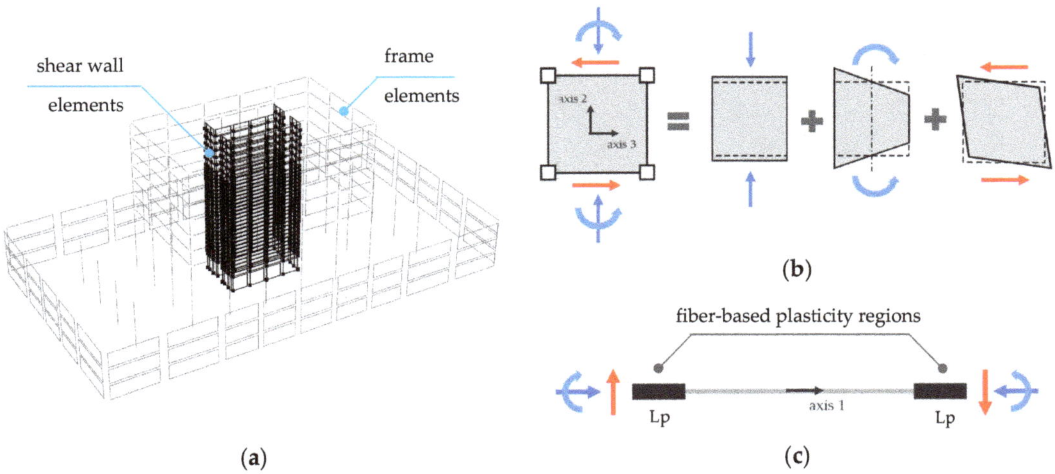

Figure 6. Perform-3D modeling. (**a**) 3D archetype; (**b**) shear-wall element; (**c**) frame element.

The 'shear wall' macro-element integrated three models to simulate the RC wall behavior: (1) a fiber-type section model, including NL steel and concrete fibers, which simulated the in-plane axial-flexural response; (2) a uniform shear layer, with a one-dimensional NL shear model that simulated the in-plane shear response; and (3) a uniform linear-elastic plate-bending model that simulated the out-of-plane response. In the 'shear wall' element, the bending and axial effects were coupled, whereas the bending and shear effects were not. As indicated by recent modeling recommendations [53], the NL material properties of the wall cross-section fibers were defined based on the uniaxial constitutive relationship (i.e., stress vs. strain backbone curve) YULRX (Y: yielding; U: ultimate; L: loss; R: residual; X: maximum).

As recommended by engineering practice [48], the nominal reinforcing steel and concrete strengths were multiplied by 1.17 and 1.3, respectively. Furthermore, the material cyclic response (i.e., unloading and reloading stiffnesses degradation) was included by defining an energy dissipation factor for concrete material and energy dissipation and stiffness factors for reinforcing steel material [53]. The shear layer was defined by an elastic-perfectly plastic shear stress vs. shear-strain backbone curve that considers the potential nonlinear shear behavior of the walls. The expected shear strength of the walls was set to 1.5 times the nominal shear strength, as defined by ACI 318-08 [42], whereas the effective shear stiffness was set to 10% of the uncracked shear stiffness (G_c^{eff} A = 0.1 G_c A, where G_c = 0.4 E_c, A is the cross-section, and E_c is the elastic modulus) [45]. This procedure is an indirect way to consider shear cracking because the shear-wall element captures neither nonlinear shear deformations nor coupling with nonlinear flexural deformations under cyclic loading. Figure 7 presents examples of the constitutive functions of the concrete and steel fibers of the core walls of archetype Bld-07-z3-sD.

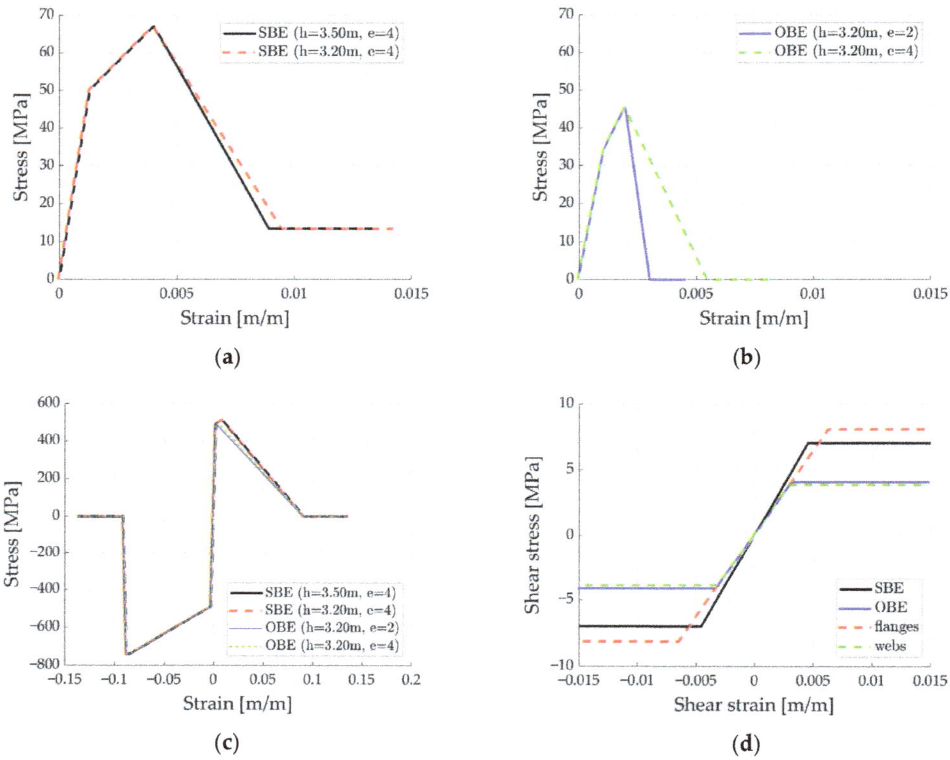

Figure 7. Archetype Bld-07-z3-sD: material models for the shear walls. (**a**) Confined concrete. (**b**) Unconfined concrete. (**c**) Steel. (**d**) Wall shear. SBE = special boundary element; OBE = ordinary boundary element; h = story height; e = number of vertical shear-wall elements per story.

It is important to note that the concrete materials (both confined and unconfined) incorporated regularization derived from the theory of constant compressive crushing energy [55]. The effect of regularized materials for confined and unconfined uniaxial YULRX curves was exemplified for the SBEs and OBEs (Figures 7a and 7b, respectively). For reinforcing steel fibers, the steel materials indirectly incorporated the buckling of steel bars for when the cover concrete material reaches the crushing point (i.e., R in YULRX curve), as shown in Figure 7c. The shear layer constitutive relationships are shown in Figure 7d. The results of the experimental tests of slender walls with different types of cross-sections (planar [56] and T-shaped [57]) were used to validate the number of shear-wall elements per story, the material regularization, and the number of fibers in the cross-sections, among other modeling aspects. This validation, not presented here for brevity, was essentially identical to that performed in previous studies [45,53,58].

Beams and columns were modeled using 'frame-type' elements (see Figure 6c) with fiber-based plasticity regions of length L_p at both ends and a linear-elastic region in between. YULRX backbone functions were also adopted to represent the uniaxial stress–strain relationships between steel and concrete materials. Length L_p was determined following recommendations found in the literature [59] (L_p = 0.5 h, where h is the depth of the element). Geometric nonlinearity was also implemented to consider the P-delta effects in the columns and walls. As indicated in previous studies [60,61], long-span post-tensioned slabs in core-wall buildings were considered so as not to substantially affect the building response. Thus, these elements were not explicitly modeled. In its place, rigid diaphragms were implemented at all floor levels and the self-weight and mass of the slabs were incorporated into the models. Lastly, energy dissipation was mostly modeled directly by the hysteretic force–deformation response of the structural components. Modal damping was set to 2.4% for all modes and the additional Rayleigh damping was set to 0.1% at 0.2 T_1 and 1.5 T_1 [43,44].

3.4. Seismic Collapse Criteria

Predictions of seismic collapse require the identification of all possible collapse modes and the evaluation of structural seismic collapse through simulated and non-simulated collapse modes. Figure 8 schematizes the collapse criteria adopted in the analyses.

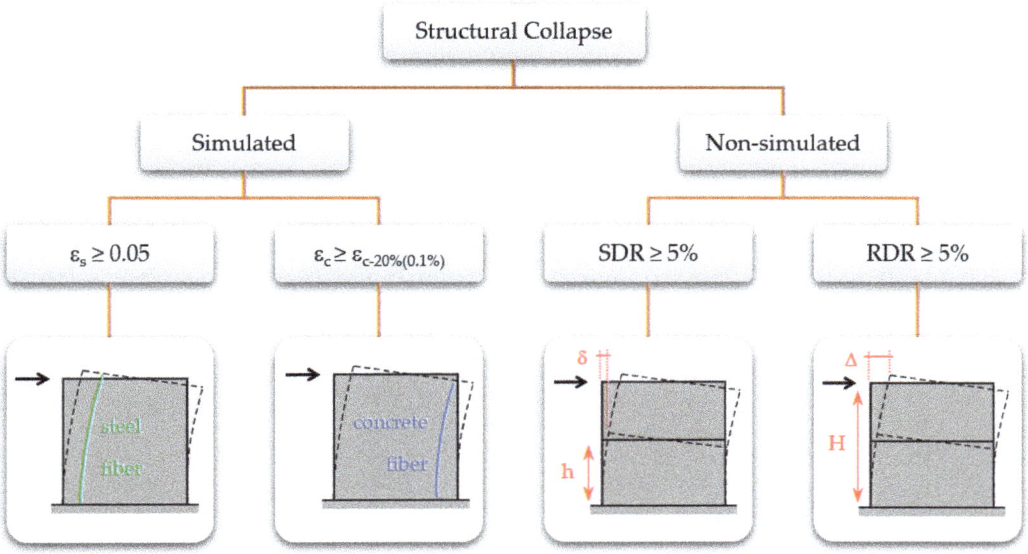

Figure 8. Structural collapse criteria.

The simulated failure criteria considered the local response parameters derived from the axial-bending demands on the walls. Collapse due to fracture of the reinforcement bars was assumed to occur when the tensile strains in the longitudinal reinforcement exceed 0.05, as indicated by Gogus and Wallace [17]. Moreover, collapse due to longitudinal reinforcement buckling and concrete crushing was assumed to occur when the concrete compressive strains exceed the crushing limit (the point where the post-peak descending branch of the concrete stress–strain curve reaches 20% of the peak stress of the confined concrete or 0.1% of the peak stress of the unconfined concrete [17]). In the case of non-simulated criteria, on the other hand, collapse is related to either axial wall failure or slab-column failure. As indicated by Kim and Foutch [62], slab-column failure was assumed to occur when a story drift ratio (SDR) exceeds 5% at any story. In addition, axial wall failure was assumed to occur when the roof drift ratio (RDR) exceeds 5% [17]. Other collapse measures, such as numerical instability or excessive increases in story drift demands for small increases of the ground motion IM ($S_a(T_1)$), were not detected.

3.5. Pushover Analyses

Initially, pushover analyses (nonlinear static analyses) were implemented to evaluate the seismic response of each archetype building in terms of strength, stiffness, and the type and location of the damage. First, the gravitational loads (100% of dead loads plus 25%

of live loads) were applied. Second, a first-mode lateral force pattern, as recommended by ASCE 41-17 [63], was applied in the transverse direction (i.e., the same direction along which the NLRH analyses were performed), both positive and negative. Figure 9 shows the pushover curves of the archetype buildings, along with some points of interest.

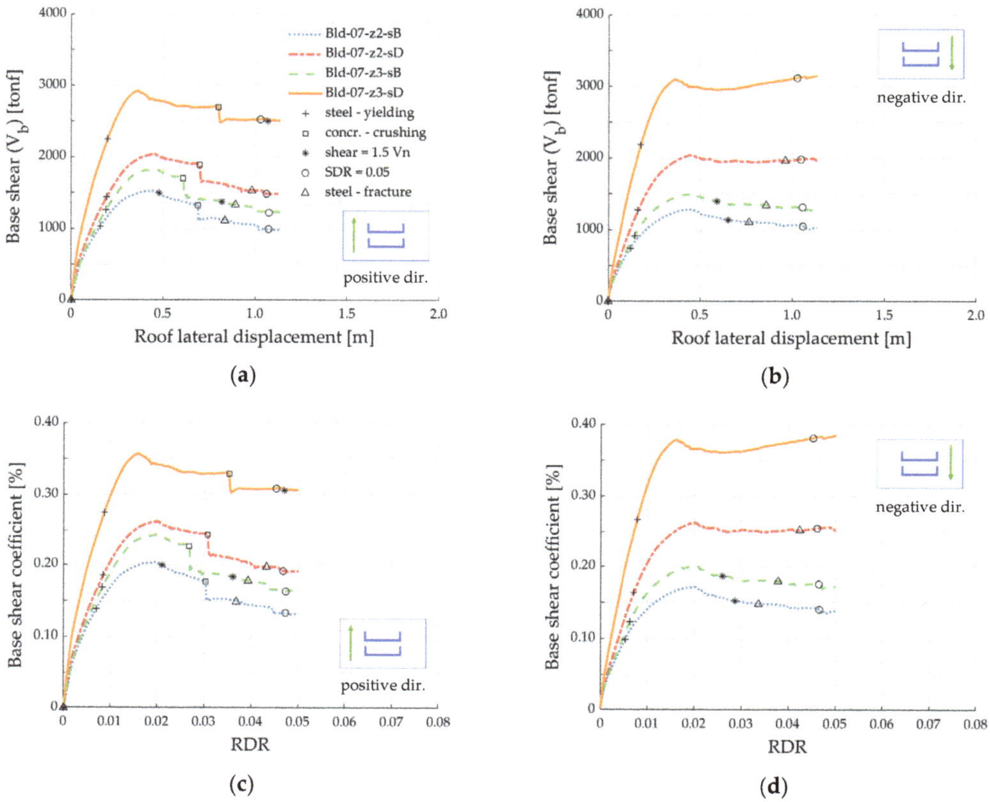

Figure 9. Capacity curves (pushover). (**a**,**b**) Base shear vs. roof displacement. (**c**,**d**) Base-shear coefficient versus roof-drift ratio.

In more detail, Figure 9a,b show the pushover curves in terms of the base shear versus the roof displacement, whereas Figure 9c,d present the same curves in terms of the base0shear coefficient (i.e., base shear divided by the structural weight) versus the RDR. The base shear V_b was evaluated at the ground level (see Figure 3) and normalized by the seismic weight of the stories above. The capacity curves in the positive direction were not equal to those in the negative direction due to the asymmetric plan layout of the core walls (see Figure 3a). Moreover, Figure 9 depicts some points of interest: (1) the first fiber at any wall that reached steel yielding, concrete crushing, shear strength, and steel fracture; and (2) the SDR reaching 5% at any story.

The pushover curves show some interesting features. Firstly, archetype Bld-07-z3-sD had considerably more stiffness and strength than the other archetypes, likely because of the increased cross-sections of the shear walls, beams, and columns. Secondly, in all cases, the maximum base-shear capacity (V^{max}) was reached at roughly the same value of the RDR (≈1.5~2.0%). Finally, there was always a clear drop in strength after the first concrete crushing when the seismic loads were applied in the positive direction. It is interesting to note that this first concrete crushing occurred at the SBEs of the walls located at the first story, where the shear and moment demands were expected to be more significant, and at RDR values higher than 2.5% (i.e., excessive drift demands).

The overstrength factor (Ω), i.e., the ratio of V^{max} to the design base shear V^{des}, was calculated. Table 3 summarizes the pushover analysis results, where the ratios of V^{des} (also above ground level) and V^{max} to the seismic weight are also shown. Along the positive transverse direction, the Ω^{+} values ranged from 2.4 to 4.4, with an average value of 3.3. Along the negative transverse direction, on the other hand, the Ω^{-} values ranged from 2.5 to 3.7, with an average value of 3.0. Again, the high values of V^{max} for archetype Bld-07-z3-sD (2918 tonf for V^{max+}, and 3149 tonf for V^{max-}) were consistent with the larger cross-section dimensions of the structural members, resulting in an archetype that had more strength.

Table 3. Summary of pushover analyses.

Archetype	V^{des} (tonf)	V^{des}/W (%)	V^{max+} (tonf)	V^{max+}/W (%)	Ω^{+}	V^{max-} (tonf)	V^{max-}/W (%)	Ω^{-}
Bld-07-z2-sB	348	4.7	1524	20.4	4.4	1283	17.2	3.7
Bld-07-z2-sD	826	10.7	2036	26.3	2.5	2038	26.3	2.5
Bld-07-z3-sB	464	6.2	1827	24.4	3.9	1495	20.0	3.2
Bld-07-z3-sD	1231	15.1	2918	35.7	2.4	3149	38.5	2.6

Referring to the location of the damage, the first nonlinear incursions of the concrete and steel fibers of the shear-wall elements appeared at the stories immediately above and below ground level. As expected, these stories were among the stories where the seismic code requires the inclusion of SBEs. As previously mentioned, when pushover was applied in the positive direction (i.e., boundary elements of the walls in compression), the first concrete crushing in the boundary elements was observed to occur at the first story, where the shear and moment demands were expected to be more significant. An example (archetype Bld-07-z2-sB) of this behavior can be seen in Figure 10, where the concrete fiber strains of SBEs are shown for two RDR values (1.5% in Figure 10a and 3.0% in Figure 10b). In addition, a schematic representation of the uniaxial constitutive relationship 'YULRX' is depicted in Figure 10c, where the colors represent the aforementioned regions of interest.

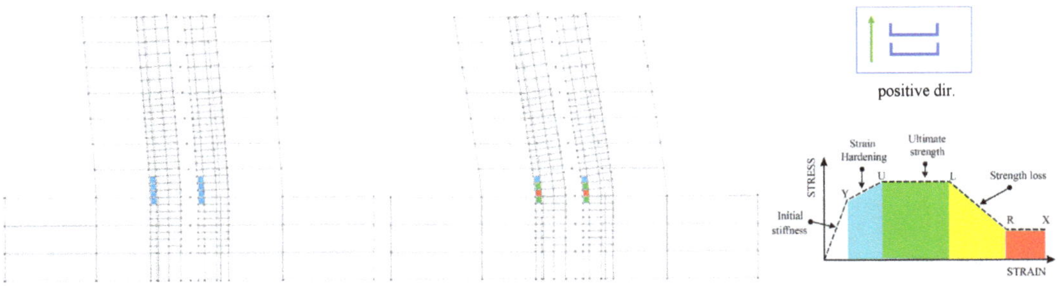

(a) (b) (c)

Figure 10. Strains at the boundary elements of archetype Bld-07-z2-sB (pushover along the positive transverse direction) (**a**) at V_b^{max} (RDR ≈ 1.5%), and (**b**) at the first concrete crushing point R (RDR = 3.0%). (**c**) Schematic YULRX curve: color scale.

4. Seismic Hazard Analyses and Selection of Subduction Ground Motions

4.1. Seismic Hazard Analyses

Probabilistic seismic hazard analysis (PSHA) was used to integrate the rupture scenarios defined in a recent Chilean seismic source model (SSM) with current Chilean ground-motion models (GMMs). The seismicity model defined by Poulos et al. [64] and the GMMs proposed by Montalva et al. [65] and Idini et al. [66] were integrated into the computational platform SeismicHazard [67] to assess the PHSA. Calculations were performed incorporating the GMM epistemic uncertainties using a logic tree with equal weights of

1/2. Figure 11a depicts the calculated seismic hazard curves in terms of the $S_a(T_1)$ versus the mean annual frequency of exceedance, $\lambda_{Sa(T_1)}$, and Figure 11b depicts the $S_a(T_1)$ versus the mean return period, Tr. Horizontal lines are also drawn to represent the traditional λ_{Sa} hazard levels related to return periods of 2475, 475, and 72 years, i.e., the maximum considered earthquake (MCE), design-basis earthquake (DBE), and service-level earthquake (SLE), respectively.

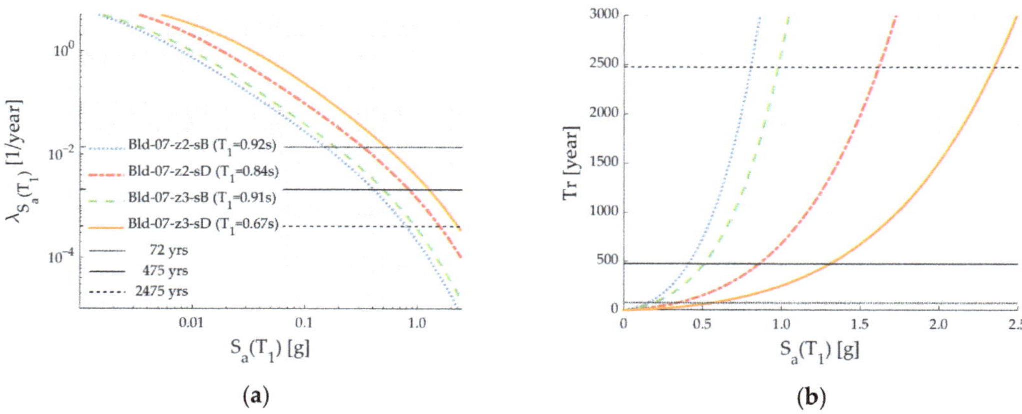

Figure 11. Seismic hazard curves plotted as a function of (**a**) the annual probability of exceedance, $\lambda_{Sa(T1)}$, and (**b**) the return period, Tr.

The $S_a(T_1)$ values for the different hazard levels are presented in Table 4, along with the values of both the Chilean code elastic design spectra, $S_a(T_1)^{des,e}$, and the reduced design spectra, $S_a(T_1)^{des,red}$, (see Figure 4). It is noted that there were relatively small differences in terms of the $S_a(T_1)_{SLE}$, $S_a(T_1)_{DBE}$, and $S_a(T_1)_{MCE}$ values between the archetypes on soil type B (i.e., Bld-07-z2-sB and Bld-07-z3-sB), which had very similar values of T_1. On the other hand, the $S_a(T_1)$ values for the archetype Bld-07-z3-sD were almost 1.5 times higher than those for archetype Bld-07-z2-sD, indicating a considerable increase in seismic demand. It is interesting to note that the values of $S_a(T_1)^{des,e}$ were different from those of $S_a(T_1)_{DBE}$. These differences were expected because the elastic design spectra of the Chilean seismic design code NCh 433 [8,9] are not uniform-hazard spectra. In addition, the Chilean code does not explicitly define the SLE and MCE hazard levels. These levels were defined as indicated in ASCE 7-22 [10].

Table 4. Summary of IM = $S_a(T_1)$.

Archetype	$S_a(T_1)_{SLE}$ (g)	$S_a(T_1)_{DBE}$ (g)	$S_a(T_1)_{MCE}$ (g)	$S_a(T_1)^{des,red}$ (g)	$S_a(T_1)^{des,e}$ (g)	P(%) in 50 y	Tr (y)
Bld-07-z2-sB	0.16	0.41	0.81	0.040	0.261	24	180
Bld-07-z2-sD	0.34	0.86	1.63	0.155	0.943	8	600
Bld-07-z3-sB	0.19	0.52	0.99	0.054	0.348	22	200
Bld-07-z3-sD	0.55	1.31	2.36	0.233	1.377	9	530

P(%) in 50 y = probability of exceedance in 50 years (as a percentage); Tr = mean return period.

For comparison purposes, Table 4 also shows the probability of exceedance in 50 years and the related return period of the spectral ordinates $S_a(T_1)^{des,e}$ (i.e., spectral ordinates of the Chilean code elastic design spectra). For the archetypes on soil type D, the probabilities of exceedance in 50 years were nearly 10% (Tr ~475 years), which is a worldwide standard for the DBE hazard level. For the archetypes on soil type B, on the other hand, the probabilities of exceedance in 50 years were higher than 22% (Tr ~200 years). In other words, the hazard level associated with the elastic design NCh 433 spectra for soil type B was smaller than that associated with the DBE. These observations are consistent with the previous results shown in Figure 5.

4.2. Selection of Ground Motions

Special care was taken to obtain sets of Chilean subduction ground motions consistent with the IM-based seismic hazard at different sites of different seismicity levels and different soil types. First, from the hazard curves obtained in the preceding section, uniform hazard spectra (UHS) were calculated at different levels of seismic hazard. A single UHS represents the acceleration spectral values for the same probability of exceedance (uniform) in a given exposure time (50 years, commonly used). For each site–soil case, the UHS with a 10% probability of exceedance in 50 years (DBE hazard level) was constructed based on the hazard curves obtained from the 'SeismicHazard' platform [67] (see Figure 5). Then, the UHS with a 2% probability of exceedance in 50 years was constructed. For each archetype, a Conditional Spectrum (CS) was rigorously constructed and defined as a target spectrum in this study. The approach proposed by Baker [68] was adopted to define the hazard-consistent target $S_a(T)$ distribution (with conditional mean and conditional standard deviation at each period within the range of interest). The CS calculations considered the target $S_a(T_1)$ for the 2%—50-year hazard level and the mean causal magnitude M, distance R, and epsilon ε obtained from the PSHA deaggregation, where each GMM was considered separately. The mean and standard deviation of logarithmic $S_a(T)$ also considered the correlation model proposed by Candia et al. [69] for the Chilean subduction zone, where the correlations are generally higher than those for other subduction zones such as Japan. Finally, single GMM calculations were combined using assumed logic-tree weights, following method 2, which was suggested by Lin et al. [70], to obtain the composite CS of each case. Although logic-tree weights are not rigorously correct, it was considered a convenient approximation for this investigation.

Using the Chilean strong motion database 'SIBER-RISK' [71], for each archetype, a hazard-consistent ensemble of 44 subduction ground motions (a typical number of records used in research [41]) was selected and scaled to match the target mean, variance, and correlations of the spectral acceleration values, $S_a(T)$, at a period between 0.2 T_1 and 2.0 T_1, following the procedure proposed by Baker and Lee [72]. Since NLRH analyses of the 3D archetypes were performed only in the transverse direction, each ground motion was a horizontal component. Following current selection guidelines [16,73], these components met the selection criteria and the objectives of consistency, representativeness, and statistical sufficiency to permit the statistical evaluation of the RTR variability in the structural response. To avoid bias in the probability of collapse when $S_a(T_1)$ was used as the IM, the amplitude scaling factor that modifies the ground motions to achieve the desired intensity level was limited to a maximum value of 5.0, as suggested by recent studies [74,75] where the spectral shape was appropriately accounted for in the selection process.

Following this procedure, 4 different sets of 44 GM records were selected and scaled (i.e., a set for each archetype), which were consistent with the respective deaggregated M, R, ε, and soil type. Figure 12 shows the spectra, mean, and dispersion of the four ground motion sets (one for each archetype building) and the respective target composite CS. The dispersion of the spectral ordinates, $S_a(T)$, was higher for archetypes on soil type D (Figure 12b,d) than for those on soil type B (Figure 12a,c). This variability was inherited from the Chilean GMMs used in the PSHA and may impact the variability of the structural response.

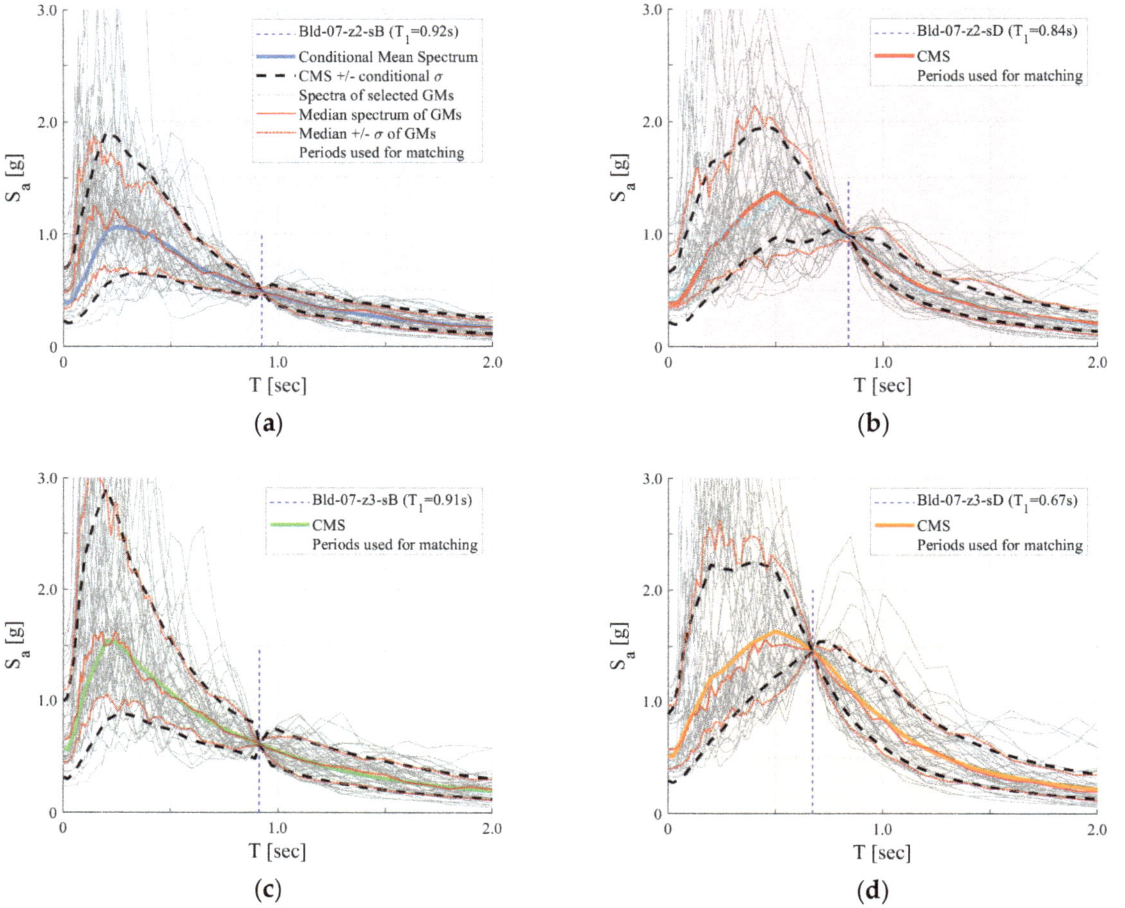

Figure 12. Ground-motion selection. (**a**) Bld-07-z2-sB; (**b**) Bld-07-z2-sD; (**c**) Bld-07-z3-sB; (**d**) Bld-07-z3-sD. Note: the 3rd, 4th, 5th and 6th series in the legend of subfigure (**a**) also apply for the other subfigures.

5. Response and Collapse Assessment Results

5.1. IDAs, Collapse Fragility Functions, P(C | $S_a(T_1)$MCE), and CMR

The results from the structural analyses are summarized in Figure 13, which shows the median Peak Story Drift Ratios (PSDRs) considering different intensity levels. As a reference, the three main intensity levels (i.e., MCE, DBE, and SLE) are explicitly indicated in the legends with the corresponding values of the $S_a(T_1)$ level. As expected, the median PSDRs increased with an increasing return period (i.e., increasing hazard level). For comparison purposes, the response for a ~50,000-year return period is also presented. Although this is an extremely large return period, the corresponding $S_a(T_1)$ values were close to the median $S_a(T_1)$ collapse values (units of gravity, g) that are presented later.

For each archetype, the $S_a(T_1)$ values that triggered a collapse and the collapse mode (i.e., either simulated or non-simulated) were identified and recorded for each ground motion. The collapse modes were found to depend on the archetypes. For instance, more than 60% of collapses of archetypes Bld-07-z2-sB, Bld-07-z3-sB, and Bld-07-z2-sD were due to local collapse criteria. Concrete crushing and, at the same time, steel buckling at the wall boundary elements when the buildings were loaded in the positive transverse direction (i.e., boundary elements in compression) were mostly observed. On average,

40% of collapses occurred in the negative transverse direction (i.e., boundary elements in tension) due to the global collapse criteria (SRD ≥ 5%). On the other hand, 70% of the collapses of archetype Bld-07-z3-sD were due to the global collapse criteria (in either the negative or positive transverse direction).

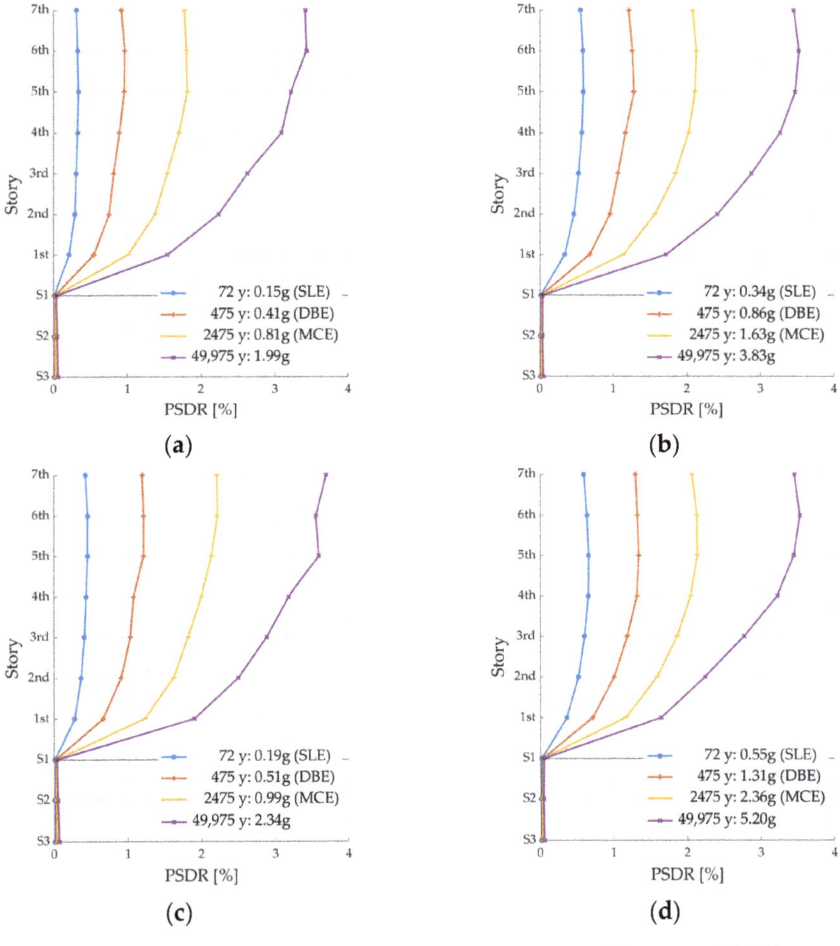

Figure 13. Median values of Peak Story Drift Ratios (PSDR). (**a**) Bld-07-z2-sB; (**b**) Bld-07-z2-sD; (**c**) Bld-07-z3-sB; (**d**) Bld-07-z3-sD.

Figure 14 shows the collapse assessment results. For brevity, the IDA curves present information about one of the EDPs (i.e., RDR) as an indicator of the global structural seismic response. In more detail, Figure 14a–d show the IDA results for each building, plotted as the RDR values against the $S_a(T_1)$ values. The figures also present the 50th collapse percentile (median), as well as the 16th and 84th collapse percentiles (equal to one logarithmic standard deviation below and above the mean when a lognormal distribution is assumed). Moreover, Figure 14e–h shows the estimated lognormal collapse fragility curves and the values of the $S_a(T_1)$ collapse intensity.

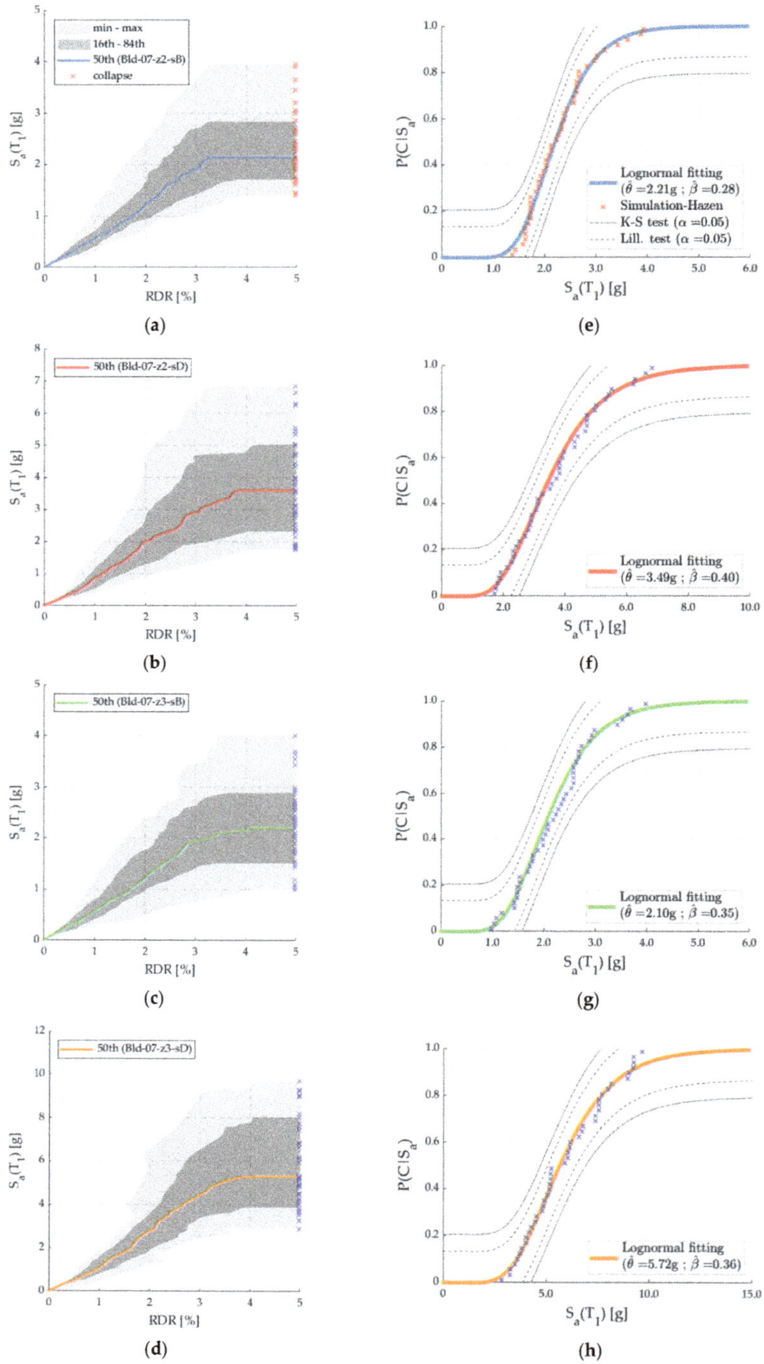

Figure 14. IDA results: (**a**) Bld-07-z2-sB; (**b**) Bld-07-z2-sD; (**c**) Bld-07-z3-sB; (**d**) Bld-07-z3-sD. Collapse simulation results and estimated collapse fragility functions: (**e**) Bld-07-z2-sB; (**f**) Bld-07-z2-sD; (**g**) Bld-07-z3-sB; (**h**) Bld-07-z3-sD. Note: the 1st, 2nd and 4th series in the legend of subfigure (**a**) also apply for subfigures (**b–d**); and the 2nd, 3rd and 4th series in the legend of subfigure (**e**) also apply for subfigures (**f–h**).

These figures also exhibit both the estimated median ($\hat{\theta}$) and dispersion ($\hat{\beta}$) values (obtained using the MLM) and the results of the K–S and Lilliefors goodness-of-fit tests, both at a 5% significance level. The fragility functions passed both tests, indicating a proper representation of the $S_a(T_1)$ collapse intensity data. In more detail, seismic zone 2 Bld-07-z2-sB had $S_a(T_1)$ collapse values that ranged from 1.40 g to 3.95 g, with $\hat{\theta}$ = 2.21 g, whereas Bld-07-z2-sD had $S_a(T_1)$ collapse intensity values ranging from 1.75 g to 6.85 g, with a greater $\hat{\theta}$ of 3.49 g. On the other hand, seismic zone 3 Bld-07-z3-sB exhibited $S_a(T_1)$ collapse values that ranged from 1.00 g to 4.00 g, with $\hat{\theta}$ = 2.10 g, whereas Bld-07-z3-sD exhibited $S_a(T_1)$ collapse intensities ranging from 2.90 g to 9.70 g, with a significantly greater $\hat{\theta}$ of 5.72 g. In particular, the $\hat{\theta}$ results were consistent with the seismic design code strength requirement (i.e., $\hat{\theta}$ = 2.21 g for Bld-07-z2-sB (V_b^{des} = 0.05 W) was smaller than $\hat{\theta}$ = 5.72 g for Bld-07-z3-sD (V_b^{des} = 0.10 W)).

In terms of dispersion, the archetype buildings had quite similar values of $\hat{\beta}$, which were less than 0.40. Dispersion values of 0.28, 0.40, 0.35, and 0.36 were estimated for archetypes Bld-07-z2-sB, Bld-07-z2-sD, Bld-07-z3-sB, and Bld-07-z3-sD, respectively. Although the dispersion of the $S_a(T_1)$ collapse values seemed to be rather high, the estimated values of $\hat{\beta}$ were either equal to or smaller than those recommended by FEMA P-58 [13].

The estimated values of $\hat{\theta}$ and $\hat{\beta}$ as well as the results of the goodness-of-fit tests, are presented in Table 5. In terms of the effect of the seismic zone, the $\hat{\theta}$ value of archetype Bld-07-z2-sB (stiff soil, moderate seismic activity) was slightly higher than that of archetype Bld-07-z3-sB (high seismic activity, also stiff soil), whereas the $\hat{\theta}$ value of archetype Bld-07-z2-sD (moderate seismic activity, moderately stiff soil) was smaller than that of archetype Bld-07-z3-sD (high seismic activity, moderately stiff soil). In other words, there was no clear relationship between the level of seismic activity and $\hat{\theta}$. The effect of the soil type was more relevant on the archetype buildings located in seismic zone 3 ($\hat{\theta}$ = 2.10 g for Bld-07-z3-sB and $\hat{\theta}$ = 5.72 g for Bld-07-z3-sD) than on those located in seismic zone 2 ($\hat{\theta}$ = 2.21 g for Bld-07-z2-sB and $\hat{\theta}$ = 3.49 g for Bld-07-z2-sD).

Table 5. Summary of collapse fragility analysis.

Archetype	$\hat{\theta}$ (g)	$\hat{\beta}$	K–S Test?	Lilliefors Test?	$P(C \mid S_a(T_1)_{MCE})$ (%)	CMR
Bld-07-z2-sB	2.21	0.28	Pass	Pass	0.02	2.7
Bld-07-z2-sD	3.49	0.40	Pass	Pass	2.78	2.1
Bld-07-z3-sB	2.10	0.35	Pass	Pass	1.58	2.1
Bld-07-z3-sD	5.72	0.36	Pass	Pass	0.66	2.4

Based on the results shown in Table 5 it can be stated that (i) the values of $\hat{\beta}$ for archetype buildings located on soil type D were greater than those for archetype buildings located on soil type B, and (ii) there was no clear relationship between the $\hat{\beta}$ values and the seismic zone. Additionally, the effect of the soil type seemed to be more relevant than that of the seismic zone. Higher values of $\hat{\beta}$ for archetype buildings located on soil type D could be the result of a greater dispersion of the CS spectral ordinates, as mentioned in Section 4.2.

The collapse fragility curves of each archetype building (plotted as a function of $S_a(T_1)$) are presented in Figure 15a and it can be observed that the fragility curves of archetypes Bld-07-z2-sB and Bld-07-z3-sB are quite similar to each other, whereas those of archetypes Bld-07-z2-sD and Bld-07-z3-sD are situated far to the right. At first glance, this observation could suggest that the latter archetype buildings have a superior seismic collapse performance, but it is important to note that for a specified $S_a(T_1)$ value, the corresponding $\lambda_{Sa(T1)}$ value may differ significantly for different archetype buildings depending on the fundamental period, seismic zone, and soil type. Thus, a direct comparison of the fragility functions expressed in terms of $S_a(T_1)$ can be misleading [38]. Consequently, Figure 15b shows the collapse fragility curves but plotted as a function of the corresponding $\lambda_{Sa(T1)}$ values, where vertical lines at different hazard levels (i.e., probabilities of exceedance equal to 2%,

10%, and 50% in 50 years, or, in other words, return periods of 72, 475 and 2475 years, respectively) are also shown.

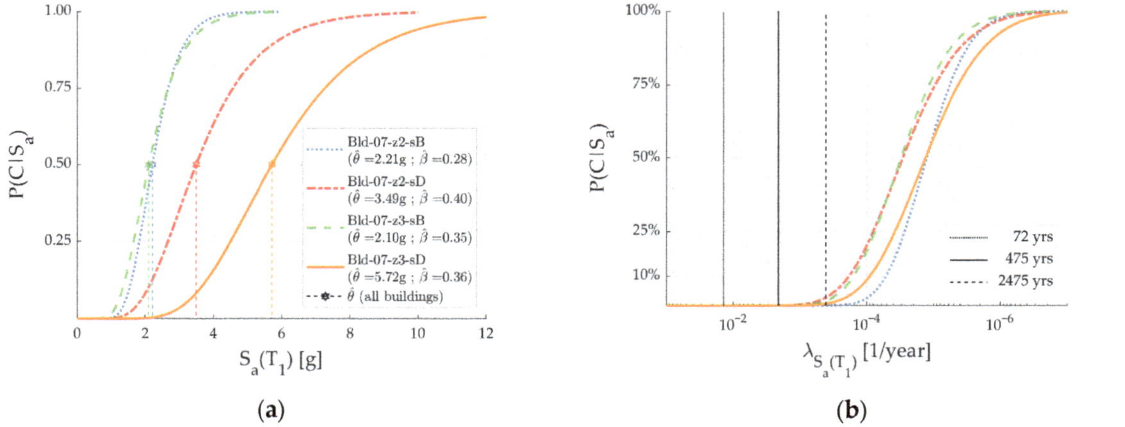

Figure 15. Collapse fragility curves plotted as a function of (**a**) $S_a(T_1)$; (**b**) $\lambda_{Sa(T1)}$.

By looking at Figure 15a,b it is clear why a direct comparison of the collapse fragility functions plotted as functions of $S_a(T_1)$ was misleading. For instance, the $S_a(T_1)$ collapse fragility curve of archetype Bld-07-z2-sB was located to the left of the remaining curves in Figure 15a (which may have been misunderstood as being inferior performance), whereas the corresponding $\lambda_{Sa(T1)}$ collapse fragility curve was located to the right of the other curves, which indicates a superior collapse performance (relative to that of the archetypes on soil type D). Although fragility functions plotted as a function of $\lambda_{Sa(T1)}$ may not provide sufficient information to quantitatively rank the earthquake-induced collapse performance of different buildings (as opposed to the information provided by λ_c, for example), this analysis provides a useful tool to compare different buildings to each other.

Figure 15b shows that each archetype building had negligible collapse probability values at hazard levels of 10% and 50% in 50 years (i.e., SLE and DBE earthquakes, respectively). Likewise, at the MCE earthquake (i.e., hazard level of 2% in 50 years) archetypes Bld-07-z2-sB and Bld-07-z3-sD had negligible collapse probability values. Instead, archetypes Bld-07-z2-sD and Bld-07-z3-sB had small but non-negligible collapse probability values (2.8% and 1.6%, respectively). Table 5 shows the collapse probability value of each archetype building at their respective MCE intensities, and it can be observed that each archetype building met the 10% conditional probability target indicated by ASCE 7-22 [10]. The latter was consistent with the satisfactory collapse prevention performance exhibited by modern Chilean RC buildings in recent earthquakes [3,6,7]. In terms of the effect of the soil type and seismic zone, the results shown in Figure 15b do not show a clear pattern. Lastly, the CMR values are also shown in Table 5 and it can be seen that these values ranged from 2.1 to 2.7 (average of 2.3) and were higher (i.e., lower collapse risk) than those indicated in FEMA P695 for RC structures, whose design seismic response coefficients and collapse uncertainties are similar to those of the archetype buildings assessed in this study [16].

5.2. *Values of λ_c and $P_c(50)$*

The estimated values of λ_c and $P_c(50)$ are presented in Table 6 and it can be observed that these values ranged from 2.17×10^{-5} to 7.24×10^{-5} and from 0.11% to 0.36%, respectively. The estimated values of $P_c(50)$ were small and consistent with the seismic response of modern Chilean RC buildings empirically observed in recent earthquakes. In addition, the target maximum probability of collapse of 1% in 50 years indicated by ASCE 7-22 [10] was achieved by each archetype building and, therefore, post-2010 Chilean RC mid-rise dual wall-frame buildings are expected to reach the collapse prevention limit state (at least at the locations and soil types considered in this study).

Table 6. Summary of values of λ_c and $P_c(50)$.

Archetype	λ_c (1/Year)	$P_c(50)$ (%)	$\lambda_c(\lambda_{Sa(T1)} = 10^{-4})/\lambda_c$ (%)
Bld-07-z2-sB	2.17×10^{-5}	0.11	5.4
Bld-07-z2-sD	7.24×10^{-5}	0.36	42.1
Bld-07-z3-sB	6.31×10^{-5}	0.31	32.6
Bld-07-z3-sD	3.56×10^{-5}	0.18	25.7

There was no clear pattern in terms of the influence of the soil type and seismic zone on the $P_c(50)$ values. For soil type B, the $P_c(50)$ value for seismic zone 3 was higher than that for seismic zone 2. Specifically, the $P_c(50)$ values for archetypes Bld-07-z3-sB and Bld-07-z2-sB were 0.31% and 0.11%, respectively. In contrast, for soil type D, the $P_c(50)$ value for seismic zone 3 was smaller than that for seismic zone 2. Specifically, values of $P_c(50)$ for archetypes Bld-07-z3-sD and Bld-07-z2-sD were 0.18% and 0.36%, respectively. Regarding the influence of the soil type, the $P_c(50)$ values for archetypes Bld-07-z2-sB and Bld-07-z2-sD were 0.11% and 0.36%, respectively, whereas those for archetypes Bld-07-z3-sB and Bld-07-z3-sD $P_c(50)$ were 0.31% and 0.18%, respectively. Although the effect of the seismic zone and soil type on the $P_c(50)$ values may appear counter-intuitive (even chaotic), it is important to highlight that Chilean seismic design codes are mostly prescriptive and do not include explicit PBEE design targets. Given that normalized design base shears, V^{des}/W, for the archetype buildings on soil type D were higher than those for the archetype buildings on soil type B (see Table 3), higher $P_c(50)$ values for the soil type D archetype buildings seemed to indicate that the difference between the actual demand and the design demand was greater on soil type D than on soil type B. Since current Chilean design codes are prescriptive and lack PBEE design targets (such as uniform collapse risk on different soil types), the substantial differences observed in the $P_c(50)$ values may be expected (which does not make them less unacceptable).

5.3. Deaggregation Values of λ_c

The deaggregation curves of λ_c based on the $S_a(T_1)$ intensity values are shown in Figure 16a, and it can be observed that the areas below the curves for archetypes Bld-07-z2-sD and Bld-07-z3-sB are significantly greater than those for archetypes Bld-07-z2-sB and Bld-07-z3-sD, which was expected because these areas represent the values of λ_c that were summarized in Table 6. Moreover, the deaggregation curve of archetype Bld-07-z3-sB is located more to the left, which indicates that the contribution of small $S_a(T_1)$ values to λ_c was greater for this archetype building than for the remaining buildings. Consequently, archetype Bld-07-z3-sB was more susceptible (in terms of collapse during its lifetime) to small/medium $S_a(T_1)$ intensities. In contrast, the deaggregation curve of archetype Bld-07-z3-sD is located more to the right, which indicates that the contribution of high $S_a(T_1)$ values to λ_c was greater for this archetype building than for the remaining buildings. Consequently, archetype Bld-07-z3-sD was more susceptible (in terms of collapse during its lifetime) to high $S_a(T_1)$ intensities. Figure 16a also shows three sets of lines that indicate the values of $\hat{\theta}$, values of $S_a(T_1)$ at 50% of λ_c, and values of $S_a(T_1)$ at 75% of λ_c.

It is worth mentioning that the values of $\hat{\theta}$ were always higher than the $S_a(T_1)$ intensity values at 50% of λ_c but were always smaller than the $S_a(T_1)$ intensity values at 75% of λ_c. For instance, for archetype Bld-07-z2-sD, $\hat{\theta}$ = 3.49 g was 28% higher than the $S_a(T_1)$ value at 50% of λ_c (= 2.73 g), whereas for archetype Bld-07-z3-sB, $\hat{\theta}$ = 2.10 g was 17% higher than the $S_a(T_1)$ value at 50% of λ_c (= 1.80 g). Although the precise characterization of the collapse fragility functions is necessary for the entire range of the $S_a(T_1)$ intensity values, these observations indicate that this characterization is needed more at $S_a(T_1)$ intensities smaller than $\hat{\theta}$ because these $S_a(T_1)$ values contribute the most to λ_c (and, as a result, also to $P_c(50)$), which is in agreement with previous studies [22,38].

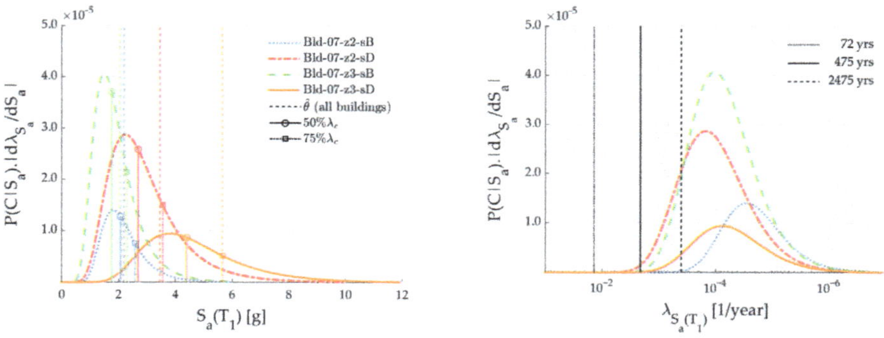

(a) **(b)**

Figure 16. Deaggregation of λ_c plotted as a function of (**a**) $S_a(T_1)$; (**b**) $\lambda_{Sa(T1)}$.

As shown previously, a direct comparison of the collapse fragility functions for the different archetype buildings (with different soil types and fundamental periods) may be misleading, and this reflection can be extended to the assessment of the λ_c deaggregation curves for different archetype buildings. The deaggregation curves of λ_c as a function of the corresponding $\lambda_{Sa(T_1)}$ values are shown in Figure 16b, where vertical lines indicate the previously defined $\lambda_{Sa(T_1)}$ hazard levels. At the SLE level, each archetype building presents small (almost negligible) values of deaggregated λ_c and this is in agreement with the observed seismic performance of Chilean RC dual wall-frame buildings in recent earthquakes. Regarding the DBE level, apart from archetype Bld-07-z2-sD, the remaining archetype buildings exhibited negligible values of deaggregated λc, whereas at the MCE level, only the archetype Bld-07-z2-sB showed negligible values of deaggregated λ_c. As seen in Figure 16b, the deaggregation curves of archetypes Bld-07-z2-sD and Bld-07-z3-sB are located more to the left, which shows that the influence of high $\lambda_{Sa(T_1)}$ values on λ_c was greater for these archetype buildings than for the remaining ones, indicating that the former archetype buildings were more susceptible (in terms of collapse during their lifetime) to more frequent ground motions (small and medium intensities). It can also be observed that, again, the influence of the seismic zone and soil type on the deaggregated λ_c curves is unclear. For example, at $\lambda_{Sa(T_1)} = 10^{-4}$, the area under the deaggregation curve is 42.1% of λ_c for archetype Bld-z2-sD but only 25.7% of λ_c for archetype Bld-z3-sD (see Table 6). On the other hand, at $\lambda_{Sa(T_1)} = 10^{-4}$, the area under the deaggregation curve is 5.4% of λ_c for archetype Bld-z2-sB but 32.6% of λ_c for archetype Bld-z3-sB (see Table 6). Therefore, it cannot be concluded that archetype buildings located in a high seismicity zone are less susceptible to collapse during their lifetime (e.g., 50 years) to more recurrent ground motions than those in a moderate seismicity zone. As a summary, Figure 17 shows a bar plot, where the λ_c and $P_c(50)$ values are plotted for the four archetype buildings.

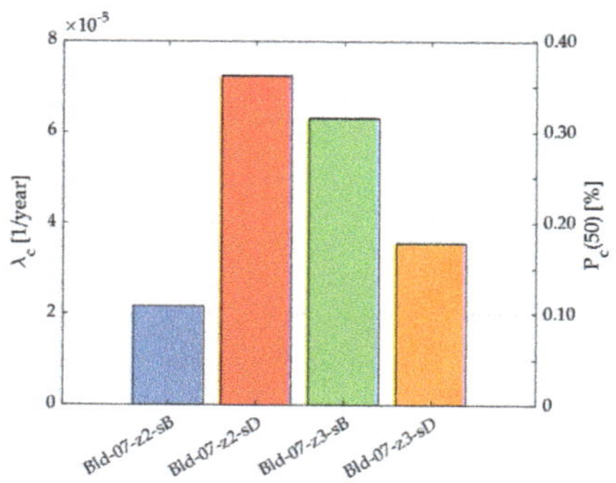

Figure 17. Bar graph of λ_c and $P_c(50)$.

6. Summary and Closing Remarks

This paper assesses the seismic collapse performance of a group of four mid-rise RC dual wall-frame archetype Chilean office buildings subjected to Chilean subduction ground motions. To the best of the authors' knowledge, quantitative characterizations of the seismic collapse performance are absent from the literature. The archetype buildings (representative of Chilean mid-rise office buildings) are code-conforming buildings that meet the minimum requirements imposed by current Chilean seismic design codes, including the amendments introduced after the M_w 8.8 2010 Chilean earthquake. This group of archetype buildings is characterized by two site locations (i.e., a high seismic zone and a moderate seismic zone, which are denoted as seismic zones 3 and 2, respectively), two soil types (B and D, where the latter is less stiff than the former), and one building height (i.e., 7 stories). The archetype buildings use the naming convention of Bld-07-zX-sY, where X is the seismic zone (i.e., 2 or 3) and Y is the soil type (i.e., B or D). The assessment of the collapse performance was obtained by implementing the latest advances in PBEE proposed by the PEER Center, following the well-established FEMA P-58 methodology. The seismic collapse assessment was evaluated through 3D nonlinear finite-element models subjected to IDAs using 44 carefully chosen and scaled Chilean subduction ground-motion records. The collapse performance of the archetype buildings was characterized by the estimation of (1) the collapse fragility functions; (2) the probability of collapse at the Maximum Considered Earthquake (MCE) intensity, $P(C \mid S_a(T_1)_{MCE})$; (3) the collapse margin ratio, CMR; (4) the mean annual frequency of collapse, λ_c; (5) the probability of collapse in 50 years, $P_c(50)$; and (6) the deaggregation of λ_c. In summary, the following conclusions can be drawn:

- The lognormal distribution is an adequate representation of the collapse fragility function of the archetype buildings since each collapse fragility function passed the Kolmogorov–Smirnov and Lilliefors goodness-of-fit tests, both at a 5% significance level. Other distribution functions can be tested in further studies.
- The estimated median collapse ($\hat{\theta}$) and logarithmic dispersion ($\hat{\beta}$) values were 2.21 g, 3.49 g, 2.10 g, and 5.72 g, and 0.28, 0.40, 0.35, and 0.36, for archetype buildings Bld-07-z2-sB, Bld-07-z2-sD, Bld-07-z3-sB, and Bld-07-z3-sD, respectively.
- The analysis and comparison of the collapse fragilities based on $\lambda_{S_a(T_1)}$ (as an alternative to $S_a(T_1)$) generated meaningful information since adequate direct comparisons of the conditional collapse probability can be obtained at different hazard levels such as the SLE, 50%—50 years; DBE, 10%—50 years; and MCE, 2%—50 years. Specifically, all archetype buildings had a negligible collapse probability at the SLE and DBE levels. At the MCE level, the archetype buildings Bld-07-z2-sD and Bld-07-z3-sB had non-negligible collapse probabilities (2.78% and 1.58%, respectively). However, all these MCE collapse probabilities were smaller than the 10% target defined in ASCE 7-22. Consequently, current Chilean seismic design standards seem to provide adequate levels of collapse prevention, which is in agreement with the observed performance of modern RC buildings in recent earthquakes.
- The calculated values of the CMR (i.e., $\hat{\theta}$ divided by the MCE intensity) ranged from 2.1 to 2.7, with a mean of 2.3. This mean value was higher (i.e., lower collapse risk) than the values stated in previous studies on RC frame buildings designed using US seismic codes and subjected to crustal ground motions. No clear influence of the seismic zone and soil type on the CMR was identified.
- The estimated values of λ_c and $P_c(50)$ were 2.17×10^{-5}, 7.24×10^{-5}, 6.31×10^{-5}, and 3.56×10^{-5}, and 0.11%, 0.36%, 0.31%, and 0.18% for archetypes Bld-07-z2-sB, Bld-07-z2-sD, Bld-07-z3-sB, and Bld-07-z3-sD, respectively. The values of $P_c(50)$ largely fulfilled the maximum 1% target collapse probability in 50 years stated by ASCE 7-22, indicating, once more, the adequate level of collapse prevention provided by current Chilean seismic design standards.
- Non-negligible differences were found between the values of $P_c(50)$ for the different archetype buildings but no clear influence of the seismic zone and soil type on $P_c(50)$ was observed. Although Chilean seismic design codes are mostly prescriptive and do

not include explicit PBEE design targets, the collapse risk should nevertheless be more uniform, suggesting that current Chilean seismic design codes (particularly the design spectra) might require a revision.

- Lastly, the deaggregation of λ_c (as a function of $\lambda_{Sa(T_1)}$, which allows adequate direct comparisons between the deaggregation functions) showed that the values of $\hat{\theta}$ were always higher than the $S_a(T_1)$ values at 50% of λ_c. In other words, in terms of the collapse risk, the contribution of the $S_a(T_1)$ values at the $S_a(T_1) < \hat{\theta}$ range was greater than that of the $S_a(T_1)$ values at the $S_a(T_1) > \hat{\theta}$ range. This observation means that, contrary to intuition, the accurate characterization of collapse fragilities is more important at the $S_a(T_1) < \hat{\theta}$ range than at the $S_a(T_1) > \hat{\theta}$ range.

In summary, although current Chilean seismic design codes for buildings appear to provide a satisfactory collapse prevention level for mid-rise RC dual wall-frame buildings, the results indicate that the collapse risk is not uniform. This lack of uniformity may indicate that a more thorough calibration of the current Chilean seismic design code is necessary. These results should be used with caution, as further research is required to evaluate the effects of different construction materials, structural systems, plan layouts, or building heights (i.e., low- and high-rise buildings) on the collapse capacity.

Author Contributions: Conceptualization, M.F.G., G.A.-L., D.L.-G. and P.F.P.; methodology, M.F.G., G.A.-L., D.L.-G. and P.F.P.; software, M.F.G. and P.F.P.; validation, M.F.G., G.A.-L., D.L.-G. and P.F.P.; formal analysis, M.F.G., G.A.-L., D.L.-G. and P.F.P.; investigation, M.F.G., G.A.-L., D.L.-G. and P.F.P.; resources, G.A.-L. and D.L.-G.; data curation, M.F.G.; writing—original draft preparation, M.F.G., G.A.-L. and D.L.-G.; writing—review and editing, G.A.-L. and D.L.-G.; visualization, M.F.G.; supervision, G.A.-L., D.L.-G. and P.F.P.; project administration, G.A.-L.; funding acquisition, G.A.-L. All authors have read and agreed to the published version of the manuscript.

Funding: This research was funded by ANID Doctorado Nacional 2018 Folio 21181157, FONDECYT de Iniciación—Grant 11191194, and the Research Center for Integrated Disaster Risk Management (CIGIDEN) ANID FONDAP 1522A0005. The APC was funded by FONDECYT de Iniciación—Grant 11191194.

Data Availability Statement: The data that support the findings of this study are available from the corresponding authors, [G.A.-L.; D.L.-G.], upon reasonable request.

Acknowledgments: Valuable comments on the Chilean seismic design practice were provided by Ian Watt (VMB Structural Engineering) and Mario Lafontaine (Rene Lagos Engineers). The strong motion database was provided by the SIBER-RISK (Simulation-Based Earthquake Risk and Resilience of Interdependent Systems and Networks) project.

Conflicts of Interest: The authors declare no conflict of interest. The funders had no role in the design of the study; in the collection, analyses, or interpretation of data; in the writing of the manuscript, or in the decision to publish the results.

References

1. Ruiz, S.; Madariaga, R. Historical and recent large megathrust earthquakes in Chile. *Tectonophysics* **2018**, *733*, 37–56. [CrossRef]
2. Massone, L.M.; Bonelli, P.; Lagos, R. Seismic design and construction practices for RC structural wall buildings. *Earthq. Spectra* **2012**, *28*, S245–S256. [CrossRef]
3. EERI. *The Mw 8.8 Chile Earthquake of 27 February 2010*; Earthquake Engineering Research Institute: Oakland, CA, USA, 2010.
4. Zareian, F.; Krawinkler, H. Assessment of probability of collapse and design for collapse safety. *Earthq. Eng. Struct. Dyn.* **2007**, *36*, 1901–1914. [CrossRef]
5. Tena-Colunga, A.; Sanchez-Ballinas, D. The collapse of Alvaro Obregon 286 building in Mexico City during the September 19, 2017 earthquake. A case study. *J. Build. Eng.* **2022**, *49*, 104060. [CrossRef]
6. EERI. *Mw 8.2 Iquique, Chile Earthquake and Tsunami: Preliminary Reconnaissance Observations*; Earthquake Engineering Research Institute: Oakland, CA, USA, 2014.
7. GEER. *Geotechnical Reconnaissance of the 2015 M8.3 Illapel, Chile Earthquake*; Geotechnical Extreme Events Reconnaissance (GEER) Association, Report No. GEER-043; GEER, 2015.
8. INN. *Norma Chilena Oficial NCh433 Of. 1996 Mod. 2009 Diseño Sismico de Edificios*; Instituto Nacional de Normalizacion: Santiago, Chile, 2009. (In Spanish)
9. MINVU. *DS 61 Diseño Sismico de Edificios*; Ministerio de Vivienda y Urbanismo: Santiago, Chile, 2011. (In Spanish)

10. ASCE. *Minimum Design Loads and Associated Criteria for Buildings and Other Structures ASCE/SEI 7-22*; American Association of Civil Engineers: Reston, VA, USA, 2022.
11. MINVU. *DS 60 Requisitos de Diseño y Calculo Para el Hormigon Armado*; Ministerio de Vivienda y Urbanismo: Santiago, Chile, 2011. (In Spanish)
12. Lagos, R.; Lafontaine, M.; Bonelli, P.; Boroschek, R.; Guendelman, T.; Massone, L.; Saragoni, R.; Rojas, F.; Yañez, F. The quest for resilience: The Chilean practice of seismic design for reinforced concrete buildings. *Earthq. Spectra* **2021**, *37*, 26–45. [CrossRef]
13. FEMA. *Seismic Performance Assessment of Buildings FEMA P-58*; Federal Emergency Management Agency: Hyattsville, MD, USA, 2018.
14. Araya-Letelier, G.; Parra, P.F.; Lopez-Garcia, D.; Garcia-Valdes, A.; Candia, G.; Lagos, R. Collapse risk assessment of a Chilean dual wall-frame reinforced concrete office building. *Eng. Struct.* **2019**, *183*, 770–779. [CrossRef]
15. Cando, M.A.; Hube, M.A.; Parra, P.F.; Arteta, C.A. Effect of stiffness on the seismic performance of code-conforming reinforced concrete shear wall buildings. *Eng. Struct.* **2020**, *219*, 110724. [CrossRef]
16. FEMA. *Quantification of Building Seismic Performance Factors FEMA P695*; Federal Emergency Management Agency: Hyattsville, MD, USA, 2009.
17. Gogus, A.; Wallace, J.W. Seismic safety evaluation of reinforced concrete walls through FEMA P695 methodology. *J. Struct. Eng.* **2015**, *141*, 04015002. [CrossRef]
18. Arabzadeh, H.; Galal, K. Seismic collapse risk assessment and FRP retrofitting of RC coupled C-shaped core walls using the FEMA P695 methodology. *J. Struct. Eng.* **2017**, *143*, 04017096. [CrossRef]
19. Dabaghi, M.; Saad, G.; Allhassania, N. Seismic collapse fragility analysis of reinforced concrete shear wall buildings. *Earthq. Spectra* **2019**, *35*, 383–404. [CrossRef]
20. Terzic, V.; Kolozvari, K.; Saldana, D. Implications of modeling approaches on seismic performance of low- and mid-rise office and hospital shear wall buildings. *Eng. Struct.* **2019**, *189*, 129–146. [CrossRef]
21. Marafi, N.A.; Ahmed, K.A.; Lehman, D.E.; Lowes, L.N. Variability in seismic collapse probabilities of solid- and coupled-wall buildings. *J. Struct. Eng.* **2019**, *145*, 04019047. [CrossRef]
22. Raghunandan, M.; Liel, A.; Luco, N. Collapse risk of buildings in the Pacific northwest region due to subduction earthquakes. *Earthq. Spectra* **2015**, *31*, 2087–2115. [CrossRef]
23. Chandramohan, R.; Baker, J.; Deierlein, G. Impact of hazard-consistent ground motion duration in structural collapse risk assessment. *Earthq. Eng. Struct. Dyn.* **2016**, *45*, 1357–1379. [CrossRef]
24. Medalla, M.; Lopez-Garcia, D.; Zareian, F. Seismic characterization of steel special moment frames subjected to megathrust earthquakes. *Earthq. Spectra* **2020**, *36*, 2033–2057. [CrossRef]
25. Marafi, N.A.; Makdisi, A.J.; Eberhard, M.O.; Berman, J.W. Impacts of an M9 Cascadia subduction zone earthquake and Seattle basin on performance of RC core wall buildings. *J. Struct. Eng.* **2020**, *146*, 04019201. [CrossRef]
26. Vamvatsikos, D.; Cornell, C.A. Incremental dynamic analysis. *Earthq. Eng. Struct. Dyn.* **2002**, *31*, 491–514. [CrossRef]
27. Monjardin-Quevedo, J.G.; Valenzuela-Beltran, F.; Reyes-Salazar, A.; Leal-Graciano, J.M.; Torres-Carrillo, X.G.; Gaxiola-Camacho, J.R. Probabilistic assessment of buildings subjected to multi-level earthquake loading based on the PBSD concept. *Buildings* **2022**, *12*, 1942. [CrossRef]
28. Kostinakis, K.; Fontara, I.; Athanatopoulou, A. Scalar structure-specific ground motion intensity measures for assessing the seismic performance of structures: A review. *J. Earthq. Eng.* **2018**, *22*, 630–665. [CrossRef]
29. Rong, X.; Yang, J.; Jun, L.; Zhang, Y.; Zheng, S.; Dong, L. Optimal ground motion intensity measure for seismic assessment of high-rise reinforced concrete structures. *Case Stud. Constr. Mater.* **2023**, *18*, e01678. [CrossRef]
30. Shome, N. Probabilistic Seismic Demand Analysis of Nonlinear Structures. Ph.D. Thesis, Stanford University, Stanford, CA, USA, 1999.
31. Ibarra, L.F.; Krawinkler, H. *Global Collapse of Frame Structures under Seismic Excitations*; Technical Report 152; John A. Blume Earthquake Engineering Center: Stanford, CA, USA, 2005.
32. Eads, L.; Miranda, E.; Krawinkler, H.; Lignos, D. An efficient method for estimating the collapse risk of structures in seismic regions. *Earthq. Eng. Struct. Dyn.* **2013**, *42*, 25–41. [CrossRef]
33. Baker, J.W. Efficient analytical fragility function fitting using dynamic structural analysis. *Earthq. Spectra* **2015**, *31*, 579–599. [CrossRef]
34. Massey, F.J., Jr. The Kolmogorov-Smirnov test for goodness of fit. *J. Am. Stat. Assoc.* **1951**, *46*, 68–78. [CrossRef]
35. Lilliefors, H. On the Kolmogorov-Smirnov test for normality with mean and variance unknown. *J. Am. Stat. Assoc.* **1967**, *62*, 399–402. [CrossRef]
36. INE. *Base de Datos de Permisos de Edificación*; Instituto Nacional de Estadística: Santiago, Chile, 2020. (In Spanish)
37. IBC. *2021 International Building Code*; International Code Council: Country Club Hills, IL, USA, 2021.
38. Araya-Letelier, G. Design of Building Structural Systems and Enhanced Partition Walls to Improve the Life Cycle Costs Associated with Risk of Earthquake Damage. Ph.D. Thesis, Stanford University, Stanford, CA, USA, 2014.
39. Porcu, M.C.; Vielma Pérez, J.C.; Pais, G.; Osorio Bravo, D.; Vielma Quintero, J.C. Some issues in the seismic assessment of shear-wall buildings through code-compliant dynamic analyses. *Buildings* **2022**, *12*, 694. [CrossRef]
40. CSI. *ETABS v16 User's Guide Manual*; Computers & Structures: Berkeley, CA, USA, 2016.
41. Akcelyan, S.; Lignos, D.G. Seismic Assessment and retrofit of pre-Northridge high rise steel moment resisting frame buildings with bilinear oil dampers. *Buildings* **2023**, *13*, 139. [CrossRef]

42. ACI. *Building Code Requirements for Structural Concrete and Commentary ACI 318-08*; American Concrete Institute: Farmington Hills, MI, USA, 2008.
43. CSI. *PERFORM 3D 7.0.0 Manual*; Computer & Structures: Berkeley, CA, USA, 2017.
44. NIST. *Nonlinear Structural Analysis for Seismic Design: A Guide for Practicing Engineers NIST GCR 10-917-5*; National Institute of Standards and Technology: Gaithersburg, MD, USA, 2010.
45. ATC. *Modeling and Acceptance Criteria for Seismic Design and Analysis of Tall Buildings PEER/ATC 72-1*; Applied Technology Council: Redwood City, CA, USA, 2010.
46. NIST. *Recommended Modeling Parameters and Acceptance Criteria for Nonlinear Analysis in Support of Seismic Evaluation, Retrofit, and Design NIST GCR 17-917-45*; National Institute of Standards and Technology: Gaithersburg, MD, USA, 2017.
47. NIST. *Guidelines for Nonlinear Structural Analysis for Design of Buildings Part I—General, NIST GCR 17-917-46v1*; National Institute of Standards and Technology: Gaithersburg, MD, USA, 2017.
48. ACHISINA. *Alternative Procedure for the Seismic Analysis and Design of Tall Buildings*; Asociacion Chilena de Sismologia e Ingenieria Antisismica: Santiago, Chile, 2017. (In Spanish)
49. TBI. *Guidelines for Performance-Based Seismic Design of Tall Buildings*; PEER Report 2017-06; Pacific Earthquake Engineering Research Center: Berkeley, CA, USA, 2017.
50. LATBSDC. *An Alternative Procedure for Seismic Analysis and Design of Tall Buildings Located in the Los Angeles Region*; Los Angeles Tall Buildings Structural Design Council: Los Angeles, CA, USA, 2020.
51. Ugalde, D.; Lopez-Garcia, D. Analysis of the seismic capacity of Chilean residential RC shear wall buildings. *J. Build. Eng.* **2020**, *31*, 101369. [CrossRef]
52. Massone, L.M.; Bedecarratz, E.; Rojas, F.; Lafontaine, M. Nonlinear modeling of a damaged reinforced concrete building and design improvement behavior. *J. Build. Eng.* **2021**, *41*, 102766. [CrossRef]
53. Lowes, L.N.; Lehman, D.E.; Baker, C. Recommendations for modeling the nonlinear response of slender reinforced concrete walls using PERFORM-3D. In Proceedings of the SEAOC Convention, Maui, HI, USA, 12–15 October 2016.
54. Kolozvari, K.; Arteta, C.; Fischinger, M.; Gavridou, S.; Hube, M.; Isakovic, T.; Lowes, L.; Orakcal, K.; Vásquez, J.; Wallace, J. Comparative study of state-of-the-art macroscopic models for planar reinforced concrete walls. *ACI Struct. J.* **2018**, *115*, 1637–1657. [CrossRef]
55. Coleman, J.; Spacone, E. Localization issues in force-based frame elements. *J. Struct. Eng.* **2001**, *127*, 1257–1265. [CrossRef]
56. Dazio, A.; Beyer, K.; Bachmann, H. Quasi-static cyclic tests and plastic hinge analysis of RC structural walls. *Eng. Struct.* **2009**, *31*, 1556–1571. [CrossRef]
57. Thomsen, J.H.; Wallace, J.W. *Displacement-Based Design of Reinforced Concrete Structural Walls: An Experimental Investigation of Walls with Rectangular and T-Shaped Cross-Sections*; Tech. Rep. No. CU/CEE-95/06; Department of Civil Engineering, Clarkson University: Postdam, NY, USA, 1995.
58. Ugalde, D.; Parra, P.F.; Lopez-Garcia, D. Assessment of the seismic capacity of tall wall buildings using nonlinear finite element modeling. *Bull. Earthq. Eng.* **2019**, *17*, 6565–6589. [CrossRef]
59. Paulay, T.; Priestley, M.J.N. *Seismic Design of Reinforced Concrete and Masonry Buildings*; Wiley: Hoboken, NJ, USA, 1992.
60. Yang, T.Y.; Moehle, J.P.; Bozorgnia, Y.; Zareian, F.; Wallace, J.W. Performance assessment of tall concrete core-wall building designed using two alternative approaches. *Earthq. Eng. Struct. Dyn.* **2012**, *45*, 1515–1531. [CrossRef]
61. Odabasi, O.; Kohrangi, M.; Bazzurro, P. Seismic collapse risk of reinforced concrete tall buildings in Istanbul. *Bull. Earthq. Eng.* **2021**, *19*, 6545–6571. [CrossRef]
62. Kim, T.; Foutch, D.A. Application of FEMA methodology to RC shear wall buildings governed by flexure. *Eng. Struct.* **2007**, *29*, 2514–2522. [CrossRef]
63. ASCE. *Seismic Evaluation and Retrofit of Existing Buildings ASCE/SEI 41-17*; American Association of Civil Engineers: Reston, VA, USA, 2017.
64. Poulos, A.; Monsalve, M.; Zamora, N.; de la Llera, J.C. An updated recurrence model for Chilean subduction seismicity and statistical validation of its Poisson nature. *Bull. Seismol. Soc. Am.* **2019**, *109*, 66–74. [CrossRef]
65. Montalva, G.A.; Bastías, N.; Rodriguez-Marek, A. Ground-motion prediction equation for the Chilean subduction zone. *Bull. Seismol. Soc. Am.* **2017**, *107*, 901–911. [CrossRef]
66. Idini, B.; Rojas, F.; Ruiz, S.; Pasten, C. Ground motion prediction equations for the Chilean subduction zone. *Bull. Earthq. Eng.* **2017**, *15*, 1853–1880. [CrossRef]
67. Candia, G.; Macedo, J.; Jaimes, M.A.; Magna-Verdugo, C. A new state-of-the-art platform for probabilistic and deterministic seismic hazard assessment. *Seismol. Res. Lett.* **2019**, *90*, 226–2275. [CrossRef]
68. Baker, J.W. Conditional mean spectrum: Tool for ground motion selection. *J. Struct. Eng.* **2011**, *137*, 322–331. [CrossRef]
69. Candia, G.; Poulos, A.; de la Llera, J.C.; Crempien, J.; Macedo, J. Correlations of spectral accelerations in the Chilean subduction zone. *Earthq. Spectra* **2020**, *36*, 788–805. [CrossRef]
70. Lin, T.; Harmsen, S.C.; Baker, J.W.; Luco, N. Conditional spectrum computation incorporating multiple causal earthquakes and ground motion prediction models. *Bull. Seismol. Soc. Am.* **2013**, *103*, 1103–1116. [CrossRef]
71. Castro, S.; Benavente, R.; Crempien, J.G.F.; Candia, G.; De la Llera, J.C. A consistently processed strong-motion database for Chilean earthquakes. *Seismol. Res. Lett.* **2022**, *93*, 2700–2718. [CrossRef]

72. Baker, J.W.; Lee, C. An improved algorithm for selecting ground motions to match a conditional spectrum. *J. Earthq. Eng.* **2018**, *22*, 708–723. [CrossRef]
73. NIST. *Selecting and Scaling Earthquake Ground Motions for Performing Response-History Analyses NIST GCR 11-917-15*; National Institute of Standards and Technology: Gaithersburg, MD, USA, 2011.
74. Davalos, H.; Miranda, E. Evaluation of bias on the probability of collapse from amplitude scaling using spectral-shape-matched records. *Earthq. Eng. Struct. Dyn.* **2019**, *48*, 970–986. [CrossRef]
75. Davalos, H.; Miranda, E. Evaluation of the scaling factor bias influence on the probability of collapse using Sa(T1) as the intensity measure. *Earthq. Spectra* **2019**, *35*, 679–702. [CrossRef]

buildings

MDPI

Article

Efficiency of an Improved Grouted Corrugated Duct (GCD) Connection Design for Precast Concrete Bridge Pier: Numerical and Parametric Study

Zhiqiang Wang, Chengjun Wu, Hongya Qu * and Wei Xiao

Department of Bridge Engineering, Tongji University, 1239 Siping Road, Shanghai 200092, China
* Correspondence: hqu@tongji.edu.cn

Abstract: In this study, finite element analysis (FEA) has been conducted for an improved grouted corrugated duct (GCD) connection design with a reserved recess in bridge footing. This study aims to understand the damage progression mechanism and to evaluate the contribution of each component in the improved GCD connection design. Numerical model based on the experimental results are first created, validated and calibrated. It is found that the confining effect (support and friction force) provided by recess sidewall keeps the connection in good integrity. It also prevents early deformation and early development of transverse cracks along the connection interface, which further avoids the damage concentration at connection joint, transfers the plastic hinge region. Parametric study is then carried out by considering different recess depths, cushion thicknesses, recess diameters, and mortar strengths. The effect of recess details on mechanical behavior is thus studied. Recess depth can be designed as 6–20% of the column section size to ensure a higher upper limit of overall strength and ductility, and it also influences the stress distribution area of the joint local. The stiffness and strength of recess control the local damage, while has limited impact on the overall performance. In addition, preliminary suggestions on the GCD design of recess depth, thickness of mortar cushion, recess diameter, the strength of mortar are proposed.

Keywords: precast pier; grouted corrugated duct connection; finite element analysis; parametric analysis; plastic hinge

Citation: Wang, Z.; Wu, C.; Qu, H.; Xiao, W. Efficiency of an Improved Grouted Corrugated Duct (GCD) Connection Design for Precast Concrete Bridge Pier: Numerical and Parametric Study. *Buildings* **2023**, *13*, 227. https://doi.org/10.3390/buildings13010227

Academic Editor: Giuseppina Uva

Received: 14 December 2022
Revised: 9 January 2023
Accepted: 11 January 2023
Published: 13 January 2023

1. Introduction

With the development of prefabrication technology, grouted corrugated duct (GCD) connection has attracted wide attention in several practical engineering application, such as HfL projects in Iowa [1] and Washington [2], as well as S6, S7 and G320 Highways in Shanghai [3]. There are mainly three types of GCDs, including plastic duct, galvanized metal duct, and improved cold formed duct [4]. GCD connection takes advantage of anchoring force of high strength mortar and transfers the force from inserted rebar to the duct. The anchorage of GCDs has been studied in various cases: with large-diameter rebar [5], with additional stainless energy dissipation bars [6], with bundled bars [7], under low temperature [8], or under cycling loading [9]. Formulas on anchorage length and several anchoring measures were also proposed, to prevent pull-out failures. Quasistatic tests [3,10–13] and shaking table tests [14] further confirmed the reliability of GCD connection.

For a precast bridge pier, the GCDs can be placed in cap beam or footing, accommodating the protruding rebars of pier column. Under earthquake input, the GCDs need to make sure internal force can be effectively transferred along the designed path, and failure of connection do not occur prior to the appearance of plastic hinge, which is emulative of cast-in-place (CIP). Experimental studies found the precast pier with GCD connections had more concentrated plastic hinge region than CIP pier. Pang et al. pointed out a remarkable difference between the precast and CIP columns from deformation distribution [10]. The plastic length of GCD was found to be approximately equal to half the column section depth and that for CIP was approximately equal to section depth [3,11,13]. In addition, an improved GCD connection design with shallow recess in the footing was developed for better integrity [4], and the spalling height was approximately equal to column section depth, which coincided with precast pier hinge length with socket connection [12,15] and CIP hinge length. While this showed that such design improved the overall seismic performance and durability, it is difficult to probe the mechanism from the experimental results alone.

Finite element analysis (FEA) can capture certain characteristics of some hard-to-observe parameters from the test. FEA was carried out for the quasistatic test of conventional reinforced concrete piers and reinforced concrete column with shape steel, respectively [16,17]. Li et al. studied the shear resistance of precast piers with different connection joint under quasistatic loading by FEA [18]. More studies utilized FEA to understand the performance of posttensioned segmental piers. Zhong et al. studied the seismic resilience of precast composite link beam connected to bridge piers (Seismic fragility and resilience assessment of bridge columns with dual-replaceable composite link beam under near-fault GMs) [19]. Ou et al. studied the seismic performance of segmental precast unbonded posttensioned concrete bridge columns, considering the joint contact, geometric nonlinearity, and confine effect on concrete by a discrete reinforcement model [20]. Finite element models were used for comparison and calibration of the proposed simplified analysis model. Dawood et al. conducted a pushover analysis on segmental unbonded posttensioned pier, using a confined concrete stress–strain model [21]. It was found that the numerical analysis could simulate the early damage development and the general behavior, but postyield strength was not well tracked. Nikbakh et al. modeled the segmental posttensioned pier, compared with CIP model, and studied the energy absorption effect by shape memory alloy or mild steel through the segment interface [22,23]. Leitner and Hao numerically analyzed the performance of fiber-reinforced-polymer wrapped rocking pier [24]. Zhong et al. simulated the hysteretic behavior, damage states and seismic demands of aging bridge piers [25]. The strength degradation behavior, energy dissipation capacity and residual displacement were well simulated by FEA, but the strength and stiffness could not be simulated simultaneously. Compared to the simplified or fiber models, the three-dimensional (3D) solid finite element model is able to represent the failure mechanism more accurately and to simulate the nonlinear shear effects more effectively [26].

The methodology of FEA simulating the experimental results followed by further parameter sensibility analysis was adopted by some studies, including the previously mentioned studies [21,23,24]. Moon et al. conducted FEA of the cast-in-steel-shell pile to reinforced concrete pier connection [27]. A monolithic pushover loading was performed. Compared with cycling test results, the stiffness and peak strength of model were similar. The parametric study analyzed five parameters from 40 models, and the weight and dispersion of the influence of each parameter were compared. Design suggestion was proposed following the comparison with existing design code.

Many studies have been developed for the rocking pier, while the simulation of precast piers that are emulative performance of CIP columns with plastic damage under earthquakes are scarce. To the authors' knowledge, the analysis on the effect of the recess design details on overall performance of precast pier with GCD connection has never been performed. FEA is thus needed to better quantify the improved design via parametric analysis. Therefore, this paper presents 3D finite element models of the improved GCD connection design, based on physical experimental data. Section 2 briefly introduced the experiment. The models are first calibrated and validated based on the results of experimental investigations in Section 3, to explain the mechanism that improved design had a better performance found in the experiment. The effects of recess depth and diameter, mortar cushion thickness, and mortar strength are then parametrically investigated in Section 4, to provide better references for such new design.

2. Experimental Investigation

In this section, the experimental investigation (Wang et al., 2020 [4]) is briefly introduced. The test consists of two 1/2.8-scale bridge columns: one with the improved GCD connection design (#1) and its CIP reference #0. The two column specimens are with the same segment design in terms of dimension and reinforcement arrangement, as shown in Figure 1. The pier segment is 500 mm in diameter and 2700 mm in height. The total height from the applied lateral loading is 2900 above the column-to-footing interface, yielding the aspect ratio of 5.8. The longitudinal reinforcements are 12 mm-diameter hot-rolled ribbed bars with nominal yield strength of 400 MPa (HRB400), and the hoops are 8 mm-diameter hot-rolled plain bars with nominal yield strength of 300 MPa (HPB300). Compressive strength of concrete is averaged to be 42.2 MPa, and that of high-strength mortar is 66 MPa. Joint detail is also given in Figure 1, and there is a 50 mm deep recess added to the connection, along with the new GCD connecters. The new GCD connector design is given in Figure 2, which improves the material strength and the effectiveness of grouting.

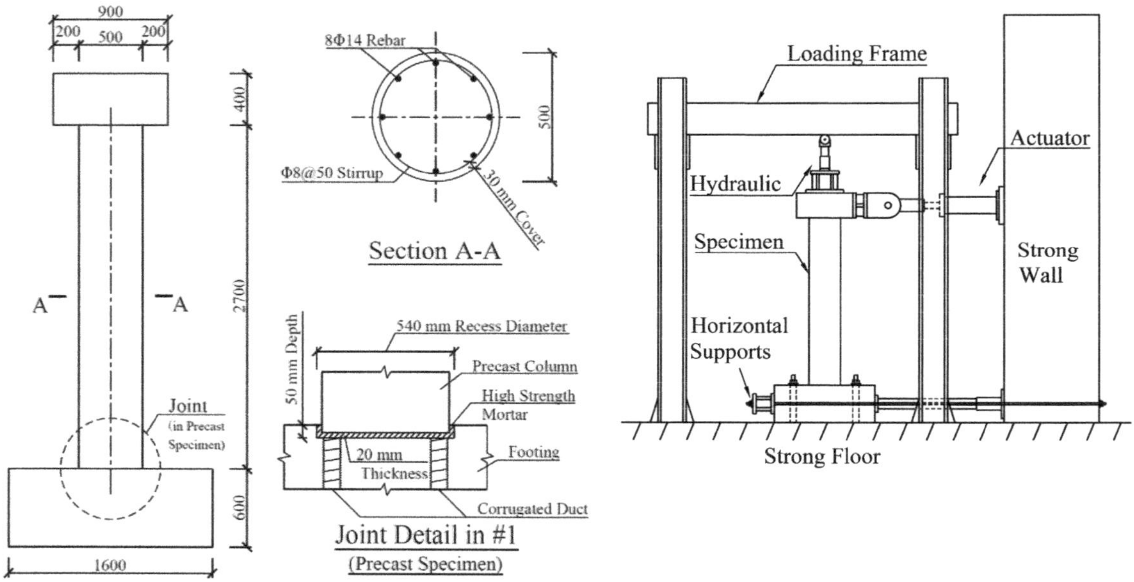

Figure 1. Test (**a**) specimen design (unit: mm); (**b**) setup.

In terms of test setup, the footing of specimens is fixed to the strong floor. Vertical load of 370 kN was applied with a 50-ton hydraulic jack to simulate superstructure mass. Lateral load was applied with the MTS 793 actuator. The specimens were test by a predefined load protocol under displacement control. The displacement level is designed as 2 mm, 5 mm, 10 mm, 15 mm, 20 mm, 25 mm, 30 mm, 40 mm and then 60 mm, 80 mm, etc. with the increment of 20 mm until the specimen felled. Each displacement level consisted of two cycles of current loading displacement and a cycle of previous displacement.

Based on the comparison of drift ratios, ductility values and residual displacements of the two column specimens, it is found that the deformation mechanism, energy dissipation potential and self-centering capability are similar between the precast and CIP specimens. This concludes that the new connection design forms an effective confinement to the column segment, and the precast column with such design is as good as, if not better than, the CIP reference.

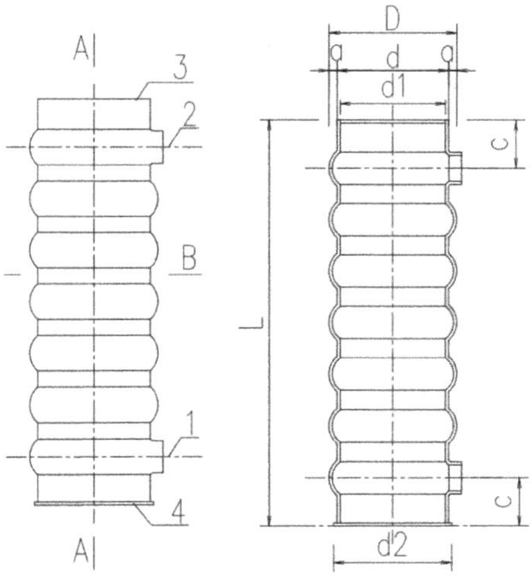

Figure 2. The new GCD connection design.

3. Numerical Simulation

3.1. Finite Element Model

3D solid finite element models are created using ANSYS to ensure the effectiveness of simulating the (1) crushing and cracking behavior of concrete and (2) mechanical behavior of the recess. The models are based on the experimental results conducted by Wang et al. [4]. Here, monotonic pushover loading is applied to the finite element model instead of cyclic loading, due to the reason that solid finite element models can hardly converge under cyclic loading in this case (especially with crushing and cracking simulation), and therefore monotonic pushover is currently the only choice to conduct the following parametric study to balance computational feasibility and analytical accuracy. The major focus of seismic behavior in practical engineering is the pushover curve, which would not lose necessary information if satisfactorily simulated for design references.

To be specific, Solid65 element is used for the concrete and mortar of pier column and upper layer (close to the recess) of footing, for crushing and cracking simulation. Solid45 element is used for cap beam and lower layer of footing. Link180 element, which can transmit uniaxial tension and compression, is used for the stimulation of longitudinal rebars. Conta173 and Target170 elements can be used to define pair-based contact, presenting the state of contact or sliding between surfaces. The interface of column and footing is thus stimulated by Conta173 and Target170.

The material property of concrete is defined according to Mander's concrete stress–strain model (Mander et al., 1988) as shown in Figure 3a, with key parameters obtained from sample tests. In addition, multilinear kinematic hardening (KINH) material model with the associated von Mises yield criterion is adopted, and Willam and Warnke's criterion for failure is used to account for cracking and crushing. Mander's model is also utilized for high-strength mortar simulation, due to similar stress–strain relation. Sliding and opening at the joint interface have been observed from the tests, and therefore detailed approach needs to be taken for simulating the joint behavior. Pair-based contact is used to simulate the axial and friction forces at the interface, where Conta173 is for the concrete column surface and Targe170 belongs to the upper surface of mortar. The Mohr–Coulomb model is used for regulating the contact behavior, and coefficient of the isotropic friction (MU) is 0.6. Maximum friction stress (TUMAX) is equal to the concrete shear strength of $\sigma_y/\sqrt{3} = 24$ MPa, and the cohesion (COHE) is set to 0 as default.

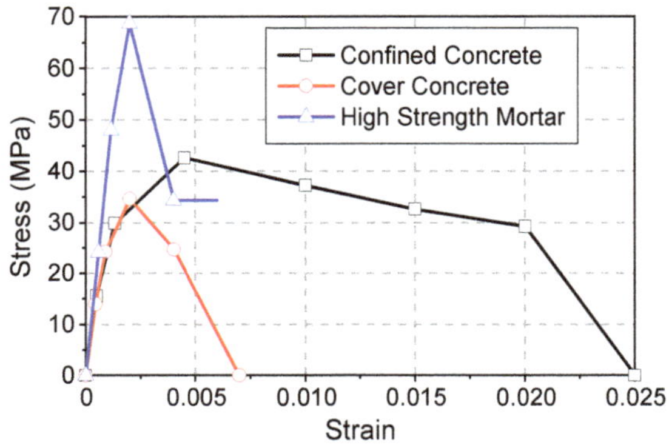

(a)

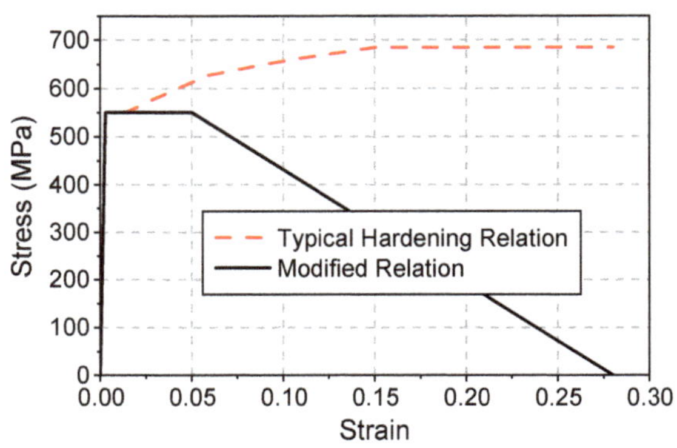

(b)

Figure 3. Stress–strain relation used in the FE model: (**a**) concrete and mortar; (**b**) rebar.

The obtained axial yield and ultimate strength of longitudinal bars are 550.5 MPa and 684.4 MPa, respectively. Longitudinal rebars are subjected to cycling loading during the test, with plastic development (Bauschinger effect) and buckling. However, only monotonic loading is applied to the finite element model due to the consideration of computational efficiency and convergence. Rebars need to be simulated with possible causes of specimen softening. Low cycle fatigue leads to plastic accumulation of the local and plastic capacity reduction of rebar, which is represented by shorter plastic plateau and earlier descending strength. Buckling of the reinforcement and the softening of the concrete under cyclic loading further complicate the degradation of the specimen capacity. The relation is modified based on ideal elastoplastic model, as shown in Figure 3b. It is determined by iterative adjustment under two conditions: (1) the numerical force–displacement monotonic loading curve agrees with backbone curve of the test result, especially the same slope of descending region; (2) at the ultimate displacement of 160 mm, the stress of the outmost tensional longitudinal rebar reduces to zero (fracture). Such definition of the constitutive model agrees with the low-cycle fatigue of longitudinal rebars reflected in the test.

There are two specimens tested in the experiment, and they are CIP reference #0 and precast specimen #1 with the same design. The precast specimen #1 is with the improved GCD connection between precast column and footing. There exists a 20 mm thick mortar as bonding material. The finite element model created based on specimen #1 is shown in Figure 4. Mesh size in the model is 50 mm, while the thickness of the grouted duct is 2 mm.

3.2. Validation with Test Results

Numerical model (precast model with recess, PMR) of precast specimen #1 is established following the details given in Section 2, and force–displacement curves are compared in Figure 5. Prior to yielding, the numerical model is with slightly lower stiffness, and maximum difference of strength is 8.69% at 20 mm. The specimen reaches the peak load of 77.4 kN, while the numerical model is 78.5 kN, which is off by 1.4%. At 65 mm, a minor hardening appears in the model curve, which is contributed by that the concrete material cannot perfectly express the triaxial behavior.

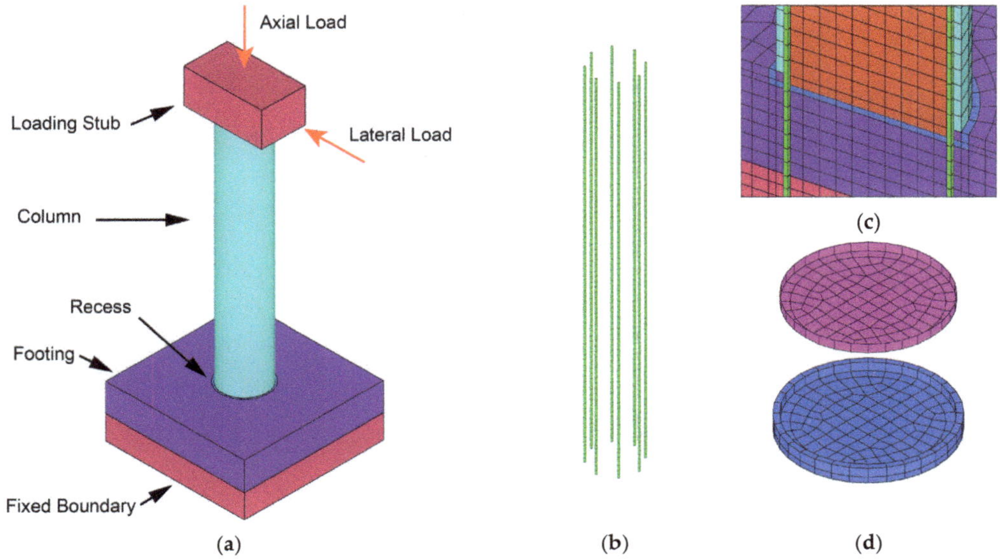

Figure 4. Numerical model of test specimen #1: (**a**) overall view; (**b**) longitudinal rebars; (**c**) recess and the mesh of profile; (**d**) contact and the mortar element.

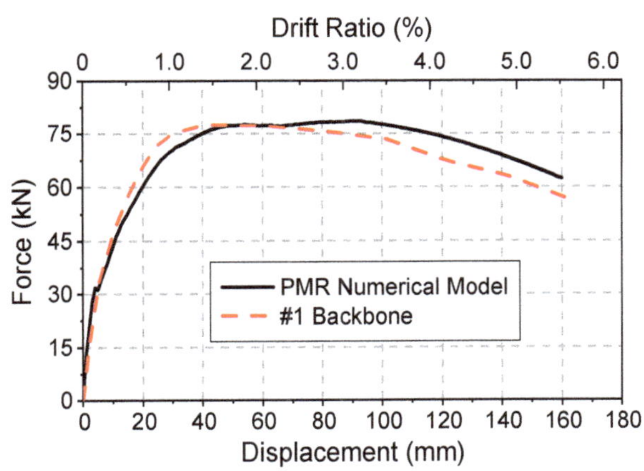

Figure 5. Force–displacement curve of PMR and the backbone curve in the test.

The contour plots of principal compressive stress and concrete cracking and crushing are shown in Figure 6. The compressive region is mainly concentrated at longitudinal reinforcement and the recess from far side of applied load. The cracks are developed along the height of 1.8 m above the footing. This is similar to the visible cracks in test specimen #1 (1.6 m), as given in Figure 6c. The crushing region covers the height of 60 cm above footing, which is higher than the test result (40 cm). As shown in Figure 7, a series of outermost rebar stress diagrams indicates that the maximum strain of longitudinal rebar is at 175 mm above the footing. It is consistent with the test results that the most severe spalling is 160 mm above the footing (Figure 6c).

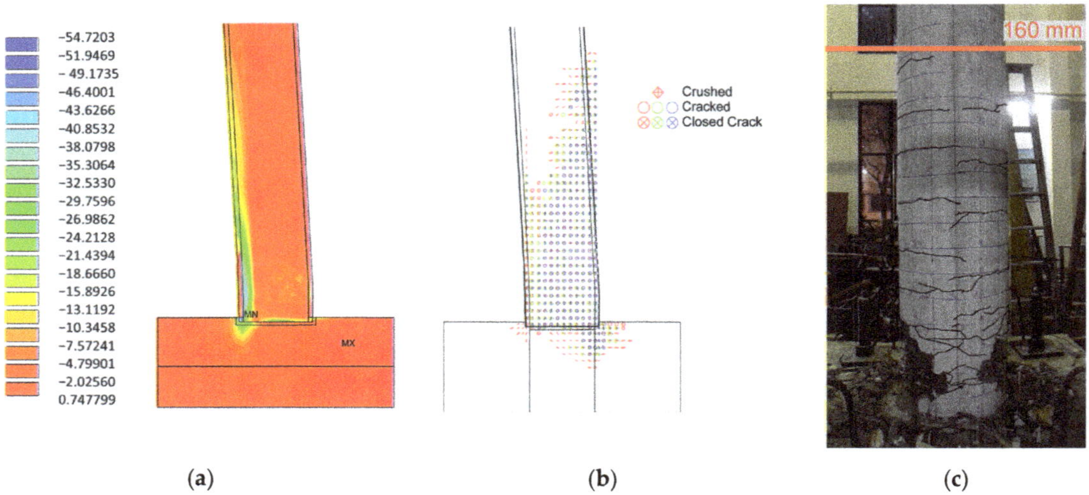

Figure 6. PMR's simulation diagram, at the final stage: (**a**) PMR's principal compressive stress (unit: MPa); (**b**) PMR's crack and crushing; (**c**) Post-test damage of #1 for comparison (cracks are intentionally marked).

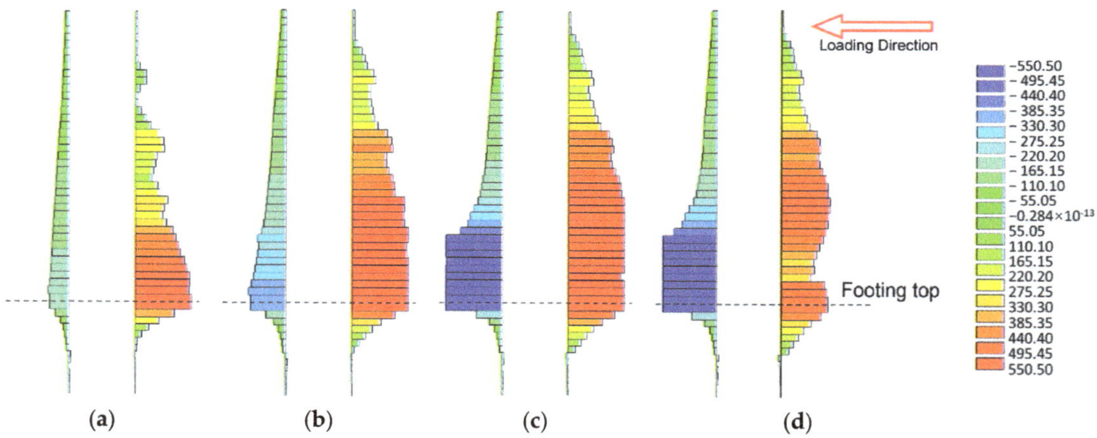

Figure 7. PMR's outmost rebar stress diagrams at the displacement of (unit: MPa): (**a**) 20 mm; (**b**) 40 mm; (**c**) 90 mm; (**d**) 120 mm.

3.3. Plastic Hinge Development

Typical precast model (TPM) without the recess design and cast-in-place model (CIPM) are also created with the same necessary elements and material as PMR, and corresponding force–displacement relations are given in Figure 8. The peak load of TPM is 70.8 kN, while those of PMR and CIPM are 78.5 kN and 80.5 kN. Postyielding capacity of TPM decreases earlier and faster than PMR, and rebar of TPM also fractures at early stage (120 mm). The yielding load values of CIPM and specimen #0 (around 50 mm) are differed by 11.4% (79.71 kN and 71.58 kN), which is larger than 8.5% of difference in concrete strength (43.9 MPa and 40.5 MPa). A more rapid decrease of force in specimen #0 is found around 100 mm.

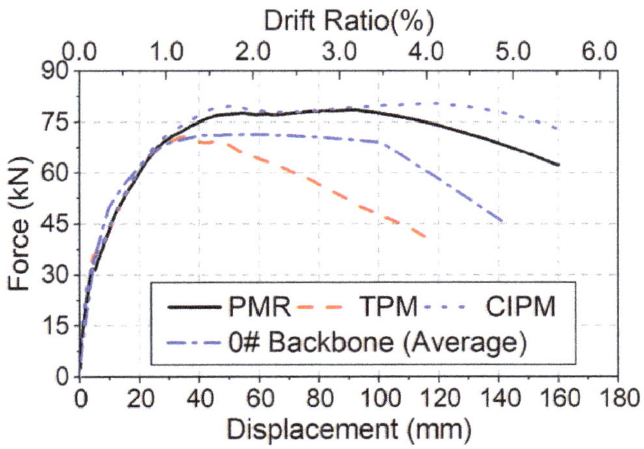

Figure 8. Force–displacement curves of TPM, CIPM and PMR.

Stress distribution of longitudinal rebar in three models are shown in Figure 9. Stress distribution in TPM is more concentrated at the column-to-footing interface, which leads to more rapid increase of strain, resulting in earlier failure than others. The maximum strain of rebar in PMR is located 175 mm above the footing, which is the same as CIPM.

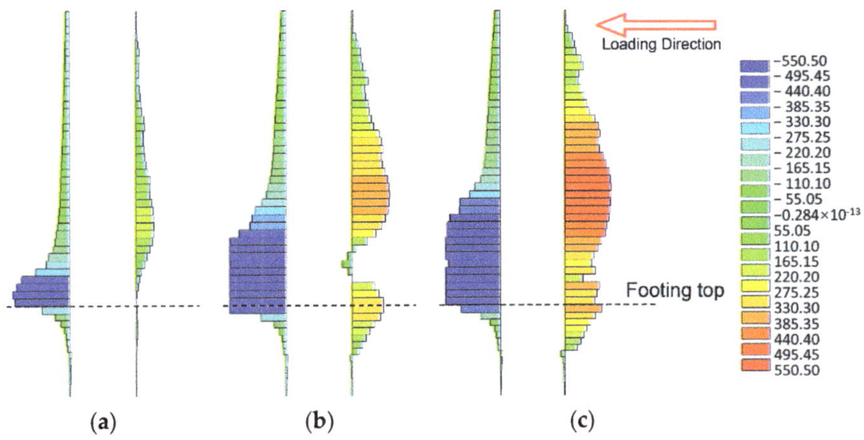

Figure 9. Rebar stress diagrams at each final state of (unit: MPa): (**a**) TPM (120 mm); (**b**) PMR (160 mm); (**c**) CIPM (160 mm).

As given in Figure 10, crushing regions of PMR and CIPM form a rectangular shape with wider distribution area, while the that of TPM is concentrated at pier bottom with triangular regions, indicating that joint of TPM is more susceptible to failure. As illustrated in Figure 11, the inclined cracks start to be formed at the neutral axis, then extended to both sides. Inclined cracks and transverse cracks together define the plastic hinge region, while vertical cracks lead to spalling. A right triangle shaped plastic hinge is formed for TPM, while the bottommost transverse cracks of PMR and CIPM extend more slowly, and still hold shear transferring capacity. The inclined cracks of PMR and CIPM are more likely to start from the end of horizontal cracks at a certain height above the column bottom instead of bottommost, which ensures more uniform development of transverse cracks along the pier column and plastic deformation to appear at the later stage.

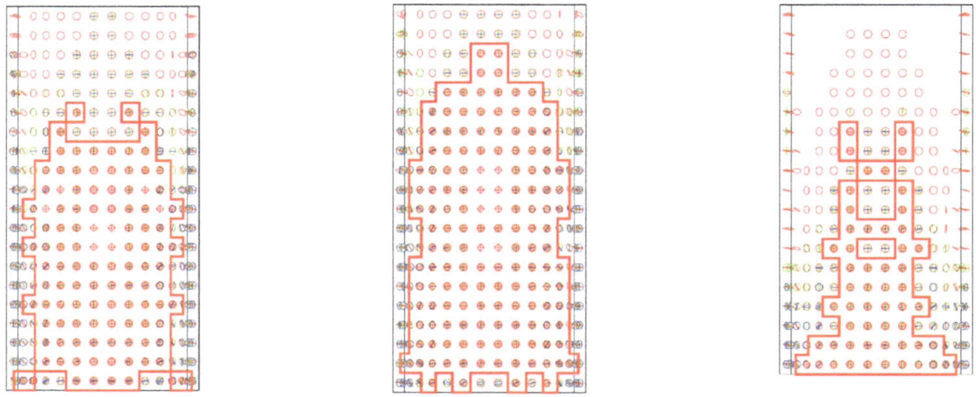

(**a**) (**b**) (**c**)

Figure 10. Crushing diagram of the cover concrete (**a**) PMR; (**b**) CIPM; (**c**) TPM.

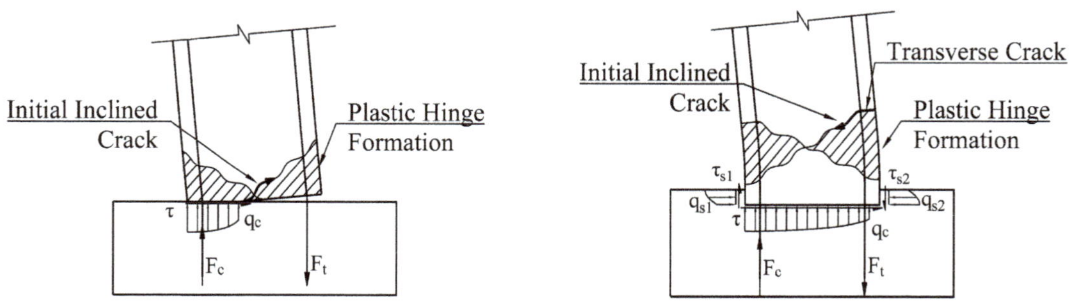

(**a**) (**b**)

Figure 11. Schematic diagram about force transferring and hinge formation in the pier (**a**) with initial horizontal defect; (**b**) without initial horizontal defect.

3.4. Local Deformation Development at Recess Joint

For the convenience of discussion, several terms are defined in Figure 12. Lifting displacement is the relative vertical displacement between the outermost node on the surface of the column tension side. Gap distance means the maximum gap distance of the contact elements at the bottom of the recess.

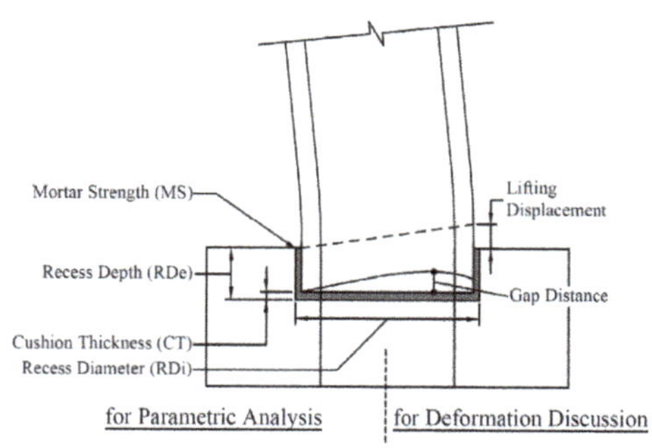

Figure 12. Illustration about joint deformation discussion (Section 3.3) and key parameter of recess. (Section 4).

Contact statuses of PMR at 50 mm and 160 mm are shown in Figure 13. Confinement effect provided by the recess keeps some elements in contact even at the displacement level of 160 mm. For the rim surface, some elements that are already in NearContact at 50 mm state return to Sticking state at 160 mm, where the upper movement of column segment makes column tip stuck to the rim surface at the near side of loading.

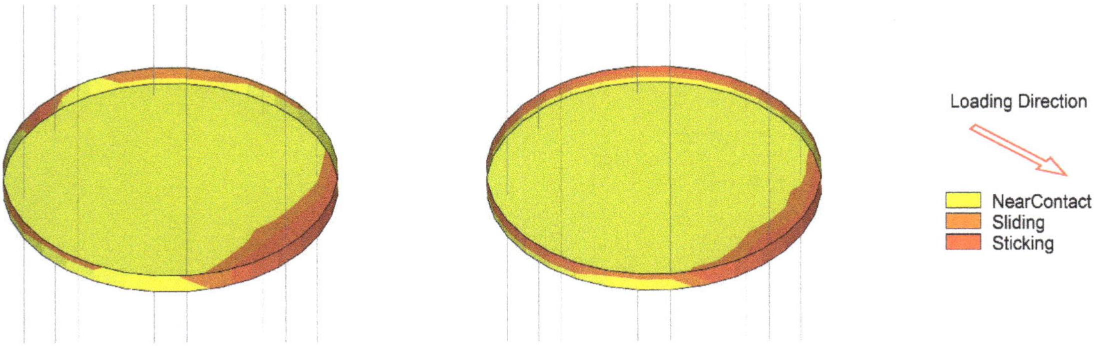

(a) (b)

Figure 13. PMR's contact status diagrams at the displacement of (**a**) 50 mm and (**b**) 160 mm.

Figure 14a shows the lifting displacement and gap distance about lateral displacement of PMR. The lifting displacement decreases with lateral resistant force at 100 mm, while gap distance keeps increasing from 0.367 mm below lifting displacement to 0.041 mm above. This indicates that the joint gap distance is mainly related to the elongation of the longitudinal rebar after crack initiation. Figure 14b shows PMR's gap distance distribution at the displacement of 160 mm, and the maximum value is 1.84 mm, where friction from the recess rim restricts the local deformation. TPM's maximum gap distance of 15.6 mm is much larger than PMR, as shown in Figure 15a. Different from PMR, and the joint opening of TPM varies almost linearly along the loading direction, due to the lack of constraints (Figure 15b).

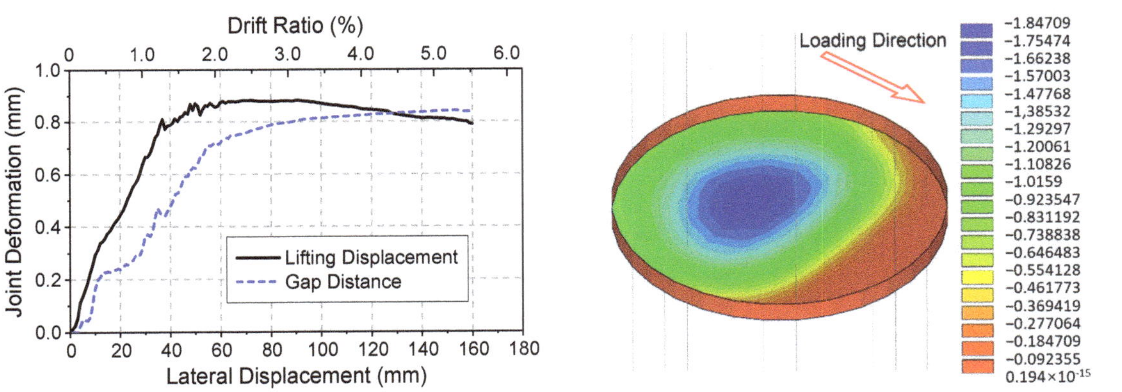

(a) (b)

Figure 14. PMR's (**a**) joint deformation curves; (**b**) gap distance at the displacement of 160 mm. (unit: mm).

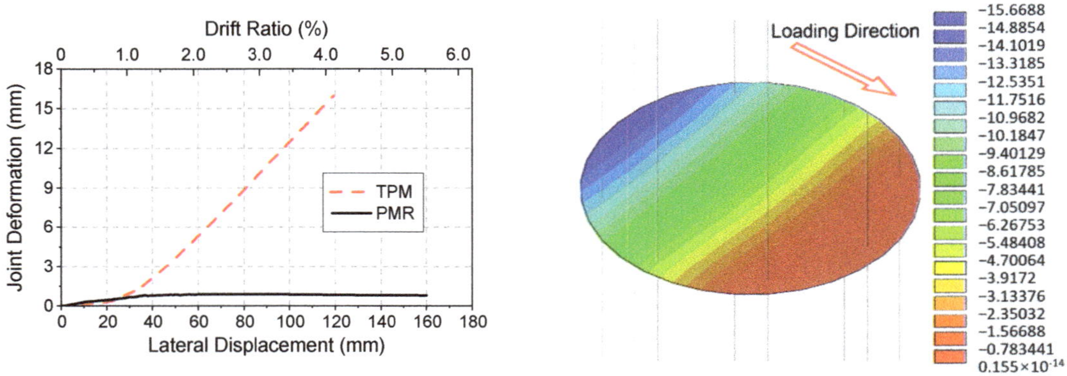

(a) (b)

Figure 15. TPM's (**a**) joint deformation curves; (**b**) gap distance of at the displacement of 120 mm. (unit: mm).

In general, the deformation development of pier with recess can be divided into two stages (Figure 16). At the first stage, initial deformation is developed by small opening of joint and the shear deformation of recess rim, where column remains elastic. After column cracking (the second stage), tension force is transferred from concrete and mortar in the recess to longitudinal reinforcement. Shear force and corresponding deformation on recess rim start to decrease, while strain of longitudinal reinforcement and gap distance continue to increase until rebars yield.

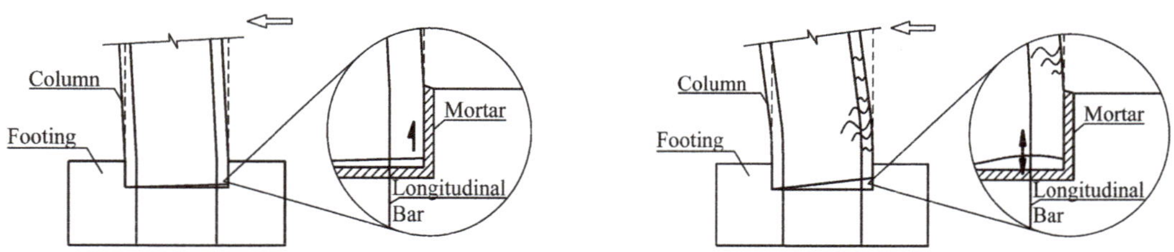

(a) (b)

Figure 16. Schematic diagram about deformation in the precast pier with recess: (**a**) Initial state; (**b**) Severely cracked.

4. Parametric Analysis

To fully understand the recess design, numerical models with different recess depths, diameters, cushion thicknesses, and mortar strengths are investigated for parametric analysis (illustrated in Figure 12). All the unstated dimensions and materials of the following models remain the same as PMR.

4.1. *Effect of Recess Depth (RDe)*

Six precast models with RDe values s of 30, 70, 100, 120, 150 and 200 mm are created and compared, along with PMR (RDe = 50 mm) and TPM (RDe = 0 mm). Force–displacement curves are compared in Figure 17along with the models of TPM and PMR. Most performance indices (e.g., initial stiffness, stiffness after cracking, yielding stiffness and yielding load) are within 10% in difference beyond RDe of 50 mm, representing enough confinement. The other two models (RDe = 0, 30 mm) are with lower peak load and shorter yield plateau, along with sudden drop of resisting force caused by excessive local stress.

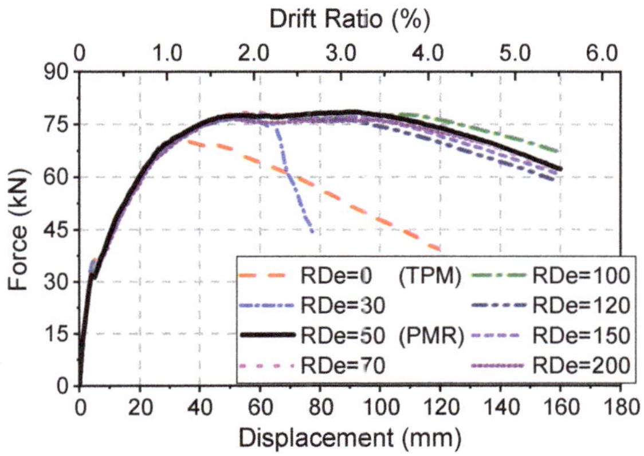

Figure 17. Force–displacement relation with different recess depths.

Ultimate displacements (the displacement when loading capacity decreases to 85% of its maximum value) of all models are listed in Table 1. Ultimate displacement of models with RDe values over 50 mm is approximately twice of models with 0 mm and 30 mm depth. Generally, the ultimate displacements of the models increase with recess depth, gradually approaching CIPM's value. Table 1 also gives the comparison of resisting forces. The lateral strength of pier keeps rising with the recess depth until 50 mm. Compared with CIPM, strengths of the models with RDe values over 50 mm are only off by 5%.

Table 1. Ultimate displacements and forces of the models with different recess depths.

Recess Depth/mm	Ultimate Displacement/mm	Difference		Maximum Lateral Loading/kN	Difference	
		CIPM	PMR		CIPM	PMR
0 (TPM)	71.79	−55.13%	−51.05%	70.86	−12.04%	−9.77%
30	68.39	−57.26%	−53.37%	76.62	−4.89%	−2.43%
50 (PMR)	146.66	−8.34%	\	78.53	−2.52%	\
70	144.91	−9.43%	−1.19%	78.43	−2.64%	−0.13%
100	160 *	0.00% *	+9.10% *	77.93	−3.26%	−0.76%
120	136.11	−14.93%	−7.19%	77.18	−4.20%	−1.72%
150	141.16	−11.78%	−3.75%	77.93	−3.26%	−0.76%
200	152.39	−4.76%	+3.91%	76.68	−4.82%	−2.36%
CIPM	160 *	\	+9.10%*	80.56	\	+2.58%

* the strength of model has not decreased to 85% of ultimate strength yet at the displacement of 160 mm.

The lifting displacement of pier at 160 mm becomes smaller when recess depth gets larger, as shown in Figure 18. Recess depth has greater influence on the lifting displacement within 70 mm, where more significant drop is found. RDe of 100 mm divide the models into two categories of joint deformations (RDe < 70 mm and RDe≥ 100 mm) (Figure 19). For RDe < 70, lifting displacement is with more gradual initial increase and subsequent decrease, and gap distance has greater impact on the lifting displacement. For RDe ≥ 100, sharp linear increase of lifting displacement is found, with more significant postpeak drop, indicating the presence of plastic hinge. No decrease of gap distance is observed in this category, and therefore the decrease of lifting displacement is more closely related to sidewall shear deformation in deeper recess.

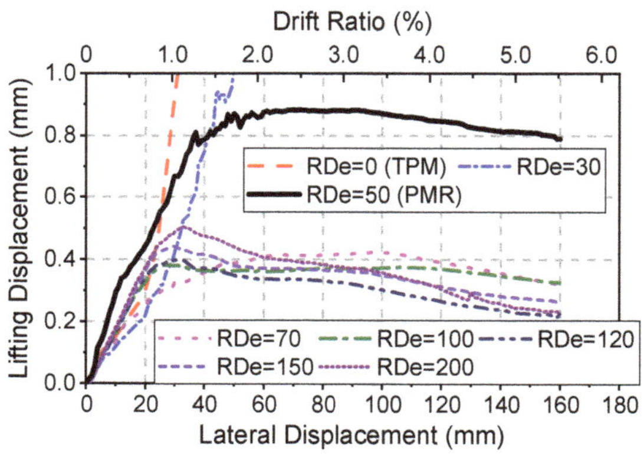

Figure 18. Lifting displacement with different recess depths.

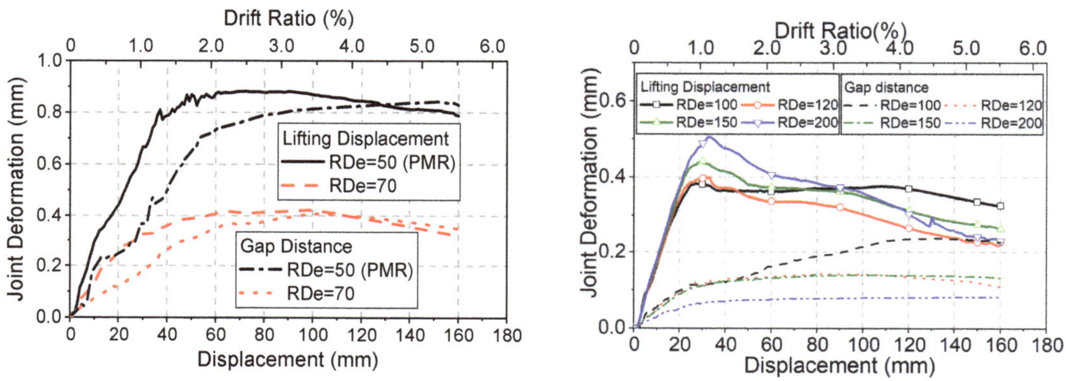

(**a**) (**b**)

Figure 19. Joint deformation with different recess depths: (**a**) RDe $\leq$ 70 mm; (**b**) RDe $\geq$ 100 mm.

The mortar stress can tell the local failure of pier column or footing. The relation of maximum principal compressive stress of mortar to loading displacement of each model is shown in Figure 20. The mortar stresses of all models are less than their ultimate strengths, and the stress level of models with over 50 mm recess depth is approximately 30% lower than that of TPM. The curves of models with RDe = 50, 100, 150, 200 mm are selected to analyze the influence of different recess depth to initial stress, as shown in Figure 21. The mortar stresses of deeper recess (RDe = 150, 200 mm) increase faster. When the mortar stresses of these two curves increase to 35 MPa, the concrete is damaged, resulting in decrease of mortar stress. In the models with shallow recess (RDe = 50, 100 mm), the mortar stresses increase more slowly with less fluctuation.

Figure 22 shows the stress distribution of the recess with various depths. In general, the stress variations of cushion mortar and sidewall mortar with the increase of recess depth are opposite, where maximum stress is at mortar cushion in shallow recess or at sidewall in deep recess. As the depth of the recess increases, the area with high stress transfers to sidewall mortar from cushion mortar. For RDe = 50 mm (PMR), the stress of mortar stays relatively small. For RDe = 70 mm, the stress of mortar cushion increases much faster, and the cushion is damaged first, following the decrease of mortar stress and supporting force. For RDe = 100 mm, the effect of cushion becomes less significant as deeper recess, so the constraint provided by sidewall is not as good as shallower recess. For RDe = 150 and 200 mm, the high stress regions are mainly at the sidewall, and the overall performances become a common socket connection.

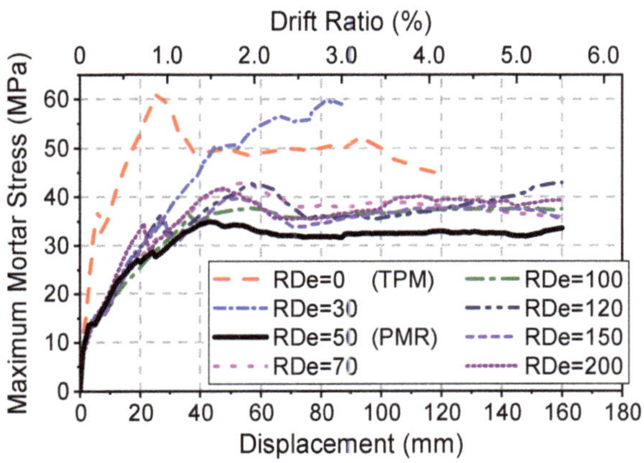

Figure 20. Mortar maximum stress with different recess depths.

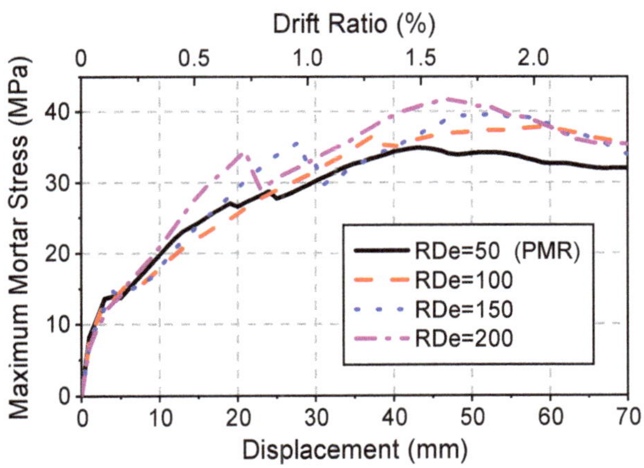

Figure 21. Mortar maximum stress with different recess depths before the displacement of 70 mm.

From the above analysis, the performance of recess is contributed by two parts: the cushion and the sidewall. When the recess is shallow, the cushion and the rebars provide the major supporting and anchoring forces to column. In deep recess cases, the column is constrained by the sidewall, and therefore the cushion is unable to fully perform its function. When the recess depth is larger than 50 mm, the overall performance of model experiences substantial improvement, and such performance becomes stable afterward.

4.2. Effect of Cushion Thickness (CT)

To investigate the effect of mortar CT on pier performance, thickness is set to 20 mm and 50 mm. Figure 23 shows two groups of force–displacement curves with different RDe levels (RDe $\leq$ 100 mm and RDe > 100 mm). The models with the same effective RDe (RDe minus CT) exhibit similar overall performance, while shallow recess is more sensitive to cushion thickness.

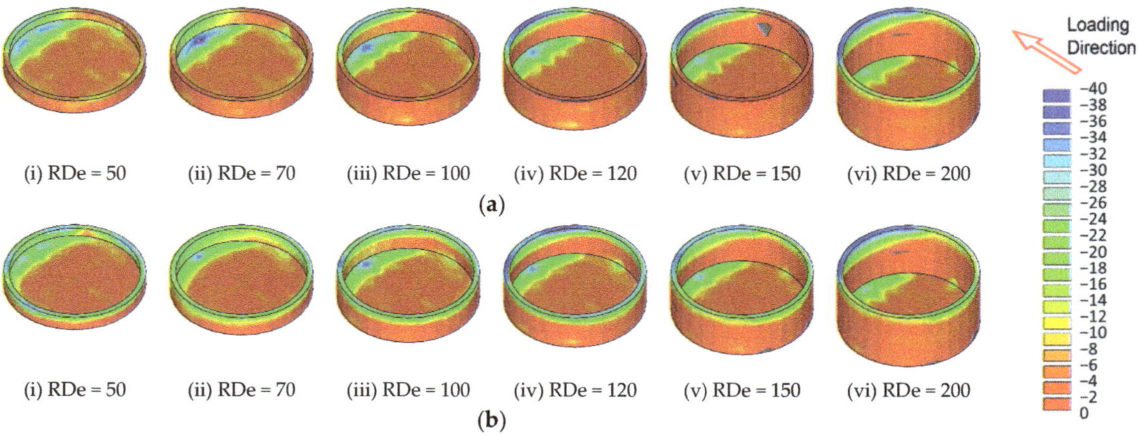

Figure 22. Mortar compressional stress with different recess depths (unit: MPa): (**a**) at the lateral displacement of 50 mm; (**b**) at the lateral displacement of 160 mm.

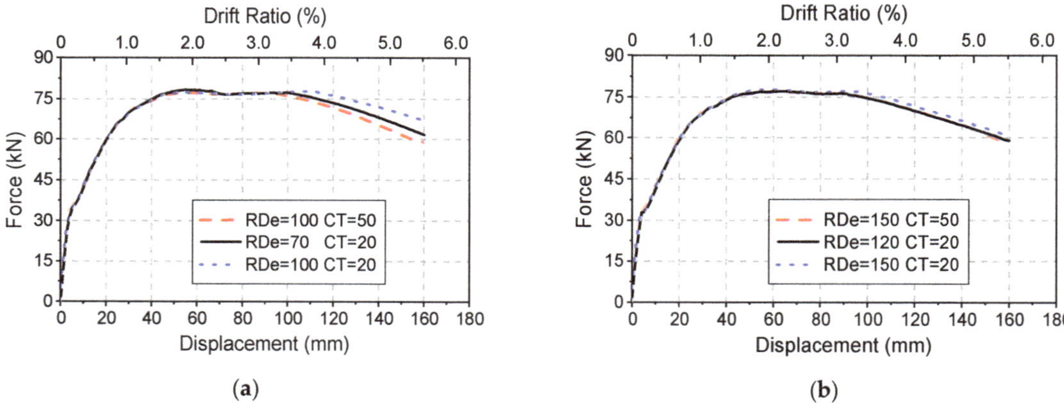

Figure 23. Force–displacement relation of different cushion thicknesses: (**a**) shallow recess; (**b**) deep recess.

Figure 24 shows the joint deformation development with various CTs but the same effective RDe. Thicker mortar cushion yields less strain penetration of rebar. Rebar deformation is concentrated above the cushion, resulting in larger gap distance. For shallower recess, the maximum differences between the two models for lifting displacement and gap distance are 0.058 mm and 0.049 mm at the displacement of 33 mm. For the deeper recess, maximum difference of gap distance is increased to 0.072 mm at 33 mm, while that of lifting displacement reaches 0.074 mm at the displacement of 53 mm. This indicates that cushion deformation in deeper recess is more difficult to occur.

Development of maximum mortar stress is given in Figure 25. For shallower recess, maximum mortar stress is decreased from 42.89 MPa to 32.57 MPa by 24%, when the thickness of cushion is increased by 30 mm. Larger CT provides more uniform distribution of stress. In addition, the two models with deeper recess become stable with limited differences in stress, as mortar cushion in deep recess is of little contribution to the distribution of mortar stress.

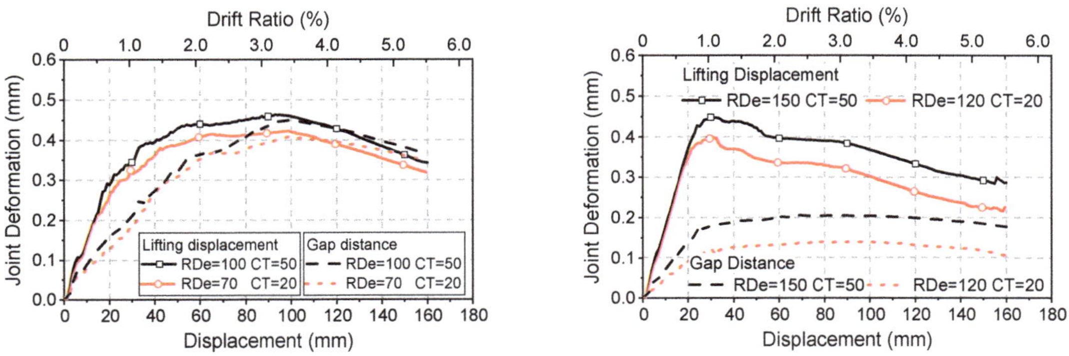

(**a**) (**b**)

Figure 24. Joint deformation curves of different cushion thicknesses: (**a**) shallow recess (effective RDe = 50 mm); (**b**) deep recess (effective RDe = 100 mm).

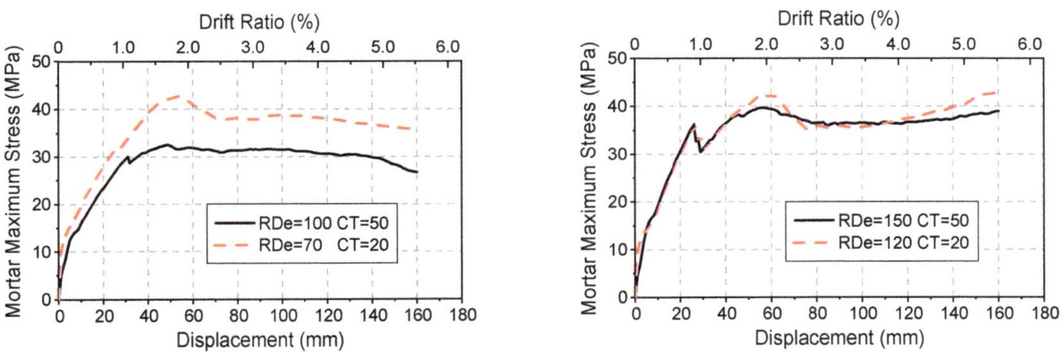

(**a**) (**b**)

Figure 25. Maximum mortar stress of different cushion thicknesses: (**a**) shallow recess; (**b**) deep recess.

The cushion acts as the medium that transfers the stress from column to footing. Though thicker cushion may slightly cause larger joint deformation and lower ductility, it is more beneficial for thicker cushion in the shallow recess in terms of local stress, to improve the robustness of joint. Considering size effect, effective RDe is more suitable than absolute depth for the design of the improved GCD connection in actual full-scale bridge. Based on the simulation results in this section, the ratio of effective RDe to the maximum column size is recommended to be within 0.06 and 0.2.

4.3. Effect of Recess Diameter (RDi)

Larger RDi brings better convenience tolerance in segment assembly, but structural integrity of precast column can be affected if it is too large. RDi in PMR is 540 mm, including a 20 mm wide gap between column and footing. Models with 600 mm RDi (gap of 50 mm) are also created for comparison. The models analyzed in this section are also divided into two groups based on recess depth (Figure 26). For depth of 50 mm, larger RDi improves the strength and ductility, where the maximum resisting force is increased from 78.53 kN to 83.20 kN. For depth of 100 mm, RDi has limited impact on the structural performance.

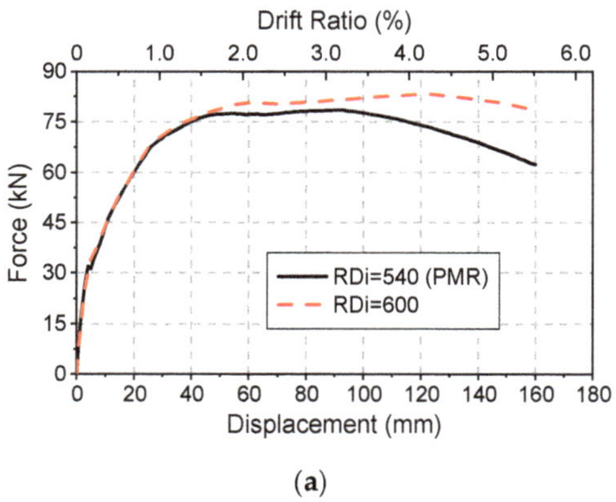

Figure 26. Force–displacement relation of different recess diameters: (**a**) shallow recess (RDe = 50 mm, CT = 20 mm); (**b**) deep recess (RDe = 100 mm, CT = 20 mm).

Larger recess diameter means longer horizontal distance between nodes and higher relative displacement under the same shear angle, and joint deformation with different RDi values are shown in Figure 27. Joint deformation of the models with shallower recess is more sensitive to RDi variation. For deeper recess, gap distance changes little with RDi increase, while lifting displacement of the model with larger RDi is higher than that with smaller RDi due to geometry change.

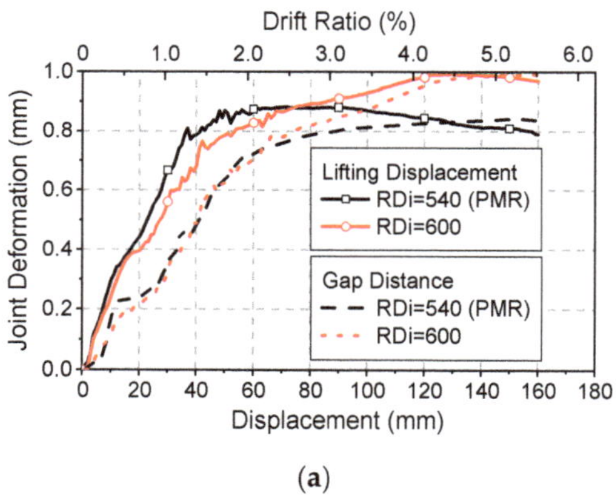

Figure 27. Joint deformation with different recess diameters: (**a**) shallow recess (RDe = 50 mm, CT = 20 mm); (**b**) deep recess (RDe = 100 mm, CT = 20 mm).

Variations of maximum mortar stress are shown in Figure 28. Little change is found with diameter increase for shallow recess, and concrete of the footing exceeds its ultimate strength at peak stress of mortar. However, stress level is reduced for deep recess by 11% for the two different RDi values in the case of deep recess, and no local compressive damage at footing concrete is found. As the strength of mortar is higher than concrete, stress is more uniformly transferred from thicker mortar, preventing concrete from local failure. The maximum stress values at displacement level of 160 mm are close (33.60 MPa and 33.44 MPa, respectively) in the case of shallow recess. Larger RDi moves compressive stress of the rim closer to the near side, giving better column support and pry-out restriction. Therefore, stiffer and higher strength boundary around the column bottom is recommended.

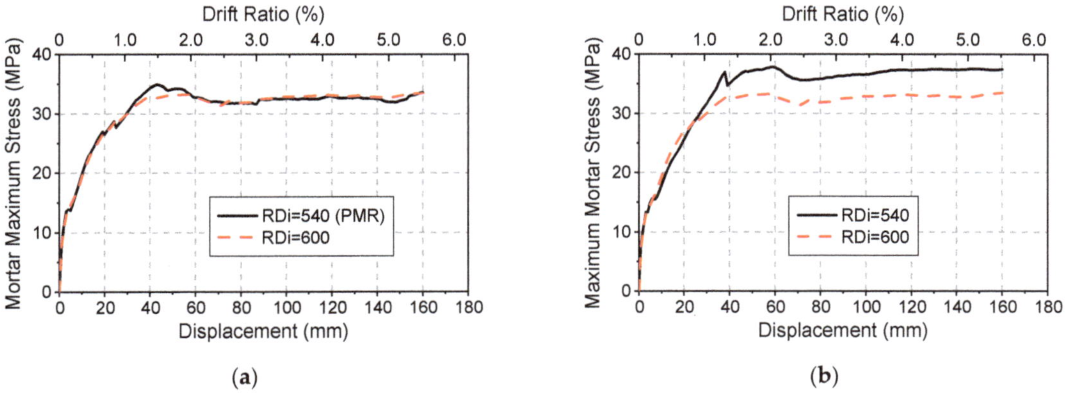

Figure 28. Maximum mortar stress curves with different recess diameters: (**a**) shallow recess (RDe = 50 mm, CT = 20 mm); (**b**) deep recess (RDe = 100 mm, CT = 20 mm).

4.4. Effect of Mortar Strength (MS)

Two models with different MSs are established and compared with PMR. PMR's mortar compressive strength is 68.71 MPa (M60). The mortar strengths of the other two models are obtained by scaling to 45.8 MPa (M40) and 91.6 MPa (M80). As shown in Figure 29, higher strength and ductility are only observed in the model with M80, while the other two models are with close variation. Figure 30 shows maximum stress variation of different mortar grades. The maximum stress values are 32.83 MPa for M40, 34.90 MPa for M60, and 38.06 MPa for M80. The larger the mortar strength, the higher stress of mortar is, while corresponding ultimate strengths are not reached in all cases. In addition, the maximum stress values are also close to each other in the three models (32.9 MP, 33.60 MPa and 34.40 MPa, respectively). Figure 31 also shows that MS has limited influence on the joint deformation, gap distance, and lifting displacement. In general, the stiffness of mortar alters the stress distribution in shallow recess, and higher stiffness improves structural integrity.

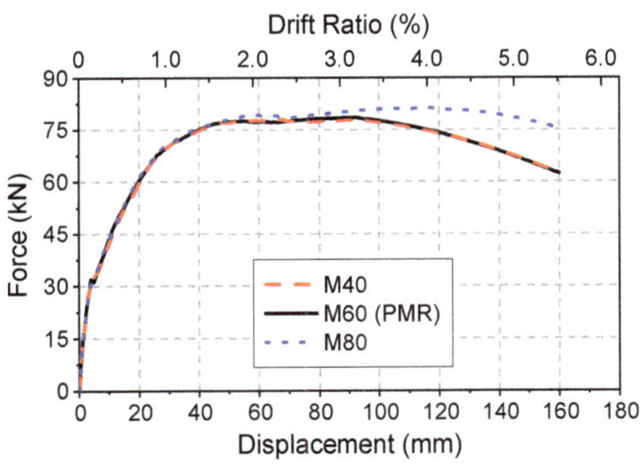

Figure 29. Force–displacement relation with different mortar grades.

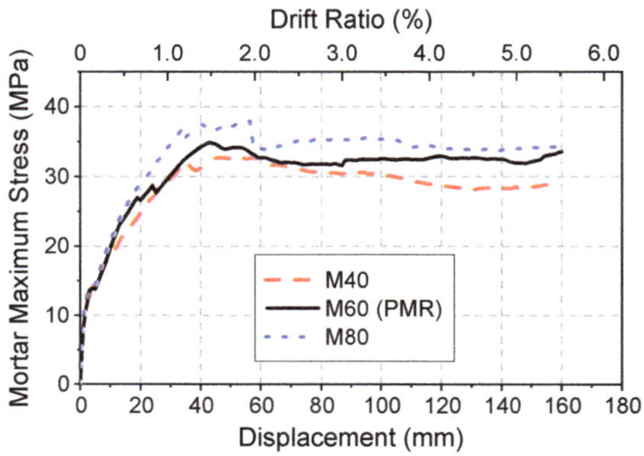

Figure 30. Maximum mortar stress with different mortar grades.

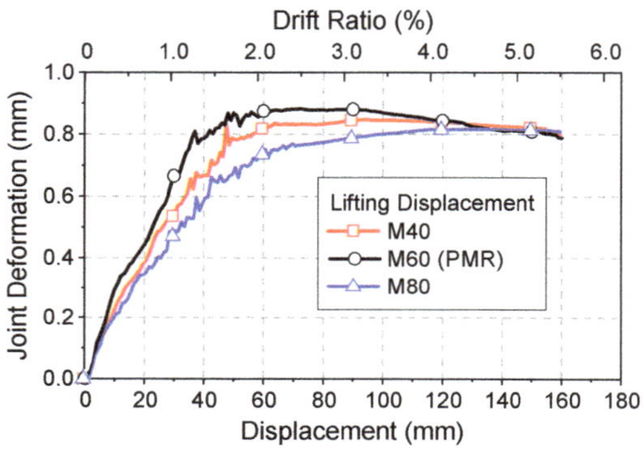

(**a**)

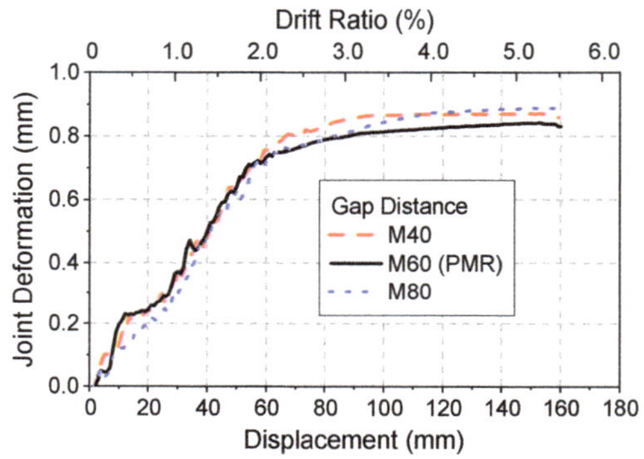

(**b**)

Figure 31. Joint deformation curves with different mortar grades: (**a**) Lifting displacement; (**b**) Gap distance.

5. Discussion

It is highly time-consuming and difficult to converge to simulate FE model with the state of high nonlinearity. In the existing articles [21,23,24], 3D solid FE analysis mainly simulates the rising region of force–displacement backbone. In this study, monotonic loading is selected, same as [18,27], to balance computational efficiency and convergence, and it may be a better option to modify the damage state by user defined material. However, the material simulation in this study is only a rough adjustment, the constitutive relationship of reinforcement needs to be further studied along with concrete material.

The FE model of test specimens with circular cross-section is mainly discussed in this study. It is worth noting that the findings can be extended to the situation with square cross-section, full scaled pier or the pier with pile foundation.

In the parametric analysis, the influence of the recess sidewall stiffness (mortar strength and sidewall thickness) on the column confinement is qualitative, and the relationship between recess stiffness and the size or material properties of column, cushion and footing is not given, so further research is needed.

6. Conclusions

Calibrated and validated by test results, finite element models are established to compare the performance differences among precast pier with recess design, typical precast pier and cast-in-place pier, thus clarifying the mechanism of this improved design's efficiency.

The parameter sensibility of recess to the pier performance is analyzed, so that the contribution and design method are discussed. Based on the study, the following conclusions can be drawn:

- The precast pier with recess (grouted ducts in the footing) can reach an equivalent seismic performance to cast-in-place pier, due to the constraint of joint deformation by the extra support and friction force from the rim. The shear resistance of connection joint does not decrease rapidly with loading, reducing the inclined cracks at column bottom. This ensures a well-distributed transverse and inclined cracks development at the plastic hinge region.
- Recess depth has the most significant influence on the performance of the pier. Above a certain depth (50 mm) of recess, enough confinement is formed, and the strength and ductility of pier can reach a value close to the cast-in-place pier. With the increase of recess depth, the lifting displacement of pier bottom gradually decreases, the components of lifting displacement redistribute, joint gap distance decreases, and sidewall shear deformation becomes more significant.
- The cushion for shallow recess and the rim for deep recess are the major part that stress is concentrated on, and the corresponding mortar should be appropriately thickened to distribute the stress. Too shallow (less than 30 mm) or too deep (greater than 100 mm) recess may produce local failure of footing. Appropriate recess depth (between 50 and 100 mm) and adequate stiffness of recess will ensure full support and confinement to the column.
- According to the study, the precast pier with recess can be designed for general case by existing code, ensuring no pullout failure of grouted duct. The detailed design for prevention of pry-up failure is achieved by the following rules:
  (1) Recess depth is suggested to be taken 6–20% of the column diameter for circular section.
  (2) Mortar is suggested to select 20 MPa higher than adjacent concrete.
  (3) Local enhancement of footing reinforcement is needed.
  (4) The recess also retains the potential to avoid the effect of cold joints of cast-in-place piers.
- The improved GCD connection design is worth studying in the condition with different section shapes or foundation constraints, and further design codes are needed for quantification with mathematical formulation.

Author Contributions: Conceptualization, Z.W. and C.W.; methodology, C.W.; validation, H.Q.; formal analysis, C.W..; investigation, W.X.; resources, Z.W.; data curation, W.X; writing—original draft preparation, C.W.; writing—review and editing, H.Q.; visualization, W.X.; supervision, H.Q.; project administration, Z.W.; funding acquisition, Z.W. All authors have read and agreed to the published version of the manuscript.

Funding: This research was funded by National Natural Science Foundation of China, grant number 51978511, 51778470 and 52008316; Natural Science Foundation of Science and Technology Commission of Shanghai Municipality, grant number No. 20ZR1461400.

Institutional Review Board Statement: Not applicable.

Informed Consent Statement: Not applicable.

Data Availability Statement: Data available on request.

Conflicts of Interest: The authors declare no conflict of interest.

References

1. Littleton, P.; Mallela, J. Iowa demonstration project: Accelerated bridge construction on US 6 over keg creek. 2013. Available online: https://rosap.ntl.bts.gov/view/dot/54259 (accessed on 9 September 2022).
2. Khaleghi, B.; Schultz, E.; Seguirant, S.; Marsh, L.; Haraldsson, O.; Eberhard, M.; Stanton, J. Accelerated bridge construction in Washington State: From research to practice. *PCI J.* **2012**, *57*, 34–49. [CrossRef]
3. Wang, Z.; Qu, H.; Li, T.; Wei, H.; Wang, H.; Duan, H.; Jiang, H. Quasi-static cyclic tests of precast bridge columns with different connection details for high seismic zones. *Eng. Struct.* **2018**, *158*, 13–27. [CrossRef]
4. Wang, Z.; Wu, C.; Li, T.; Xiao, W.; Wei, H.; Qu, H. Experimental Study on the Seismic Performance of Improved Grouted Corrugated Duct Connection (GCDC) Design for Precast Concrete Bridge Column. *J. Earthq. Eng.* **2020**, *26*, 2469–2490. [CrossRef]
5. Steuck, K.P.; Eberhard, M.O.; Stanton, J.F. Anchorage of Large-Diameter Reinforcing Bars in Ducts. *ACI Struct. J.* **2009**, *106*, 506. [CrossRef]
6. Zhou, Y.; Ou, Y.-C.; Lee, G.C. Bond-slip responses of stainless reinforcing bars in grouted ducts. *Eng. Struct.* **2017**, *141*, 651–665. [CrossRef]
7. Galvis, F.A.; Correal, J.F. Anchorage of Bundled Bars Grouted in Ducts. *ACI Struct. J.* **2018**, *115*, 415–424. [CrossRef]
8. Provost-Smith, D.; Elsayed, M.; Nehdi, M. Effect of early-age subfreezing temperature on grouted dowel precast concrete wall connections. *Constr. Build. Mater.* **2017**, *140*, 385–394. [CrossRef]
9. Elsayed, M.; Ghrib, F.; Nehdi, M. Experimental and analytical study on precast concrete dowel connections under quasi-static loading. *Constr. Build. Mater.* **2018**, *168*, 692–704. [CrossRef]
10. Pang, J.B.K.; Eberhard, M.O.; Stanton, J.F. Large-Bar Connection for Precast Bridge Bents in Seismic Regions. *J. Bridg. Eng.* **2010**, *15*, 231–239. [CrossRef]
11. Tazarv, M.; Saiidi, M.S. UHPC-filled duct connections for accelerated bridge construction of RC columns in high seismic zones. *Eng. Struct.* **2015**, *99*, 413–422. [CrossRef]
12. Mashal, M.; White, S.; Palermo, A. Quasi-static cyclic testing of emulative cast-in-place connections for Accelerated Bridge Construction in seismic regions. *Bull. New Zealand Soc. Earthq. Eng.* **2016**, *49*, 267–282. [CrossRef]
13. Mashal, M.; Palermo, A. Emulative seismic resistant technology for Accelerated Bridge Construction. *Soil Dyn. Earthq. Eng.* **2019**, *124*, 197–211. [CrossRef]
14. Shoushtari, E.; Saiidi, M.S.; Itani, A.; Moustafa, M.A. Design, Construction, and Shake Table Testing of a Steel Girder Bridge System with ABC Connections. *J. Bridg. Eng.* **2019**, *24*, 04019088. [CrossRef]
15. Haraldsson, O.S.; Janes, T.M.; Eberhard, M.O.; Stanton, J.F. Seismic Resistance of Socket Connection between Footing and Precast Column. *J. Bridg. Eng.* **2013**, *18*, 910–919. [CrossRef]
16. Si, B.; Sun, Z.; Ren, X.; Wang, D.; Wang, Q. Finite element analysis of the hysteretic behavior of RC bridge piers. *J. Harbin Inst. Technol.* **2009**, *41*, 105–109. (In Chinese)
17. Chen, Y.; Zeng, L.; Xiao, Y.; Chen, J.; Wang, B.; Gong, Q. Numerical analysis of the seismic performance of steel-reinforced con-crete columns with a t-shaped steel cross-section. *China Earthq. Eng. J.* **2017**, *39*, 196–204. (In Chinese) [CrossRef]
18. Li, T.; Qu, H.; Wang, Z.; Wei, H.; Jiang, S. Seismic performance of precast concrete bridge columns with quasi-static cyclic shear test for high seismic zones. *Eng. Struct.* **2018**, *166*, 441–453. [CrossRef]
19. Zhong, J.; Mao, Y.; Yuan, X. Lifetime seismic risk assessment of bridges with construction and aging considerations. *Structures* **2023**, *47*, 2259–2272. [CrossRef]
20. Ou, Y.-C.; Chiewanichakorn, M.; Aref, A.J.; Lee, G.C. Seismic Performance of Segmental Precast Unbonded Posttensioned Concrete Bridge Columns. *Eng. Struct.* **2007**, *133*, 1636–1647. [CrossRef]
21. Dawood, H.; ElGawady, M.; Hewes, J. Behavior of Segmental Precast Posttensioned Bridge Piers under Lateral Loads. *J. Bridg. Eng.* **2012**, *17*, 735–746. [CrossRef]
22. Nikbakht, E.; Rashid, K.; Hejazi, F.; Osman, S.A. A numerical study on seismic response of self-centring precast segmental columns at different post-tensioning forces. *Lat. Am. J. Solids Struct.* **2014**, *11*, 864–883. [CrossRef]
23. Nikbakht, E.; Rashid, K.; Hejazi, F.; Osman, S.A. Application of shape memory alloy bars in self-centring precast segmental columns as seismic resistance. *Struct. Infrastruct. Eng.* **2014**, *11*, 297–309. [CrossRef]
24. Leitner, E.J.; Hao, H. Three-dimensional finite element modelling of rocking bridge piers under cyclic loading and exploration of options for increased energy dissipation. *Eng. Struct.* **2016**, *118*, 74–88. [CrossRef]
25. Zhong, J.; Zheng, X.; Wu, Q.; Jiang, L.; He, M.; Dang, X. Seismic fragility and resilience assessment of bridge columns with dual-replaceable composite link beam under near-fault GMs. *Structures* **2023**, *47*, 412–424. [CrossRef]
26. Delgado, R.; Delgado, P.; Pouca, N.V.; Arêde, A.; Rocha, P.; Costa, A. Shear effects on hollow section piers under seismic actions: Experimental and numerical analysis. *Bull. Earthq. Eng.* **2008**, *7*, 377–389. [CrossRef]
27. Moon, J.; Lehman, D.E.; Roeder, C.W.; Lee, H.-E.; Lee, T.-H. Analytical Evaluation of Reinforced Concrete Pier and Cast-in-Steel-Shell Pile Connection Behavior considering Steel-Concrete Interface. *Adv. Mater. Sci. Eng.* **2016**, *2016*, 4159619. [CrossRef]

buildings

MDPI

Article

Seismic Risk Mitigation and Management for Critical Infrastructures Using an RMIR Indicator

Alon Urlainis [1] and Igal M. Shohet [2,3,*]

[1] Department of Civil Engineering, Ariel University, Ariel 40700, Israel
[2] Department of Civil and Environmental Engineering, Faculty of Engineering Sciences, Ben-Gurion University of the Negev, Beer-Sheva 84105, Israel
[3] Department of Civil and Construction Engineering, Chaoyang University of Technology, Taichung 413, Taiwan, China
* Correspondence: igals@bgu.ac.il

Abstract: Recent earthquake events have highlighted the importance of critical infrastructure (CI) resilience, as a strong correlation was found between economic loss and severity of CI damage. CIs are characterized by a complex structure composed of sub-components that are essential for the continuous performance of the system. CI owners and governments allocate ample resources to retrofitting and upgrading CI systems and components to increase the resilience of CIs and reduce risk in case of seismic events. Governments and decision makers must manage and optimize the retrofitting efforts to meet budget and time constraints. This research presents a probabilistic methodology for CI seismic risk mitigation and management. The risk expectancy is appraised according to an FTA-based stochastic simulation. The simulation includes the development of exclusive fragility curves for the CI and an examination of the expected damage distribution as a function of earthquake intensity and fragility uncertainty of the components. Furthermore, this research proposes a novel RMIR (risk mitigation to investment ratio) indicator for the priority setting of seismic mitigation alternatives. The RMIR is a quantitative indicator that evaluates each alternative's cost-effectiveness in terms of risk expectancy mitigation. Following the alternative's RMIR value, it is possible to prioritize the alternatives meeting budget and time constraints. This paper presents the implementation of the proposed methodology through a case study of a generic oil pumping station. The case study includes twelve mitigation alternatives examined and evaluated according to the RMIR indicator.

Keywords: critical infrastructures; earthquake; risk mitigation; risk management

Citation: Urlainis, A.; Shohet, I.M. Seismic Risk Mitigation and Management for Critical Infrastructures Using an RMIR Indicator. *Buildings* **2022**, *12*, 1748. https://doi.org/10.3390/buildings12101748

Academic Editors: Xiaowei Wang, Chao Li, Jian Zhong, Weiping Wen and Suiwen Wu

Received: 21 September 2022
Accepted: 17 October 2022
Published: 20 October 2022

1. Introduction

Damage or disruption to critical infrastructures (CIs) can have a significant adverse effect on the economy, safety, and well-being of the public and private sector [1]. Recent earthquake events have highlighted the importance of critical infrastructure resilience, as a strong correlation was found between economic loss and the severity of CI damage [2,3]. Furthermore, along with the development of CIs, gradual increasing of essential services has depended on the continuous performance of multiple critical infrastructures such as energy, power supply, water supply, communications, etc. Typical CIs are nuclear power plants, desalination plants, bridges, security, and governance facilities characterized by complex systems architecture with a need to combine robustness, resilience, and redundancy in the design for continuous performance [4–11].

CI systems are characterized by a complex structure that is composed of various essential components (e.g., building, pumps, electro-mechanical equipment, and power supply in an oil pumping station) and subcomponents (e.g., building and pump foundations). The full functionality of the system requires a continuous performance of all components. Subsequently, the CI resilience is derived from the resilience, robustness, and redundancy of its core components and subcomponents. Studies show that there is a consistent interdependency across sectors of CI systems, as most of the CIs are connected and dependent on each other, and damage to one critical infrastructure will, most likely, lead to other CI failures [12–14]. Consequently, a failure of a single CI component can lead to a series of propagating disruptions of other CIs and affects a wide range of consumers from different sectors. The CI systems' interdependent structure increases vulnerability for cascading and rippling effects that increase the impact and the magnitude of each damage or disruption by initiating multi-hazard events [15]. The growing dependency on CIs, interdependencies between different infrastructures, and the growing number of infrastructures significantly increase CI seismic risk. Therefore, there is a vital need to protect and ensure the continuous performance and resilience of CI systems and assets after extreme seismic events.

In general, risk is defined as a measure of the probability and severity of adverse effects [16,17]. In the case of seismic risk, it should reflect the value of the potential consequences resulting from possible earthquakes throughout a defined duration of time [18]. Since the risk is calculated for CI in a specific location, the occurrence probability of possible earthquakes should be represented as the exceedance probability for a certain severity of IM in the location of interest. The deterministic approach focuses on a single earthquake, a small number of earthquakes, or a specific ground motion value [19–21]. This approach is useful for worst-case scenario analysis and for particular seismic scenarios. Several studies for seismic risk assessment were carried out based on deterministic approach [22–24]. However, the deterministic approach does not consider the uncertainties of the time, location, and magnitude of possible earthquakes. Moreover, targeting the retrofit efforts based on deterministic risk assessment may mislead the decision makers due to potential ignoring of possible events and subsequently avoiding the optimal alternative. Probabilistic Seismic Hazard Analysis (PSHA) is aimed at considering all possible earthquake scenarios and ground-motion levels that can occur in the system's location. The PSHA process produces a hazard curve that presents the annual rate of exceedance for any value for IM [25,26]. The probabilistic approach has been widely used to assess risk and develop seismic hazard maps [27–31].

Consequences are the outcome and the effects of an earthquake event. An examination of previous earthquakes reveals inconsistency in the severity of post-earthquake consequences [1]. For a specific earthquake event, there is a wide range of damage levels observed in similar types of structures and infrastructures in the same place. Many parameters can influence actual consequences, such as integrated maintenance and seismic retrofit frameworks [32–34], quality of materials reducing partial seismic capacity due to poor materials, quality of construction, degree of supervision during construction, and more. In [35], the authors presented the influence of different levels of corrosion on the seismic performance of concrete bridges.

Fragility curves are traditional damage functions to evaluate the expected damage distribution of CIs due to earthquake events [36–46]. A variety of generic fragility curves for CI systems and components are presented [47–50]. However, generic fragility curves do not necessarily reflect the actual system layout and components. In [51], the authors presented a comprehensive methodology for developing exclusive fragility curves for CIs by decomposing the system into subcomponents and a fault tree analysis to determine the system's failure mechanisms. Moreover, the condition of CI components will affect the seismic performance of the CI.

CI private sector owners and the public sector (governments) allocate ample resources in retrofitting and upgrading CI systems and components to improve the resilience of CIs and reduce risk in case of seismic events. However, governments and decision makers have to consider several possible mitigation strategies and choose the best solution to reduce risk under budget constraints, i.e., the optimal mitigation strategy/alternative. Several fundamental questions must be addressed in this process: how many earthquakes and what intensities should be considered for risk assessment, what are the exceedance probabilities of a certain intensity in a specific location, what are the expected consequences for a given earthquake, how to assess the effectiveness of a mitigation alternative, and more. However, no comprehensive and universal framework offers a systematic decision support tool for CI seismic risk assessment and risk mitigation.

There is a lack of risk-based key performance indicators in the literature. Several studies have offered approaches to measure risk management performance and risk management indices [52]. The model proposed by [53] evaluates and quantifies the performance indicators by the opinion of local experts. Hence, the values are based on expert opinions and are not fully objective parameters. In [54], prioritizing risk reduction is proposed according to the disaster risk management index (DRMi), physical risk factors, and aggravating coefficient. However, in their study, the DRMi was also evaluated by a survey of experts and not by fully objective values and parameters. Furthermore, Ref. [55] proposes a scenario-based model to evaluate the effectiveness of earthquake emergency management by simulations of possible earthquake disaster scenarios. However, this model is scenario-based, that is, it does not cover all possible seismic threats and therefore may present a limited risk assessment that depends on the selected scenario. In [56], the authors developed a technological platform for resource allocation under budget limitations in order to achieve the optimal seismic risk mitigation.

This paper presents a comprehensive and efficient framework for CI seismic risk assessment and management. The proposed framework intends to address three key issues: (1) seismic risk assessment; (2) quantification of mitigation alternative effectiveness; and (3) prioritization of alternative mitigation strategies.

2. Methodology

A four-step methodology is proposed: (1) determination of the seismic scenarios; (2) calculation of CI system vulnerability; (3) quantitative assessment of risk; (4) implementation of risk-mitigation alternatives and prioritization of risk-mitigation alternatives.

2.1. Determination of the Seismic Scenarios

The first step of the risk appraisal process is the definition of the threat scenarios that critical infrastructure (CI) components are exposed to. In our case, this is an occurrence of an earthquake event and its subsequent effects on the CI. An earthquake can occur at various locations and with different intensities. However, the on-site ground motion will determine the impact on a specific CI system after the earthquake. In this research, the seismic scenarios are defined by a hazard curve. The seismic hazard curve is derived using a PSHA [57], and it determines the annual probability of exceeding a peak ground motion in a specific location. Theoretically, the hazard curve represents possible seismic scenarios and their occurrence probabilities. The hazard curve can be derived based on the probabilities of exceedance of 10%, 5%, and 2% in 50 years or by producing a complete PSHA process.

2.2. Definition of System Seismic Vulnerability

The seismic vulnerability of the system is represented by an exclusive fragility curve. The fragility curves express the probability of reaching or exceeding different damage states for a given level of ground intensity motion (e.g., PGA, PGV, and PGD). The exclusive fragility curve allows for a customized, in-depth risk analysis and later examination of the effectiveness of various retrofit alternatives. The exclusive curves are derived following the methodology presented in [51].

The fragility curves for CI systems are formed as a lognormal cumulative distribution function (CDF) that expresses the probability of reaching or exceeding a certain damage state (DS) for a given level of ground motion intensity (e.g., PGA, PGV, and PGD). This fragility function is defined by the median capacity to resist the damage state i (θ_i) and the standard deviation of the capacity (β_i), as formulated in Equation (1). Then, Equation (2) calculates the probability of exceeding a specific damage state:

$$P[DS \geq ds|IM = x] = \Phi\left(\frac{\ln(x/\theta_{ds})}{\beta_{ds}}\right);ds \in \{1,2,\ldots N_{DS}\} \quad (1)$$

$$P(DS = ds_i|IM) = \begin{cases} 1 - P(DS \geq ds_i|IM) & i = 0 \\ P(DS \geq ds_i|IM) - P(DS \geq ds_{i+1}|IM) & 1 \leq i \leq N_{DS} - 1 \\ P(DS \geq ds_i|IM) & i = N_{DS} \end{cases} \quad (2)$$

where P stands for a conditional probability of being at or exceeding a damage state (DS) for a given seismic intensity, and x is defined by the earthquake intensity measure (IM). Furthermore:

DS—Damage state of a particular component {0, 1,... N_{DS}}.
ds_i—A particular value of DS.
N_{DS}—Number of possible damage states.
IM—Uncertain excitation, the ground motion intensity measure (i.e., PGA, PGD, or PGV).
x—A particular value of IM.
Φ—Standard cumulative normal distribution function.
θ_{ds}—The median capacity of the component to resist damage state DS measured in terms of IM.
β_{ds}—The logarithmic standard deviation of the uncertain capacity of the component to resist damage state DS.

2.3. *Quantitative Assessment of Risk*

The product of this step is a seismic risk curve that expresses the expected annual risk for any given value of IM. Since risk represents the potential impact and loss and is defined as the product of the occurrence probability and the expected consequences, this curve is constructed by multiplying the annual rate of the exceedance curve by the direct damage curve by matching between the IM values in both curves and correlating the expected consequence and its probability to occur. This matching produces a curve that correlates the annual risk expectancy and the IM value.

The total risk expectancy for a T-years lifespan TRE_T (Equation (4)) expresses the overall risk to which the system is exposed to earthquake events during the system's lifespan. The TRE_T is calculated based on possible seismic scenarios, their occurrence probability, and the expected consequences. The R_U (Equation (3)) expresses the overall consequences that are expected in case of complete damage to the system, which is expressed in terms of cost (USD).

$$R_U = \left(\sum C_R + \sum C_D\right) \cdot C_I \quad (3)$$

where:

C_R—Repair cost (USD).
C_D—Direct loss (USD).
C_I—Indirect loss coefficient.
R_U—Overall consequences (USD).

$$TRE_T = \left[\sum_{IM}\left(\sum_{i=1}^{N} P(ds_i|IM) \cdot DR_{ds_i}\right) \cdot PE_A(IM)\right] \cdot R_U \cdot T \quad (4)$$

where:

TRE_T—Total risk expectancy for T years.
DR_{ds_i}—Damage rate of damage state i.
$P(ds_i\ |IM)$—Conditional probability of being in a certain damage state i for a given IM.
T—Design life cycle.

$PE_A(IM)$—Annual rate of exceedance of a given IM.

2.4. Implementation of Risk-Mitigation Alternatives and Prioritization of the Risk-Mitigation Alternatives

Based on the derived risk curve and the total risk expectancy (TRE), mitigation alternatives are considered to find the most beneficial one. The mitigation alternatives are examined to predict their cost-effectiveness on the preparedness level of the CIs by quantifying the reduction of risk. Each alternative has different effects on the robustness and resilience of the system, which is reflected through the fragility curve parameters. The change in the fragility parameters will subsequently affect the system's level of risk; therefore, evaluating mitigation strategies according to risk reduction level is proposed. To examine the cost-effectiveness of the mitigation alternative, a novel risk mitigation to investment ratio (RMIR) performance indicator is proposed. The RIMR is a quantitative indicator that attempts to calculate the overall value of the examined alternative (Equation (5)). The RMIR is the ratio of the expected risk mitigation along T years of service life (ERMT), expressed in monetary terms, to the estimated mitigation cost (EMC). The RMIR aims to examine and prioritize the alternatives. If an alternative has an RMIR greater than 1.0, the alternative is expected to be efficient in terms of cost–benefit analysis. On the other hand, if the alternative has an RMIR lower than 1.0, the alternative is considered to be inefficient, meaning that the investment is higher than the expected benefits of risk mitigation. Moreover, it is possible to prioritize the mitigation alternatives based on their RMIR value. The higher the value of RMIR, the higher the cost-effectiveness of the mitigation alternative. Figure 1 presents the general flowchart of the risk-mitigation alternative's evaluation and prioritization process.

$$RMIR = \frac{ERM^T}{EMC} \tag{5}$$

where:

$RMIR$—Risk mitigation to investment ratio.
ERM^T—Expected risk mitigation along T years of service life.
EMC—Estimated mitigation cost for the alternative.

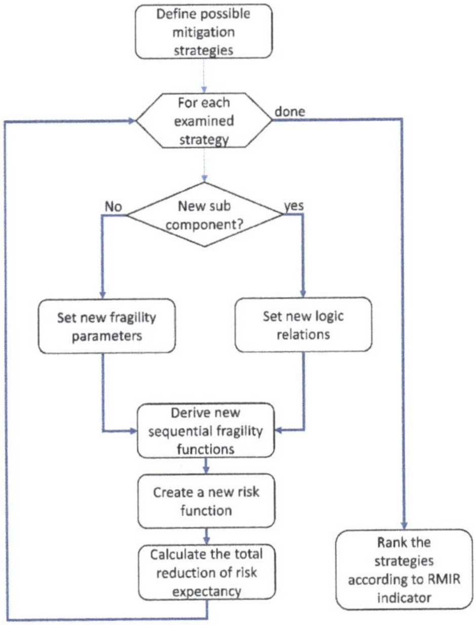

Figure 1. Critical infrastructure risk mitigation flow chart.

3. Case Study

3.1. Introduction

This section presents an implementation of the methodology through a pumping station facility case study. This case study is based on the generic pumping station presented in [51]. The generic pumping station is composed of four main subcomponents that are vital for the functionality of the pumping station: building, pumps, electro-mechanical equipment, and power supply. In this example, the original pumping station is composed of a one-story RC concrete moment frame building, one horizontal pump, mechanical and electrical equipment, and the electric power supply is based on the external commercial power grid. The derived exclusive fragility parameters and curves are as proposed by [51] presented in Figure 2. The probability of damage states (1–4) is derived from the median fragility and the standard deviation.

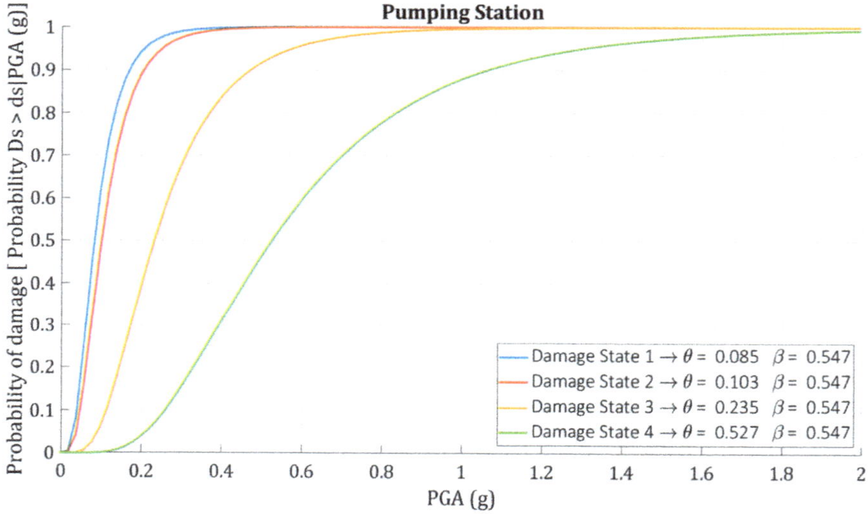

Figure 2. The exclusive fragility curves that were derived for an oil pumping station (based on [51]).

3.2. Risk Appraisal

The risk appraisal processes were carried out for two seismic zones in Israel: Be'er Sheva region and Bik'at HaYarden region. The Be'er Sheva region is considered an area with low seismic risk, while the Bik'at HaYarden region is considered an area with relatively high seismic risk. The selected ground-motion intensity measure for the pumping station is peak ground acceleration (PGA). The hazard curves for those regions (Figure 3) were approximated to exponential function according to the Geophysical Institute of Israel (GII) data of annual rate ground motion for 2%, 5%, and 10% probability of exceedance in 50 years [58]. The PGA values of 2%, 5%, and 10% probability of exceedance in 50 years for the Be'er Sheva region are 0.09 g, 0.07 g, and 0.06 g and for Bik'at HaYarden are 0.41 g, 0.30 g, and 0.23 g.

The full repair cost of the station was calculated as the cost of constructing a new station, estimated at USD 1.5 million. The direct loss is calculated according to disruption to the average daily capacity in barrels for seven consecutive days. In this scenario, the indirect loss was estimated by an indirect-damage coefficient of 2.5. Moreover, in the case of CI such as a pumping station, the facility is not occupied with stuff permanently, and there will not be human casualties.

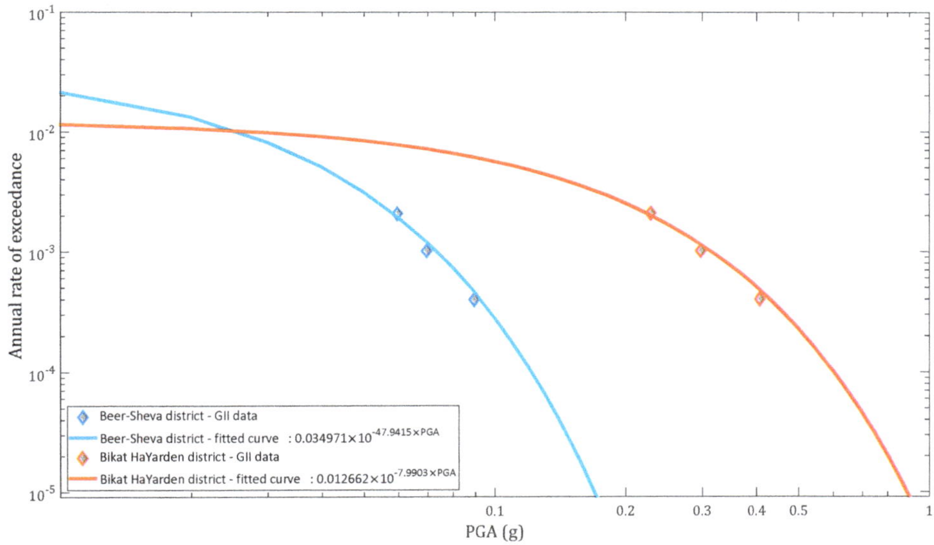

Figure 3. Hazard curve for Be'er Sheva and Bik'at HaYarden regions.

Following the methodology, the risk curves for each region were yielded (Figure 4). The yielded risk curves are the composition of the rate exceedance and expected damage of a specific value of the PGA. Afterward, according to Equation (4), the total risk expectancy for a life span of 50 years (TRE_{50}) was calculated. The TRE_{50} for the Be'er Sheva region is estimated at USD 364,721, and the TRE_{50} for the Bik'at HaYarden region it is estimated at USD 2,913,852. The difference between the values is consistent with the assumption that the seismic risk in the Be'er Sheva region is significantly lower than in the Bik'at HaYarden region.

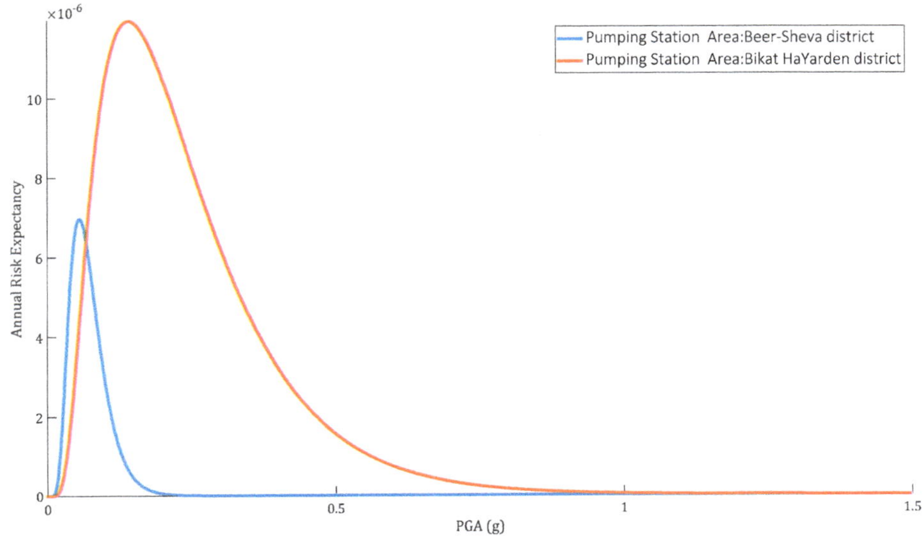

Figure 4. Risk curve for Be'er Sheva and Bik'at HaYarden regions.

3.3. Examination of Possible Mitigation Alternatives

In the risk management process, the mitigation alternatives are aimed at reducing the vulnerability of the sub-component and subsequently decreasing the vulnerability of the pumping station. The examined mitigation alternatives are focused on the sub-components of the station: pump layout (single pump or two pumps), the power supply (power grid only, power grid and diesel backup generator without seismic isolation, and power grid and a diesel backup generator with seismic isolation), and the building type (concrete moment frame building (C1L) or concrete shear walls building (C2L)). The fragility parameters for the components are based on [47,49,51].

In total, twelve alternatives were examined for each site to find the most beneficial one. The mitigation alternatives are composed of different combinations of sub-component layouts. Table 1 presents the sub-component layouts in each alternative and the estimated costs of the alternative.

Table 1. Alternative mitigation strategies.

Alternative No.	Building	Pump	Power Supply	Estimated Cost (USD)
1	C1L	Single pump	Only Grid	-
2	C1L	Single pump	Grid + Generator w/o	70,000
3	C1L	Single pump	Grid + Gen with Isolation	80,500
4	C1L	Two pumps	Only Grid	250,000
5	C1L	Two pumps	Grid + Generator w/o	320,000
6	C1L	Two pumps	Grid + Gen with Isolation	330,500
7	C2L	Single pump	Only Grid	100,000
8	C2L	Single pump	Grid + Generator w/o	170,000
9	C2L	Single pump	Grid + Gen with Isolation	180,500
10	C2L	Two pumps	Only Grid	350,000
11	C2L	Two pumps	Grid + Generator w/o	420,000
12	C2L	Two pumps	Grid + Gen with Isolation	430,500

The risk management process is intended to indicate the optimal mitigation alternative. The optimal alternative weighs the contribution of the alternative to risk-mitigating and the cost of the alternative.

3.4. Results and Discussion

For each region, a dozen proposed mitigation alternatives have been implemented and investigated. Table 2 presents the total risk expectancy for a 50-year designed life cycle (TRE_{50}), the estimated risk mitigation (ERM^{50}), and the calculated RMIR for each alternative. Figures 5 and 6 present the risk mitigation and risk mitigation to investment (RMIR) ratios. Alternative number one, the generic station, is used as a default alternative for the examination of risk reduction.

Table 2. Total risk expectancy for 50-year designed life cycle (TRE_{50}), estimated risk mitigation (ERM^{50}), and calculated RMIR for different mitigation alternatives.

Mitigation Alternative		Be'er Sheva Region			Bik'at HaYarden Region		
#	Estimated Mitigation Cost (USD)	TRE_{50}	ERM^{50} [% (USD)]	RMIR	TRE_{50}	ERM^{50} [% (USD)]	RMIR
1	-	364,721	0% (USD 0)	-	2,913,852	0% (USD 0)	-
2	70,000	292,256	19.9% (USD 72,465)	1.035	2,731,600	6.3% (USD 182,252)	2.604
3	80,500	291,842	20% (USD 72,879)	0.905	2,725,695	6.5% (USD 188,156)	2.337
4	250,000	364,692	0% (USD 30)	0	2,913,062	0% (USD 789)	0.003
5	320,000	292,208	19.9% (USD 72,514)	0.227	2,730,594	6.3% (USD 183,257)	0.573
6	330,500	291,772	20% (USD 72,949)	0.221	2,724,603	6.5% (USD 189,249)	0.573
7	100,000	267,820	26.6% (USD 96,901)	0.969	2,601,564	10.7% (USD 312,288)	3.123
8	170,000	177,057	51.5% (USD 187,665)	1.104	2,303,721	20.9% (USD 610,131)	3.589
9	180,500	176,453	51.6% (USD 188,268)	1.043	2,294,304	21.3% (USD 619,548)	3.432
10	350,000	267,810	26.6% (USD 96,911)	0.277	2,600,415	10.8% (USD 313,437)	0.896
11	420,000	176,955	51.5% (USD 187,766)	0.447	2,301,969	21% (USD 611,882)	1.457
12	430,500	176,380	51.6% (USD 188,341)	0.437	2,292,495	21.3% (USD 621,356)	1.443

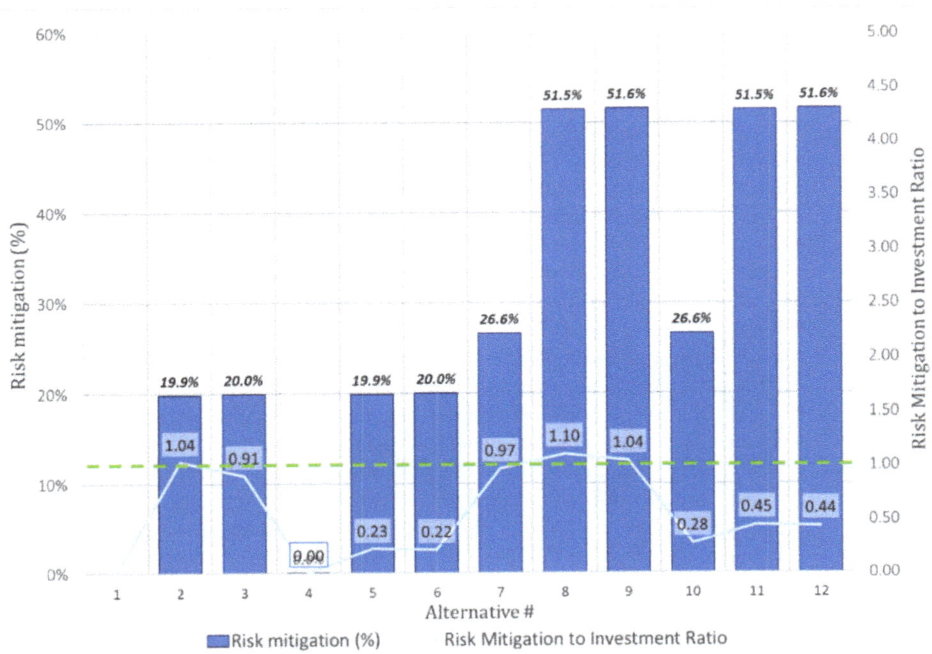

Figure 5. Analysis of the mitigation alternatives by the percentage of risk mitigation and RMIR case of the Be'er Sheva region.

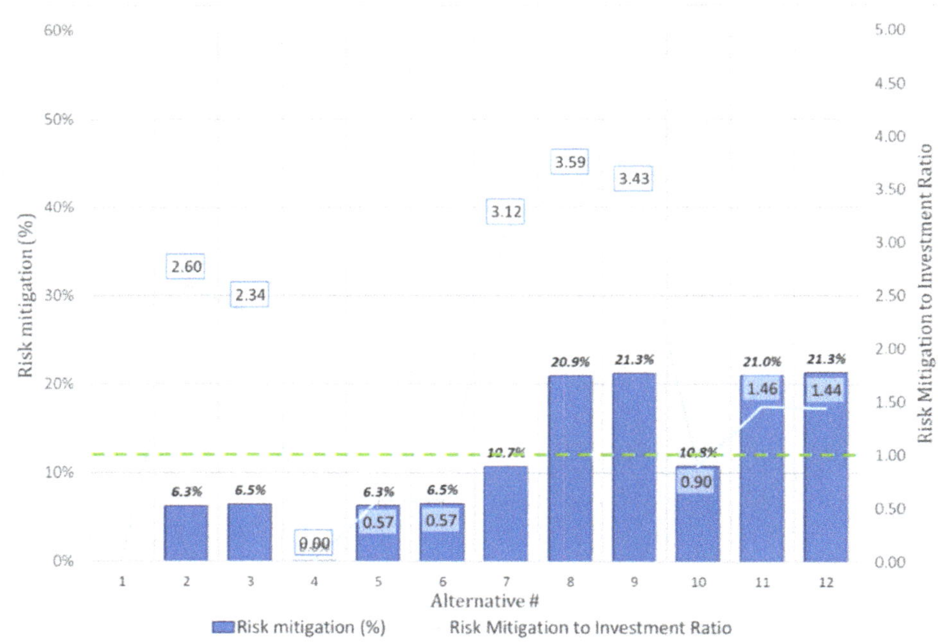

Figure 6. Analysis of the mitigation alternatives by the percentage of risk mitigation and RMIR case of the Bik'at HaYarden region.

In the case of the Be'er Sheva region, alternatives 8, 9, 11, and 12 reduce the risk by 51.5%, 51.6%, 51.5%, and 51.6%, respectively. Those alternatives include the building retrofit and an addition of a backup generator, while alternatives 11 and 12 include the addition of a redundant pump. However, the ERM change is minor, indicating that the redundant pump's marginal benefit to the risk mitigation is minor. Alternative number four, which included the addition of a backup pump, has a negligible effect on the risk mitigation and the RMIR value is zero. Accordingly, it can be concluded that the pump's influence on the station's seismic vulnerability is very low. Therefore, it is not beneficial to retrofit this component. The efficiency of the alternatives can be analyzed by the proposed RMIR indicator, which evaluates the alternative's cost-effectiveness in terms of risk expectancy mitigation. Alternatives two, seven, and nine have an RMIR value higher than 1.0 (1.035, 1.104, and 1.043, respectively). Therefore, it can be determined that only these alternatives are economically viable. However, the values of the RMIR are only slightly higher than 1.0, which makes the efficiency of the alternative uncertain. That is, it is possible that if the cost of the alternative turns out to be higher than planned, the alternative's viability will turn negative in terms of cost–benefit analysis. Therefore, in cases where the RMIR value is close to 1.0, it is advisable to perform an additional estimate of the retrofit costs. In addition, it is important to note that alternatives 11 and 12, which have a high value of ERM, are unviable according to the RMIR cost–benefit analysis.

In the case of the Bik'at HaYarden region, alternatives 8, 9, 11, and 12 have the highest values of ERM at 20.9%, 21.3%, 21.0%, and 21.3%, respectively. These results were expected since these alternatives had the highest ERM values in the Be'er Sheva region. The mitigation alternatives improve the seismic vulnerability of the station, which depends only on the components of the station. On the other hand, the alternative's effectiveness highly depends on the station location since it is analyzed according to the risk. Alternatives 2, 3, 7, 8, 9, 11, and 12 have RMIR values higher than 1.0, that is, these alternatives are economically justifiable according to RMIR cost–benefit analysis. Alternative number eight has the highest value of RMIR (3.589), meaning that this alternative achieves the best risk-mitigation percentage per money. Alternative number nine has the second-best value of RMIR (3.432), with an ERM of 21.3%, which is higher than alternative eight (20.9%). This alternative achieves higher risk mitigation but lower cost–benefit efficiency. In addition, the RMIR value enables budget-based considerations. In case of budget constraints, alternatives two and three present good values of RMIR (2.604 and 2.337, respectively) and can be considered for implementation.

4. Conclusions

This research introduces a comprehensive methodology for seismic risk appraisal and management. The proposed methodology examines the preparedness of critical infrastructures through an appraisal of the risk that CIs are exposed to in case of extreme seismic events. This research establishes a probabilistic quantitative model that assesses the total risk expectancy of CIs to extreme earthquake events and produces a decision support tool that allows decision makers to manage and analyze alternative courses of action in order to mitigate the risk considering a wide range of risk scenarios. The proposed methodology was illustrated through a case study of an oil pumping station. In this case study, three alternative mitigation strategies were examined: additional pump installation (redundancy), installation of a diesel generator with and without seismic isolation (redundancy), and a building retrofit (robustness).

Twelve possible combinations of those strategies were examined. It was found that an additional pump (pump redundancy) yields only a minor contribution to the risk mitigation, whereas the building retrofit yields the most significant impact on the risk mitigation and cost-effectiveness. The pump sub-component is not vulnerable to low accelerations; therefore, it has a low impact on the overall risk. In contrast, according to the structure's sub-component fragility curve, it is vulnerable to low-to-moderate acceleration intensities; consequently, its impact on overall risk reduction is high.

In addition, this study proposes the RMIR (risk mitigation to investment ratio) as a novel quantitative indicator in order to examine and prioritize alternative mitigation strategies. The proposed RMIR indicator evaluates alternative mitigation strategies based on cost-effectiveness of the mitigation strategies, considering integrated probabilities of all damage states and all possible seismic scenarios. This indicator is unbiased and depends on quantitative values and objective estimates of the seismic risk, CI resistance, and derived effectiveness of the mitigation alternative. Moreover, the proposed RMIR indicator covers all possible seismic scenarios due to the PSHA approach that is implemented systematically in the methodology. The RMIR examines and prioritizes the alternatives based on the risk mitigation to investment ratio. If an alternative has an RMIR greater than 1.0, the alternative is expected to be efficient in terms of risk mitigation. On the other hand, if the alternative has an RMIR lower than 1.0, the alternative is considered to be inefficient in terms of risk mitigation, meaning that the investment is higher than the expected benefits of risk mitigation. In addition, the higher the value of RMIR, the higher the cost-effectiveness of the mitigation alternative. The benefits of the proposed indicator were illustrated in the case study. The indicator allows us to examine the cost-effectiveness of the alternatives and prioritize the mitigation alternatives according to decision-maker policies. The research's novelty focuses on a synthesis of the fragility and morphology of the system into the risk expectancy and derives a coherent indicator for seismic risk mitigation in critical infrastructures. This contribution can be used for stimulating the preparedness of energy-related CI for extreme events and can be extended to risk mitigation of other hazards such as storms, floods, shocks, and blasts [10,59].

The presented methodology considers damage as a result of a single earthquake (i.e., mainshock) and does not consider sequential seismic events (aftershocks and foreshocks). Furthermore, the maintenance levels and wear and tear of the CI components are usually not reflected in the fragility curves. Moreover, adjustments will be required in order to include unique seismic retrofit solutions [32,60,61]. These issues can be addressed in follow-up studies.

Author Contributions: Conceptualization, I.M.S.; data curation, A.U.; investigation, A.U. and I.M.S.; methodology, A.U.; project administration, I.M.S.; software, A.U.; supervision, I.M.S.; validation, I.M.S.; visualization, A.U.; writing—original draft, A.U.; writing—review and editing, I.M.S. All authors have read and agreed to the published version of the manuscript.

Funding: This research received no external funding.

Conflicts of Interest: The authors declare no conflict of interest.

References

1. Urlainis, A.; Ornai, D.; Levy, R.; Vilnay, O.; Shohet, I.M. Loss and Damage Assessment in Critical Infrastructures Due to Extreme Events. *Saf. Sci.* **2022**, *147*, 105587. [CrossRef]
2. Urlainis, A.; Shohet, I.M.; Levy, R.; Ornai, D.; Vilnay, O. Damage in Critical Infrastructures Due to Natural and Man-Made Extreme Events—A Critical Review. *Procedia Eng.* **2014**, *85*, 529–535. [CrossRef]
3. Yu, J.; Cruz, A.M.; Piatyszek, E.; Lesbats, M.; Tardy, A.; Hokugo, A.; Tatano, H. A Survey of Impact on Industrial Parks Caused by the 2011 Great East Japan Earthquake and Tsunami. *J. Loss Prev. Process Ind.* **2017**, *50*, 317–324. [CrossRef]
4. Moteff, J.; Parfomak, P. *Critical Infrastructure and Key Assets: Definition and Identification*; Library of Congress: Washington, DC, USA, 2004.
5. Mendonça, D.; Lee, E.E., II; Wallace, W.A. Impact of the 2001 World Trade Center Attack on Critical Interdependent Infrastructures. In Proceedings of the 2004 IEEE International Conference on Systems, Man and Cybernetics, SMC 2004, The Hague, The Netherlands, 10–13 October 2004; Volume 5, pp. 4053–4058.
6. Mendonça, D.; Wallace, W.A. Impacts of the 2001 World Trade Center Attack on New York City Critical Infrastructures. *J. Infrastruct. Syst.* **2006**, *12*, 260–270. [CrossRef]
7. Johansson, J.; Hassel, H. An Approach for Modelling Interdependent Infrastructures in the Context of Vulnerability Analysis. *Reliab. Eng. Syst. Saf.* **2010**, *95*, 1335–1344. [CrossRef]
8. Poljanšek, K.; Bono, F.; Gutiérrez, E. Seismic Risk Assessment of Interdependent Critical Infrastructure Systems: The Case of European Gas and Electricity Networks. *Earthq. Eng. Struct. Dyn.* **2012**, *41*, 61–79. [CrossRef]
9. Ornai, D.; Elkabets, S.M.; Kivity, Y.; Ben-Dor, G.; Chadad, L.; Gal, E.; Tavron, B.; Gilad, E.; Levy, R.; Shohet, I.M. A Methodology of Risk Assessment, Management, and Coping Actions for Nuclear Power Plant (NPP) Hit by High-Explosive Warheads. *Adv. Eng. Inform.* **2020**, *46*, 101192. [CrossRef]
10. Yao, X.; Wei, H.H.; Shohet, I.M.; Skibniewski, M.J. Assessment of Terrorism Risk to Critical Infrastructures: The Case of a Power-Supply Substation. *Appl. Sci.* **2020**, *10*, 7162. [CrossRef]
11. Urlainis, A.; Shohet, I.M. Probabilistic Risk Appraisal and Mitigation of Critical Infrastructures for Seismic Extreme Events. In Proceedings of the Creative Construction Conference (CCC2018), Ljubljana, Slovenia, 30 June–3 July 2018.
12. Lam, C.Y.; Shimizu, T. A Network Analytical Framework to Analyze Infrastructure Damage Based on Earthquake Cascades: A Study of Earthquake Cases in Japan. *Int. J. Disaster Risk Reduct.* **2021**, *54*, 102025. [CrossRef]
13. Rinaldi, S.M.; Peerenboom, J.P.; Kelly, T.K. Identifying, Understanding, and Analyzing Critical Infrastructure Interdependencies. *IEEE Control Syst. Mag.* **2001**, *21*, 11–25. [CrossRef]
14. Pescaroli, G.; Alexander, D. Critical Infrastructure, Panarchies and the Vulnerability Paths of Cascading Disasters. *Nat. Hazards* **2016**, *82*, 175–192. [CrossRef]
15. Cutter, S.L. Compound, Cascading, or Complex Disasters: What's in a Name? *Environment* **2018**, *60*, 16–25. [CrossRef]
16. Shamseldin, A.; Lowrance, W.W. Of Acceptable Risk: Science and the Determination of Safety. *J. Am. Stat. Assoc.* **1978**, *73*, 686. [CrossRef]
17. Haimes, Y.Y. *Risk Modeling, Assessment, and Management*, 3rd ed.; John Wiley & Sons: Hoboken, NJ, USA, 2008; ISBN 978-0-470-28237-3.
18. Bozorgnia, Y.; Campbell, K.W. Chapter 5: Engineering Characterization of Ground Motion. In *Earthquake Engineering: From Engineering Seismology to Performance-Based Engineering*; CRC Press: Boca Raton, FL, USA, 2004; Volume 1, pp. 1–976. ISBN 9780429204968.
19. Robinson, T.R.; Rosser, N.J.; Densmore, A.L.; Oven, K.J.; Shrestha, S.N.; Guragain, R. Use of Scenario Ensembles for Deriving Seismic Risk. *Proc. Natl. Acad. Sci. USA* **2018**, *115*, E9532–E9541. [CrossRef]
20. Starita, S.; Scaparra, M.P. Optimizing Dynamic Investment Decisions for Railway Systems Protection. *Eur. J. Oper. Res.* **2016**, *248*, 543–557. [CrossRef]
21. Rusydy, I.; Idris, Y.; Mulkal; Muksin, U.; Cummins, P.; Akram, M.N. Syamsidik Shallow Crustal Earthquake Models, Damage, and Loss Predictions in Banda Aceh, Indonesia. *Geoenviron. Disasters* **2020**, *7*, 8. [CrossRef]
22. Oakes, B.D.; Mattsson, L.G.; Näsman, P.; Glazunov, A.A. A Systems-Based Risk Assessment Framework for Intentional Electromagnetic Interference (IEMI) on Critical Infrastructures. *Risk Anal.* **2018**, *38*, 1279–1305. [CrossRef]
23. Cao, X.; Lam, J.S.L. Simulation-Based Catastrophe-Induced Port Loss Estimation. *Reliab. Eng. Syst. Saf.* **2018**, *175*, 1–12. [CrossRef]
24. McDonald, M.; Mahadevan, S.; Ambrosiano, J.; Powell, D. Risk-Based Policy Optimization for Critical Infrastructure Resilience against a Pandemic Influenza Outbreak. *ASCE ASME J. Risk Uncertain. Eng. Syst. A Civ. Eng.* **2018**, *4*, 04018007. [CrossRef]
25. Sucuoglu, H.; Akkar, S. *Basic Earthquake Engineering: From Seismology to Analysis and Design*, 2014th ed.; Springer: Cham, Switzerland, 2014; Volume 1, ISBN 9783319010250.

26. Baker, J.W. Efficient Analytical Fragility Function Fitting Using Dynamic Structural Analysis. *Earthq. Spectra* **2015**, *31*, 579–599. [CrossRef]
27. Abrahamson, N.A.; Kuehn, N.M.; Walling, M.; Landwehr, N. Probabilistic Seismic Hazard Analysis in California Using Nonergodic Ground-Motion Models. *Bull. Seismol. Soc. Am.* **2019**, *109*, 1235–1249. [CrossRef]
28. Giardini, D.; Danciu, L.; Erdik, M.; Şeşetyan, K.; Demircioğlu Tümsa, M.B.; Akkar, S.; Gülen, L.; Zare, M. Seismic Hazard Map of the Middle East. *Bull. Earthq. Eng.* **2018**, *16*, 3567–3570. [CrossRef]
29. Akkar, S.; Kale, Ö.; Yakut, A.; Çeken, U. Ground-Motion Characterization for the Probabilistic Seismic Hazard Assessment in Turkey. *Bull. Earthq. Eng.* **2018**, *16*, 3439–3463. [CrossRef]
30. Peñarubia, H.C.; Johnson, K.L.; Styron, R.H.; Bacolcol, T.C.; Sevilla, W.I.G.; Perez, J.S.; Bonita, J.D.; Narag, I.C.; Solidum, R.U.; Pagani, M.M.; et al. Probabilistic Seismic Hazard Analysis Model for the Philippines. *Earthq. Spectra* **2020**, *36* (Suppl. 1), 44–68. [CrossRef]
31. Mulargia, F.; Stark, P.B.; Geller, R.J. Why Is Probabilistic Seismic Hazard Analysis (PSHA) Still Used? *Phys. Earth Planet. Inter.* **2017**, *264*, 63–75. [CrossRef]
32. Vona, M.; Flora, A.; Carlucci, E.; Foscolo, E.; Aprile, A.; Monti, G. Seismic Retrofitting Resilience-Based for Strategic RC Buildings. *Buildings* **2021**, *11*, 111. [CrossRef]
33. Menna, C.; Felicioni, L.; Negro, P.; Lupíšek, A.; Romano, E.; Prota, A.; Hájek, P. Review of Methods for the Combined Assessment of Seismic Resilience and Energy Efficiency towards Sustainable Retrofitting of Existing European Buildings. *Sustain. Cities Soc.* **2022**, *77*, 103556. [CrossRef]
34. Tena-Colunga, A.; Godínez-Domínguez, E.A.; Hernández-Ramírez, H. Seismic Retrofit and Strengthening of Buildings. Observations from the 2017 Puebla-Morelos Earthquake in Mexico City. *J. Build. Eng.* **2022**, *47*, 103916. [CrossRef]
35. Crespi, P.; Zucca, M.; Valente, M.; Longarini, N. Influence of Corrosion Effects on the Seismic Capacity of Existing RC Bridges. *Eng. Fail. Anal.* **2022**, *140*, 106546. [CrossRef]
36. Afrouz, S.G.; Farzampour, A.; Hejazi, Z.; Mojarab, M. Evaluation of Seismic Vulnerability of Hospitals in the Tehran Metropolitan Area. *Buildings* **2021**, *11*, 54. [CrossRef]
37. Dolce, M.; Prota, A.; Borzi, B.; da Porto, F.; Lagomarsino, S.; Magenes, G.; Moroni, C.; Penna, A.; Polese, M.; Speranza, E.; et al. Seismic Risk Assessment of Residential Buildings in Italy. *Bull. Earthq. Eng.* **2021**, *19*, 2999–3032. [CrossRef]
38. Flenga, M.G.; Favvata, M.J. Fragility Curves and Probabilistic Seismic Demand Models on the Seismic Assessment of RC Frames Subjected to Structural Pounding. *Appl. Sci.* **2021**, *11*, 8253. [CrossRef]
39. O'Rourke, M.J.; So, P. Seismic Fragility Curves for On-Grade Steel Tanks. *Earthq. Spectra* **2000**, *16*, 801–815. [CrossRef]
40. Razzaghi, M.S.; Eshghi, S. Probabilistic Seismic Safety Evaluation of Precode Cylindrical Oil Tanks. *J. Perform. Constr. Facil.* **2015**, *29*. [CrossRef]
41. Rosti, A.; del Gaudio, C.; Rota, M.; Ricci, P.; di Ludovico, M.; Penna, A.; Verderame, G.M. Empirical Fragility Curves for Italian Residential RC Buildings. *Bull. Earthq. Eng.* **2021**, *19*, 3165–3183. [CrossRef]
42. Rosti, A.; Rota, M.; Penna, A. Empirical Fragility Curves for Italian URM Buildings. *Bull. Earthq. Eng.* **2021**, *19*, 1–20. [CrossRef]
43. Rahmani, A.Y.; Bourahla, N.; Bento, R.; Badaoui, M. An Improved Upper-Bound Pushover Procedure for Seismic Assessment of High-Rise Moment Resisting Steel Frames. *Bull. Earthq. Eng.* **2018**, *16*, 315–339. [CrossRef]
44. Amin, J.; Gondaliya, K.; Mulchandani, C. Assessment of Seismic Collapse Probability of RC Shaft Supported Tank. *Structures* **2021**, *33*, 2639–2658. [CrossRef]
45. Belejo, A.; Bento, R. Improved Modal Pushover Analysis in Seismic Assessment of Asymmetric Plan Buildings under the Influence of One and Two Horizontal Components of Ground Motions. *Soil Dyn. Earthq. Eng.* **2016**, *87*, 1–15. [CrossRef]
46. Porter, K.; Hamburger, R.; Kennedy, R. Practical Development and Application of Fragility Functions. In Proceedings of the Research Frontiers at Structures Congress 2007, Long Beach, CA, USA, 20 June 2007.
47. Gehl, P.; Desramaut, N.; Réveillère, A.; Modaressi, H. Fragility Functions of Gas and Oil Networks. In *SYNER-G: Typology Definition and Fragility Functions for Physical Elements at Seismic Risk*; Springer: Dordrecht, The Netherlands, 2014; Volume 1, ISBN 9789400778719.
48. ALA. ALA Seismic Fragility Formulations For Water Systems. In *Alliance American Lifelines (ALA)*; ASCE-FEMA: Reston, VA, USA, 2001; p. 104.
49. NIBS. *HAZUS-MH: Users's Manual and Technical Manuals. Report Prepared for the Federal Emergency Management Agency*; National Institute of Building Sciences, Federal Emergency Management Agency (FEMA): Washington, DC, USA, 2004.
50. Rossetto, T.; D'Ayala, D.; Ioannou, I.; Meslem, A. Evaluation of Existing Fragility Curves. In *SYNER-G: Typology Definition and Fragility Functions for Physical Elements at Seismic Risk*; Pitilakis, K., Crowley, H., Kaynia, A., Eds.; Springer: Dordrecht, The Netherlands, 2014. [CrossRef]
51. Urlainis, A.; Shohet, I.M. Development of Exclusive Seismic Fragility Curves for Critical Infrastruc-Ture: An Oil Pumping Station Case Study. *Buildings* **2022**, *12*, 842. [CrossRef]
52. Carreño, M.L.; Cardona, O.D.; Barbat, A.H. A Disaster Risk Management Performance Index. *Nat. Hazards* **2007**, *41*, 1–20. [CrossRef]
53. Carreño, M.L.; Cardona, O.D.; Barbat, A.H. New Methodology for Urban Seismic Risk Assessment from a Holistic Perspective. *Bull. Earthq. Eng.* **2012**, *10*, 547–565. [CrossRef]

54. Lantada, N.; Carreño, M.L.; Jaramillo, N. Disaster Risk Reduction: A Decision-Making Support Tool Based on the Morphological Analysis. *Int. J. Disaster Risk Reduct.* **2020**, *42*, 101342. [CrossRef]
55. Zhang, Y.; Weng, W.G.; Huang, Z.L. A Scenario-Based Model for Earthquake Emergency Management Effectiveness Evaluation. *Technol. Soc. Chang.* **2018**, *128*, 197–207. [CrossRef]
56. Nuzzo, I.; Caterino, N.; Novellino, A.; Occhiuzzi, A.; Ditommaso, R.; di Cesare, A.; Shohet, I.M.; Asteris, P.G. Computer-Aided Decision Making for Regional Seismic Risk Mitigation Accounting for Limited Economic Resources. *Appl. Sci.* **2021**, *11*, 5539. [CrossRef]
57. Baker, J.; Bradley, B.; Stafford, P. *Seismic Hazard and Risk Analysis*; Cambridge University Press: Cambridge, UK, 2021; ISBN 978-1-108-42505-6.
58. Klar, A.; Meirova, T.; Zaslavsky, Y.; Shapira, A. *Spectral Acceleration Maps for Use in SI 413 Amendment No.5*; The Geophysical Institute of Israel: Jerusalem, Israel, 2011.
59. Yum, S.-G.G.; Kim, J.-M.M.; Wei, H.H. Development of Vulnerability Curves of Buildings to Windstorms Using Insurance Data: An Empirical Study in South Korea. *J. Build. Eng.* **2020**, *34*, 101932. [CrossRef]
60. Shmerling, A.; Gerdts, M. Optimization of Inelastic Multistory Structures under Seismic Vibrations Using Shape-Memory-Alloy Material. *Sci. Rep.* **2022**, *12*, 16844. [CrossRef]
61. Abbaszadeh, A.; Chaallal, O. Enhancing Resilience and Self-Centering of Existing RC Coupled and Single Shear Walls Using EB-FRP: State-of-the-Art Review and Research Needs. *J. Compos. Sci.* **2022**, *6*, 301. [CrossRef]

buildings

MDPI

Article

Experimental Study of the Seismic Performance of a Prefabricated Frame Rocking Wall Structure

Fuwen Zhang [1], Xiangmin Li [1,*], Zhuolin Wang [1], Kun Tian [1], Kent A. Harries [2] and Qingfeng Xu [1]

[1] Shanghai Key Laboratory of Engineering Structure Safety, Shanghai Research Institute of Building Sciences Co., Ltd., Shanghai 200032, China
[2] Department of Civil and Environmental Engineering, University of Pittsburgh, Pittsburgh, PA 15260, USA
* Correspondence: lixiangmin@sribs.com

Abstract: This paper proposes a prefabricated frame rocking wall (PFRW) structure system in which beams, columns, and rocking walls are all prefabricated components. The rocking wall and the frame are connected by energy-dissipating connectors, and three prestressed tendons are arranged inside the rocking wall. A quasi-static test for the PFRW structure and a conventional frame (CF) structure was conducted. The research results show that the seismic load-resisting capacity of the PFRW structure is increased by about 190% relative to the CF structure, and the energy dissipation coefficient of the PFRW structure is increased to twice that of the CF structure.

Keywords: prefabricated structure; rocking wall; experimental study; quasi-static test; seismic performance

Citation: Zhang, F.; Li, X.; Wang, Z.; Tian, K.; Harries, K.A.; Xu, Q. Experimental Study of the Seismic Performance of a Prefabricated Frame Rocking Wall Structure. *Buildings* **2022**, *12*, 1714. https://doi.org/10.3390/buildings12101714

Academic Editors: Jian Zhong, Xiaowei Wang, Suiwen Wu, Chao Li and Weiping Wen

Received: 6 September 2022
Accepted: 14 October 2022
Published: 17 October 2022

1. Introduction

A rocking wall structure is a structure in which the restraint provided at the base of the wall (connection with foundation) is relaxed or softened. Based on the restraint release and the characteristics of the resulting permitted motion, two types of structure result: In the first, the restraints in the vertical degrees of freedom at the wall base are relaxed, allowing the wall to 'lift' in a vertical direction under the action of an earthquake. The second approach is to relax the rotational degree of freedom at the wall base. While the first approach permits rotation to develop, the second does so without lifting under the action of an earthquake [1]. A rocking wall structure system can change the deformation modes of a structure under earthquake action, mitigating the likelihood of concentrated damage [2]. Rocking wall structures can also affect self-centering functionality, resulting in little or no post-earthquake lateral drift.

Various rocking wall systems have been proposed. Kurama et al. [3] used full wall-height, vertically-oriented unbonded post-tensioned tendons to anchor precast reinforced concrete walls to their foundation. The resulting system allowed the walls to rock under lateral seismic loads, although damage to the system tended to accumulate locally at the wall toes. Ajrab et al. [4] proposed a rocking wall-frame structure in which the shear wall component of the structure was a rocking wall. Ajrab et al. proposed using vertically-draped post-tensioned tendons forming an X-arrangement over the height of the wall. These 'damping cables' were shown to improve the seismic response of the structure. Hitaka et al. [5] proposed a rocking joint shear wall system in which the deformation under the action of an earthquake is mainly concentrated in the wall boundary element consisting of steel coupling beams, reinforced concrete wall limbs, and concrete-filled steel tube side columns.

Because of the nature of the rocking wall mechanism, these systems exhibit relatively low energy absorption [6]. To improve the energy absorption capacity, viscous dampers [7], metal dampers [8,9], and energy-dissipating connectors [10–12] have been proposed for use with rocking wall structures. To mitigate the concentration of damage at the wall base (primarily the wall toes) associated with rocking [13], Cui et al. [14] proposed an arrangement of steel plates at the bottom of the wall and rubber blocks at the wall toes. Tagliafierro et al. [15] proposed and evaluated the effect of a novel seismic isolation system on seismic control for a steel storage pallet racking system. To improve the self-centering performance of rocking wall structures, a disc spring self-centering device [16] and shape memory alloy self-centering devices [17] have been proposed.

Rocking wall systems have been used in practice. Wada et al. [2] report a rocking wall system used to reinforce an eleven-story building at the Tokyo Institute of Technology, which employed a V-shaped pin support between the base of the rocking wall and the foundation. The support affects an essentially single-pin unrestrained rotational degree of freedom at the wall base. Wu et al. [18] report a rocking wall system used to reinforce a hospital building. In this case, the base of the wall has a connection that limits horizontal lateral movement, while self-centering was realized through post-tensioned tendons.

Extant research on rocking walls primarily focuses on their use in retrofitting existing structures. However, as prefabricated components, rocking walls are particularly well suited for inclusion in prefabricated structures. This results in frame-rocking wall structures having a high prefabrication rate and opens the possibility for the industrialization of such building systems. This paper proposes a prefabricated frame rocking wall (PFRW) system. The beams, columns, and rocking walls are all prefabricated components, while the cast-in-place joint regions of frame components are made robust and ductile through the use of engineered cementitious composites (ECC). Finally, the connection between the rocking wall and frame is made with energy-dissipating connectors. The failure behavior, hysteretic performance, backbone curves, energy dissipation capacity, and residual deformations of the resulting PFRW structure were studied through a pseudo-static reversed-cyclic load test.

2. Experimental Program

Under the lateral load, the frame and the rocking wall in the structure belong to a parallel relationship. Namely, the reaction forces of the frame and the rocking wall in the PFRW are superimposed to resist external loads. The comparison between the single frame without any extra added walls (the CF) and PFRW can better analyze the role played by the rocking wall. Meanwhile, the deformation capacities of the CF and the PFRW are similar due to the constraint between the bottom of the rocking wall, and the base is usually released. Therefore, two concrete structures were fabricated (Figure 1).

The CF structure was entirely cast-in-place (CIP). The PFRW structure consisted of prefabricated components and CIP beam-column connections. Each half-scale frame has three 1500 mm stories and consists of three 1800 mm long bays. Frame CF is the 'control' specimen and the basis of comparison for the second PFRW frame. Precast concrete columns (Figure 1c) are 200 mm square and reinforced with eight 16 mm bars (reinforcing ratio, ρ = 0.04), and confined with 8 mm ties spaced at 100 mm. Precast beams (Figure 1d) are 200 mm $\times$ 120 mm having two 12 mm bars on top and bottom (ρ = 0.017) and 8 mm ties spaced at 100 mm. Column and beam reinforcing bars are made continuous through joint regions using grouted splice sleeves. At the end of beams and tops of columns, straight bar extensions are anchored into extensions of each joint region, as seen in Figure 1a,b. The CIP joint regions are enclosed ECC to enhance the durability and ductility of these regions.

The PFRW specimen replaces the middle frame bay with a 120 mm thickness precast rocking wall (Figure 1b,f). The wall has thirty 8 mm diameter vertical bars (ρ = 0.010) arranged in two layers through the wall thickness, 8 mm horizontal bars at spacings of 50 mm (near the base of the wall), and 100 mm (elsewhere). The reinforcing arrangement also provides eight 8 mm cross ties through the wall thickness coinciding with the horizontal bars. The vertical reinforcement is welded to a continuous steel plate at the base of the wall. Three 15.2 mm diameter unbonded post-tensioning tendons are also provided in the middle region of the wall (Figure 1f); these will provide self-centering capacity for the wall.

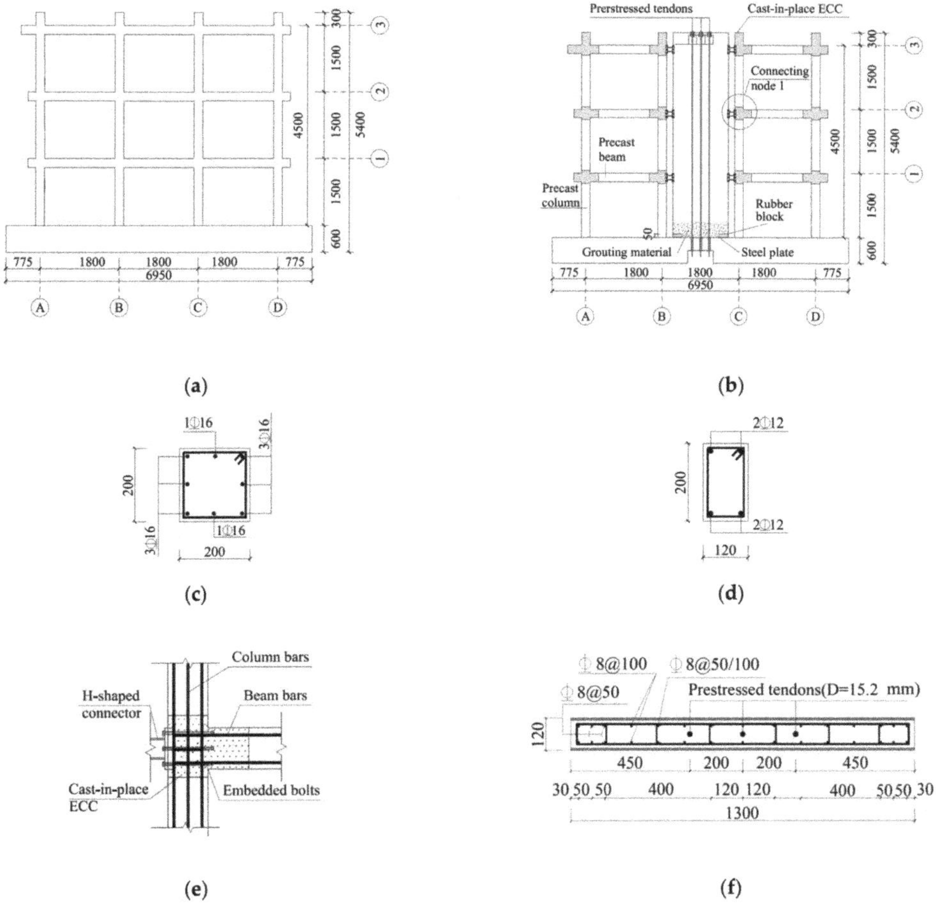

Figure 1. Test specimen dimensions and reinforcement (unit: mm). (**a**) Geometry of CF; (**b**) Geometry of PFRW; (**c**) frame column; (**d**) frame beam; (**e**) typical cast-in-place beam-column connection (connecting node 1); (**f**) rocking wall.

When placed on the foundation, 50 mm tall, 220 mm long neoprene rubber pads are placed at each wall toe. The remaining 860 mm long middle region of the connection is filled with a 50 mm thick course of high-strength cementitious grout. Once assembled, each of the three post-tensioning tendons is stressed to 0.6 f_{ptk}, resulting in a total measured axial prestress force of 442 kN.

The rocking wall is connected to the adjacent frames with the H-shaped energy-dissipating connectors, as shown in Figure 2. The connector is fabricated from a hot-rolled Grade Q235 plate and is designated H100 × 80 × 8 × 8 mm. The H-beam is fully welded to 15 mm end plates and bolted to pre-threaded 25 mm plates embedded in each column-beam joint. The connectors are inserted and secured with six 14 mm diameter bolts at each end plate. Due to the small specimen scale, high-strength cap screws were used in this study rather than high-strength structural (hex) bolts.

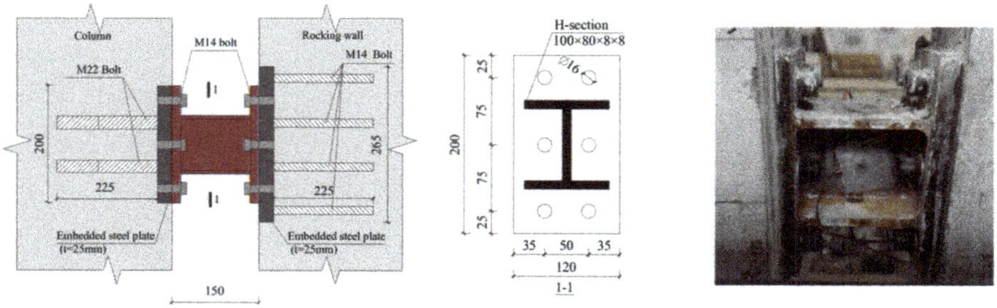

(a) (b) (c)

Figure 2. Details of energy-dissipating connector in PFRW (unit: mm). (**a**) Schematic diagram of connecting node; (**b**) Section view of 1–1; (**c**) Photo of connecting node.

2.1. Material Properties

The measured material properties of the concrete, ECC, and high-strength grout used in the specimens are given in Table 1. The ECC had a volume ratio of polyvinyl alcohol (PVA) fiber of 2.5%.

Table 1. Measured cementitious material properties.

	Compression Strength/MPa	Tension Strength/MPa	Tensile Elongation
CF precast columns and beams	38.9	-	-
PFRW columns and beams	40.8	-	-
PFRW rocking wall	43.4	-	-
wall base grout	120.4	-	-
ECC	30.5	3.91	0.025

The rubber blocks had a Shore A hardness of 59, a tensile strength of 20 MPa, a compression modulus of 4.5 MPa, and an elongation at break of 545%. Measured strengths of all reinforcing bars are given in Table 2. The nominal tensile strength of the steel tendon was f_{ptk} = 1860 MPa, and the measured tensile strength and modulus were 1960 MPa and 202 GPa, respectively. The 1000-h relaxation was measured to be 1.54%.

Table 2. Measured strength of reinforcing bar and steel plate.

	Yield Strength/MPa	Ultimate Strength/MPa
6 mm reinforcing bar	408	619
8 mm reinforcing bar	319	520
12 mm reinforcing bar	433	602
16 mm reinforcing bar	411	597
Q235 steel H-section	273	402

2.2. Specimen Loading Protocol

The loading arrangement used is shown in Figure 3. Vertical loads of 143 kN were applied at the top of each column using hydraulic rams. This load is approximately 30% of the column design capacity determined from GB 50010-2010 [19]. This load is applied through a sliding block to remain constant during horizontal load applications. The horizontal load is applied using two servo-hydraulic actuators: one actuator acts at the centerline of the third story beam (i.e., 4400 mm above the foundation block), and the other acts on a distribution beam loading the first and second floor frame beams. The distribution beam is arranged such that it loads the second story beam in a ratio of 2:1 to the first story beam. Both actuators are run at the same load level resulting in a distribution of load to the third:second:first stories = 1:0.66:0.33.

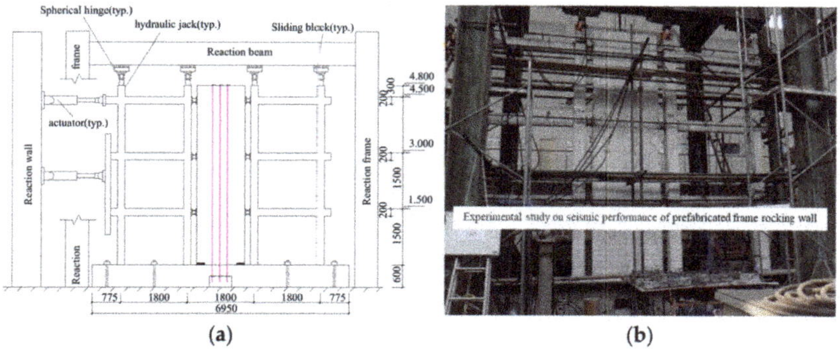

Figure 3. Constant axial load—reversed cyclic lateral load test apparatus. (a) Diagram of test loading apparatus; (b) Photo of loading apparatus.

Displacement-controlled reversed-cyclic lateral loads are applied to the frames. The third story horizontal displacement is the reference displacement, and the loading sequence is shown in Figure 4. The values in Figure 4 (and elsewhere in this paper) are given in terms of roof drift. The height of the control displacement is H = 4400 mm. In all data, positive represents the actuator "pulling" and negative is "pushing". The first excursion to each displacement level was in the positive direction.

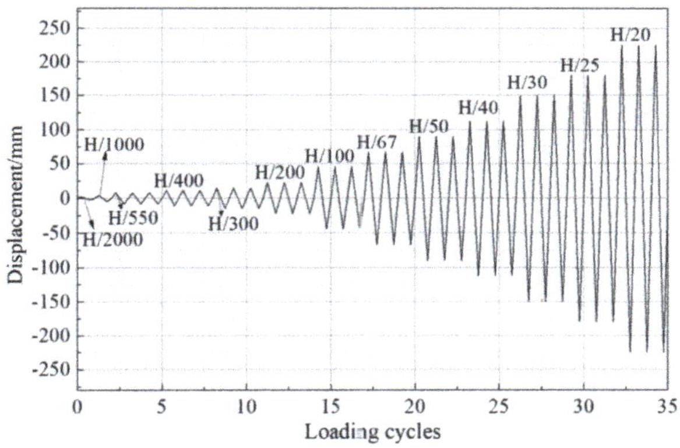

Figure 4. Loading scheme.

The initial cycles to drifts of H/2000 and H/1000 were used to capture the initial cracking state of the frames. Only one loading cycle to these displacements is performed. For the remainder of the test, three fully-reversed cycles for each roof drift were performed.

2.3. Instrumentation

An array of 18 displacement transducers (Figure 5) was used to measure (a) horizontal displacements at each story level, (b) rotation at the base of each column (using vertical transducers at either side of the column), and (c) displacements of the foundation block. For the PFRW specimen, (d) vertical displacement at both rocking wall toes, and (e) horizontal sliding at the base of the rocking wall were also measured. The force in each of the three post-tensioned tendons was also monitored using force sensors in line with the tendon anchorage. Forces in all actuators were also recorded.

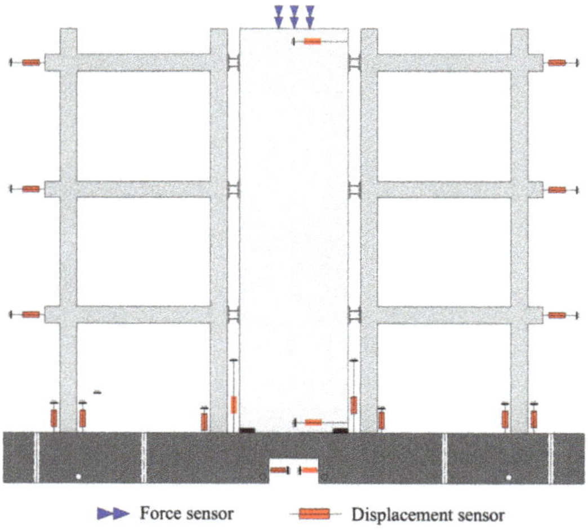

Figure 5. Instrumentation (PFRW shown).

3. Experimental Observations and Comparisons

The total applied horizontal load versus roof drift hysteresis of both frames is shown in Figure 6. The load F is the sum of the actuator forces. Thus, the applied lateral load at the roof level is $F/2$, on the second story: $F/3$, and on the first story: $F/6$.

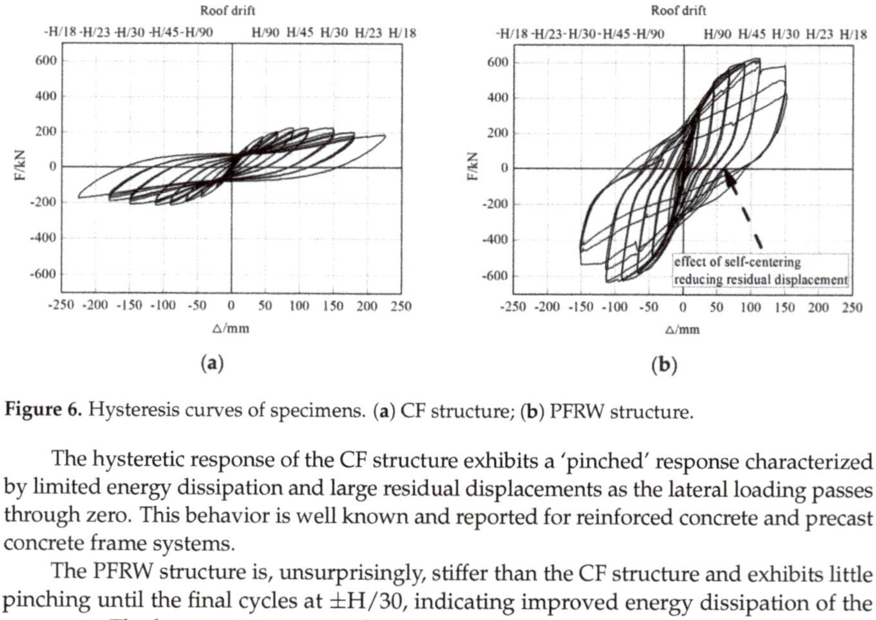

Figure 6. Hysteresis curves of specimens. (**a**) CF structure; (**b**) PFRW structure.

The hysteretic response of the CF structure exhibits a 'pinched' response characterized by limited energy dissipation and large residual displacements as the lateral loading passes through zero. This behavior is well known and reported for reinforced concrete and precast concrete frame systems.

The PFRW structure is, unsurprisingly, stiffer than the CF structure and exhibits little pinching until the final cycles at $\pm H/30$, indicating improved energy dissipation of the structure. The hysteretic response also exhibits a degree of self-centering as the lateral loading passes through zero: from the peak load, the structure initially shows little elastic rebound upon unloading. As the lateral load falls, the self-centering effect of the vertical post-tensioning is seen as a 'horizontal pinching of the hysteresis, as shown in Figure 6b.

3.1. *Behavior of CF Structure*

The lateral load-roof drift hysteresis of CF is shown in Figure 6a. At a roof drift of $\pm H/1000$, short vertical cracks developed in the tension zone at the beam ends of the first story. At a drift of ±H/400, horizontal cracks developed at the base of each column. Damage progressed in a relatively uniform manner with continued cycling. The peak lateral load capacity was observed at ±H/40, although this was relatively constant from ±H/50 to ±H/30. At a drift of ±H/25, concrete spalling initiated at the beam ends and column bases (Figure 7a,b), and evidence of buckling of the longitudinal beam bars was seen at the face of the second story exterior column A2 (Figure 7a). At a drift of ±H/20, the lateral load capacity had fallen to 85% of its peak, and the test was ended. A view of frame CF at the –H/20 is shown in Figure 7c, and the final cracking patterns are recorded in Figure 7d. The failure mechanism is dominated by flexural hinges forming in the first and second story beams and at the base of each column.

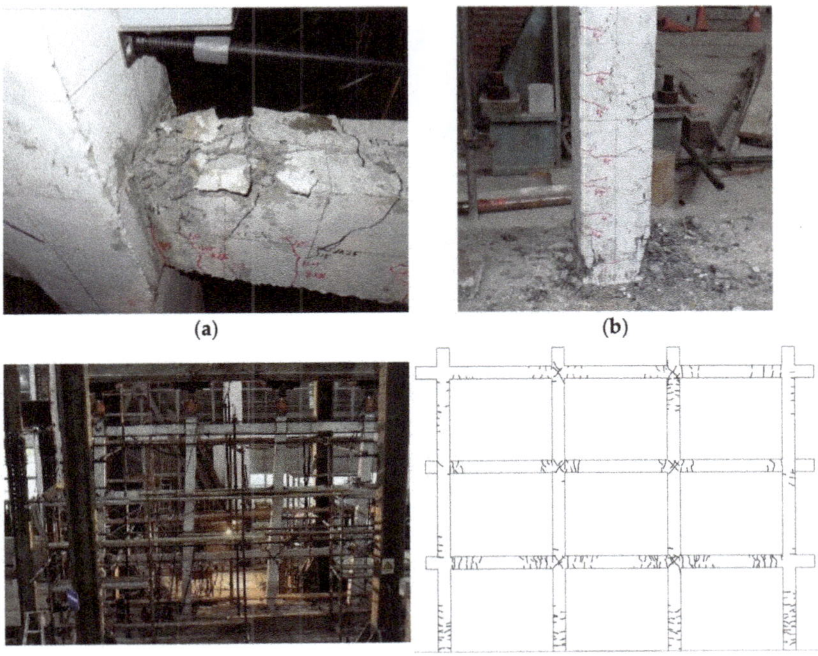

Figure 7. Damage of CF structure. (**a**) damage at second story beam end at face of column A at –H/25; (**b**) damage at base of column A at –H/25; (**c**) overall view of CF at –H/20; (**d**) final crack distribution.

3.2. *Behavior of PFRW Structure*

The lateral load-roof drift hysteresis of PFRW is shown in Figure 6b. As expected, replacing the middle bay beams with a rocking wall resulted in a considerably stiffer lateral load-resisting system. Initially, at a roof drift of −H/2000, two short vertical cracks developed at the right end (face of column B) of the beam spanning between columns A and B at the first story. At a drift of ±H/550, a crack developed at the base of column B. At a drift of ±H/100, the rocking wall lifting from the foundation at the tension toe, and the energy-dissipating connectors were obviously deformed. Horizontal cracks were evident at the upper and lower edges of the 25 mm embedded steel plates anchoring the connectors. The peak lateral load capacity was observed at ±H/40. By the time a drift of ±H/30 was achieved, all the connectors were significantly damaged (Figure 8a). The concrete at the base of the two middle columns, B and C, was crushed and severely damaged (Figure 8b); the maximum crack width of the column was 3.5 mm. The energy-dissipating connectors were seriously damaged (Figure 8c). The base of the rocking wall exhibited significant uplift, spalling, and cracks up to 0.9 mm wide at the edge of the rubber insert at the wall toe

(Figure 8d). The peak lateral load capacity was observed at ±H/30; the lateral load capacity had fallen to 85% of its peak, and the test ended. A view of frame PFRW at +H/30 is shown in Figure 8d, and the final cracking patterns are recorded in Figure 8f. Due to the similar drift levels, damage to the frame elements was similar to that observed in the CF tests. The energy-dissipating connections of the rocking wall to the frame at each story introduced additional damage at the beam column joints at which the connections were made. Damage to the wall on the second story, primarily to the rocking wall, was also observed but was relatively minor.

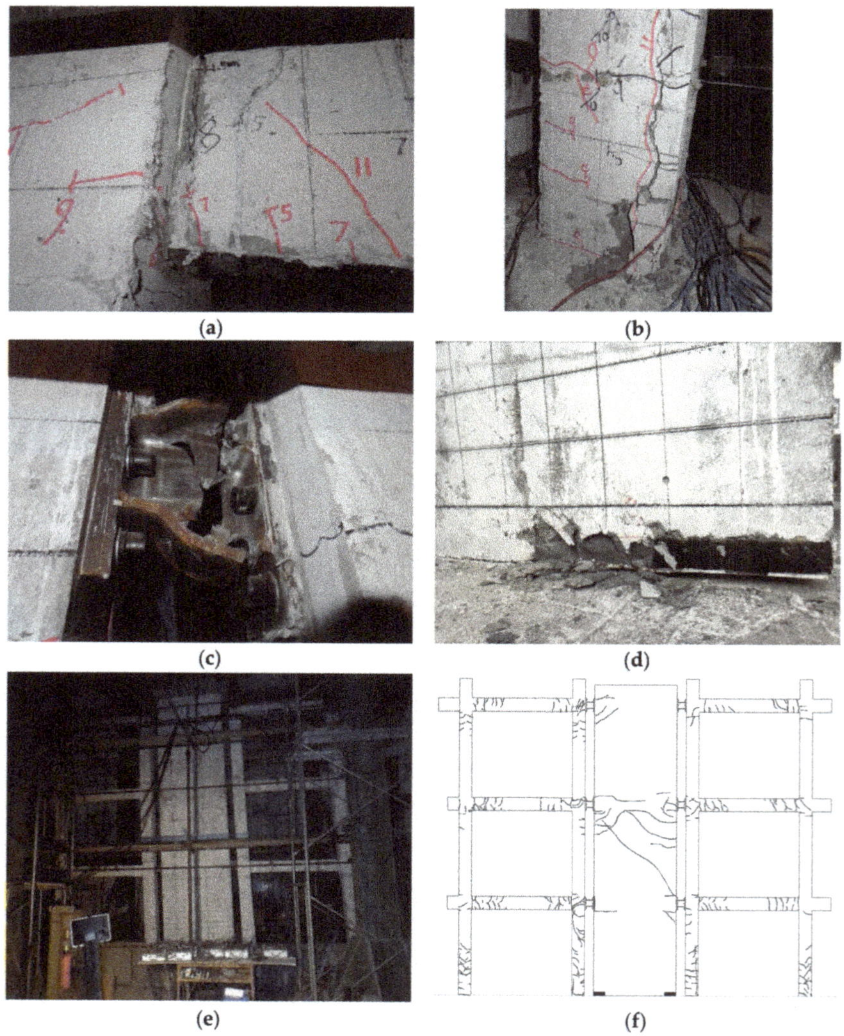

Figure 8. Damage of PFRW structure. (**a**) damage of first story beam at face of column A at +H/30; (**b**) damage at base of column C at +H/30; (**c**) damage to first story energy-dissipating connector at −H/30; (**d**) uplift and spalling at the base of the rocking wall at −H/30; (**e**) overall view of PFRW at −H/30; (**f**) final crack distribution.

3.3. Backbone Curves

A comparison of the backbone curves drawn through the first cycle peaks of frames CF and PFRW is shown in Figure 9. The corresponding key response parameters are summarized in Table 3. In Table 3, the yield point is determined according to the energy equivalent method [20], $K = F_y/\Delta_y$ is the secant stiffness of the frame defined at the yield point, and $\mu = \Delta_u/\Delta_y$ is the displacement ductility at the ultimate load—defined at 85% of the peak load attained.

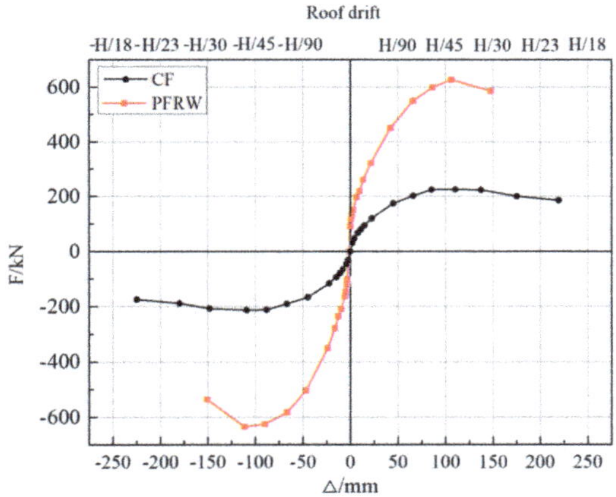

Figure 9. Backbone curves.

Table 3. Key response parameters of the backbone curves.

	Initial Cracking		Yield			Peak Load		Ultimate Load at 0.85 F_{max}		
	F_{cr} /kN	Δ_{cr} /mm	F_y /kN	Δ_y /mm	$K = F_y/\Delta_y$ /kN/mm	F_{max} /kN	Δ_{max} /mm	F_u /kN	Δ_u /mm	$\mu = \Delta_u/\Delta_y$
CF	48.3	4.5 (H/1000)	186.7	54.2 (H/80)	3.4	225.2	110.4 (H/40)	191.4	202.6 (H/20)	3.74
	−47.4	−4.5 (H/1000)	−174.4	−52.2 (H/80)	3.3	−212.7	−109.2 (H/40)	−180.8	−204.8 (H/20)	3.93
PFRW	65.1	3.0 (H/1500)	519.6	60.2 (H/70)	8.6	628.1	107.6 (H/40)	587.3	149.3 (H/30)	2.48
	−80.2	−4.5 (H/1000)	−529.1	−52.0 (H/80)	10.2	−633.1	−110.0 (H/40)	−534.6	−149.1 (H/30)	2.87

Although the stiffness of the PFRW frame increased significantly with the inclusion of the rocking wall—2.8 times based on secant stiffness at yield—the deflections at yield were similar. The rocking wall, therefore, increased the yield load 2.9 times. This observation indicates that the rocking wall was behaving as intended and served to couple the two exterior frames. Had the wall behaved as a shear wall, the stiffness would have been increased, but the yield displacement would likely have fallen due to the limited displacement capacity of conventional shear walls. The significant damage to the energy-dissipating connectors (Figure 8c) reinforces the larger displacements of the rocking wall behavior.

At the peak behavior, observed at the same drift ratio (H/40) in each frame, the capacity of frame PFRW remained about 2.9 times that of frame CF. However, the behavior of the frames near their peak load was different. CF exhibited an extended plateau near its peak load, extending from a drift ratio of approximately H/50 (88 mm) to H/30 (145 mm). In contrast, PFRW exhibited a more defined peak and more 'brittle' behavior, reaching a peak and abruptly losing capacity. The apparent ductility of CF was approximately 1.4 times greater than PFRW.

3.4. *Stiffness Degradation*

The stiffness degradation of each specimen is calculated as follows:

$$K_i = \sum\nolimits_{j=1}^{n} F_{j,max}^i / \sum\nolimits_{j=1}^{n} \Delta_j^i \quad (1)$$

where K_i is the stiffness of the cycle, $F_{j,\max}^i$ and Δ_j^i are the peak load and corresponding roof displacement, respectively, for the j-th cycle (j = 1 to 3) at the i-th load level. The initial stiffness of each specimen, K_0, is determined for each frame in its pre-cracked state. The evolution of frame stiffness and the degradation curves normalized by K_0 are shown in Figure 10a,b, respectively.

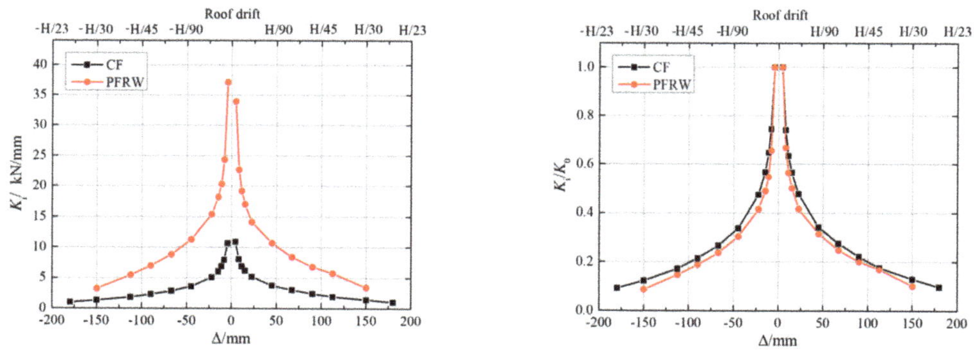

(**a**) (**b**)

Figure 10. Stiffness degradation of frames. (**a**) evolution of secant stiffness of each frame, K_i; (**b**) stiffness degradation normalized by K_0 for each frame.

Although PFRW is considerably stiffer than CF (Figure 10a), the rate of stiffness degradation is very similar in both specimens. At H/50, lateral stiffness has fallen to about 20% of the initial uncracked stiffness in each frame.

3.5. *Energy Dissipation*

The energy dissipation of the structure is measured by the area enclosed by the hysteresis curve (i.e., A_{loop}) shown in the inset of Figure 11. The equivalent elastic damping coefficient, β, given by Equation (2), is a measure of the energy absorption characteristics of the hysteresis and is normalized to permit direct comparison [21]. A larger value of β indicates a greater ability to dissipate energy. The maximum theoretical value of β, corresponding to an elastic-perfectly plastic hysteresis, is $2/2\pi = 0.318$.

$$\beta = A_{loop}/2\pi A_e \quad (2)$$

where A_{loop} is the area contained with a single hysteresis loop; and A_e is the area of the triangles defined by the equivalent elastic stiffness to the peak load and displacement of each cycle i; i.e., $A_e{}^i = 0.5\ [F_{i+}\Delta_{i+} + F_{i-}\Delta_{i-}]$.

Figure 11a plots the evolution of energy dissipation with cycling, i.e., the accumulation of A_{loop}. Figure 11b shows the evolution of β with cycling. The upper limit of each band is the first cycle at each drift ratio, and the lower limit is the third. Thus, the width of the data band indicates the deterioration of energy dissipation with cycling.

The absolute energy dissipation of PFRW clearly exceeds that of CF (Figure 11a), as does the rate of increase of energy dissipation of the PFRW structure. The equivalent elastic damping of PFRW also exceeds that of CF (Figure 11b). The equivalent elastic damping coefficient of the PFRW structure levels off at about 110 mm (H/40) lateral deflection, and the difference between the first and third cycles increases substantially. This decay reflects the rapid deterioration of the energy-dissipating connector (Figure 8c). It is also seen as the dramatic drop in capacity in the hysteresis curve of PFRW at H/30 (Figure 6b).

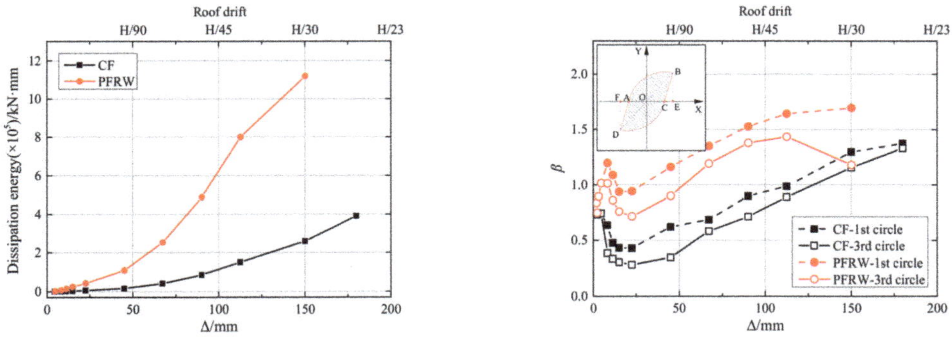

(a) **(b)**

Figure 11. Energy dissipation performance of specimens. (**a**) Cumulative energy dissipation; (**b**) equivalent elastic damping.

3.6. Wall Rocking and Residual Displacements

In the PFRW structure, the post-tensioned tendons are the only reinforcement passing across the interface at the base of the wall, connecting the wall with the foundation. Arranged in the middle third of the wall panel, these tendons provide limited resistance to overturning and therefore permit rocking of the wall. Additionally, the post-tensioning force provides a degree of elastic self-centering to the wall. Each tendon was instrumented to monitor the change in load in the tendon throughout the test (Figure 12). The initial force in each tendon was marginally different, as seen in Figure 12: from left to right, the tendons had initial forces of 147 kN, 140 kN, and 154 kN. As seen in Figure 12, these forces increased as the wall toe nearest the tendon experienced tension and eventually uplift (note that the forces for the left and right tendons are out of phase and the middle tendon exhibits less variation). At a roof drift of H/50, the maximum tendon stress has increased by about 25% (right tendon); by failure at H/30, the increase is about 32%. Nonetheless, all tendons remained below 0.8 f_{ptk}. The marginal loss is tendon force during the H/30 cycles reflects the spalling of the wall (Figure 8d) approaching the location of the strand resulting in some relaxation of the strand during the compression cycle.

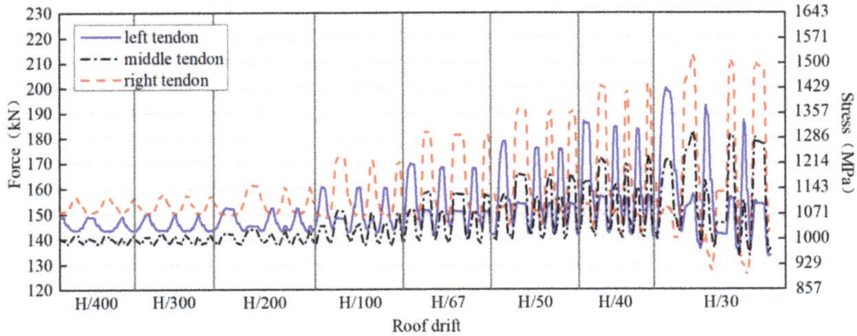

Figure 12. Internal force-time histories of post-tensioning tendons.

The tendons have the effect of self-centering the wall. However, in a wall-frame system, this is 'resisted' by the frame with no self-centering capacity. In the PFRW, the post-tensioned cables are located in the middle third of the wall, giving them a short lever arm and, therefore, a limited ability to generate restoring moments. Only once the surrounding frames deteriorate significantly, beyond a drift of about H/50, do the effects of self-centering become evident. The restoring moment has not increased; rather, the resistance to self-centering has decreased. Table 4 summarizes the greatest residual deflection observed during selected load cycles. The improved hysteretic behavior of PFRW results in more robust hysteretic loops (Figure 6) and greater relative energy dissipation (Figure 11b). However, the same behavior results in large residual deflections at zero lateral loads. Only as the structure deteriorates, at drifts exceeding H/50, does the benefit of self-centering become apparent, reducing the residual deformation by about 30% at drifts of H/30.

Table 4. Largest residual deformation observed at selected load cycles (unit: mm).

Roof Displacement Drift	H/550	H/300	H/100	H/50	H/40	H/30
CF	1.2	1.4	6.9	33.5	52.0	86.4
PFRW	3.2	5.4	11.2	38.7	45.6	61.2

This last observation has ramifications for PFRW design, indicating a tradeoff between the energy dissipation possible through rocking action and the self-centering capability of the post-tensioned wall pier. That is, efficient self-centering will place the post-tensioned tendons as far from the rocking axis as possible; however, this restrains the rocking behavior.

4. Conclusions

A hybrid precast frame-rocking wall (PFRW) structure is proposed, and a prototypical half-scale model tested under reversed cyclic pseudo-dynamic loading. This structural form was envisioned for new construction but also has the potential for a seismic upgrade retrofit of existing frame structures. To facilitate assembly and reduce the damage to the frame component of the system, engineered cementitious composite (ECC) material was used to form cast-in-place joints between precast beams and columns. The rocking wall component was connected to the frame column using energy-dissipating connectors consisting of short steel beam sections. Three vertical post-tensioning tendons, located in the middle third of the rocking wall, provided a degree of self-centering capacity to the system. The proposed PFRW system was experimentally compared to a frame having the same precast details but no rocking wall component. The following conclusions were drawn:

1. Compared with the CF structure having no rocking wall, the lateral load capacity of the PFRW structure was significantly improved. The greater stiffness of the structure, however, results in marginally reduced drift capacity and reduced displacement ductility;
2. The rocking action engages the energy-dissipating elements connecting the rocking wall to the frame. As a result, the hysteretic response of the PFRW was more robust than that of the frame alone, resulting in not only proportionally greater energy dissipation but (relatively) improved energy dissipation characteristics;
3. In the PFRW, only the rocking wall has self-centering capacity. In this study, only after significant degradation of the frame component was the self-centering evident for the structure itself. The residual displacements of both the PFRW and CF structures were comparable through roof drifts of H/50. Only beyond this—as the frame was damaged—were the residual deformations of the PFRW notably improved;
4. The combination of conclusions 2 and 3 highlights the compromise the designer must make between the energy dissipation possible through rocking action and the self-centering capability of the post-tensioned wall pier. That is, efficient self-centering will place the post-tensioned tendons as far from the rocking axis as possible; however, this restrains the rocking behavior.

Author Contributions: Formal analysis, Z.W.; investigation, F.Z.; resources, X.L.; data curation, K.T.; writing—original draft preparation, K.T.; writing—review and editing, K.A.H. and Q.X.; supervision, X.L., K.A.H. and Q.X.; project administration, F.Z. All authors have read and agreed to the published version of the manuscript.

Funding: This work was financially sponsored by Program of Shanghai Technology Research Leader (21XD1432800), State Key Laboratory of Disaster Reduction in Civil Engineering (SLDRCE20-03), and Natural Science Foundation of Shanghai (20ZR1424500).

Data Availability Statement: Not applicable.

Acknowledgments: The authors are greatly indebted to the anonymous reviewers for their valuable comments and suggestions, which helped in improving the overall quality of this manuscript greatly.

Conflicts of Interest: The authors declare no conflict of interest.

References

1. Zhou, Y.; Wu, H.; Gu, A.Q. Earthquake engineering: From earthquake resistance, energy dissipation, and isolation, to resilience. *Eng. Mech.* **2019**, *36*, 1–12. (In Chinese)
2. Wada, A.; Qu, Z.; Ito, H.; Motoyui, S.; Sakata, H.; Kasai, K. Seismic retrofit using rocking walls and steel damper. In *Improving the Seismic Performance of Existing Buildings and Other Structures*; American Society of Civil Engineers: San Francisco, CA, USA, 2009; pp. 1010–1021.
3. Kurama, Y.; Sause, R.; Pessiki, S.; Lu, L.W. Lateral load behavior and seismic design of unbonded post-tensioned precast concrete walls. *ACI Struct. J.* **1999**, *96*, 622–633.
4. Aiorab, J.; Pekcan, G.; Mander, J. Rocking wall-frame structures with supplemental tendon systems. *J. Struct. Eng.* **2004**, *130*, 895–903.
5. Hitaka, T.; Sakino, K. Cyclic tests on a hybrid coupled wall utilizing a rocking mechanism. *Earthq. Eng. Struct. Dyn.* **2008**, *37*, 1657–1676. [CrossRef]
6. Yoorasertchai, E.; Warnitchai, P. Seismic performance of precast hybrid moment-resisting frame/rocking wall systems. *Mag. Concr. Res.* **2018**, *70*, 1118–1134.
7. Wight, G.; Ingham, J.; Kowalsky, M. Shake table testing of rectangular post-tensioned concrete masonry walls. *ACI Struct. J.* **2006**, *103*, 587–595.
8. Toranzo, L.A.; Restrepo, J.I.; Mander, J.B.; Carr, A.J. Shake-Table Tests of Confined-Masonry Rocking Walls with Supplementary Hysteretic Damping. *J. Earthq. Eng.* **2009**, *13*, 882–898. [CrossRef]
9. Marriott, D.; Pampanin, S.; Palermo, A. Quasi-static and pseudo-dynamic testing of unbonded post-tensioned rocking bridge piers with external replaceable dissipaters. *Earthq. Eng. Struct. Dyn.* **2009**, *38*, 331–354. [CrossRef]
10. Nazari, M.; Sritharan, S. Seismic design of precast concrete rocking wall systems with varying hysteretic damping. *PCI J.* **2019**, *64*, 58–76. [CrossRef]
11. Sritharan, S.; Aaleti, S.; Henry, R.S.; Liu, K.-Y.; Tsai, K.-C. Precast Concrete Wall with End Columns (PreWEC) for Earthquake Resistant Design. *Earthq. Eng. Struct. Dyn.* **2015**, *44*, 2075–2092. [CrossRef]
12. Restrepo, I.; Rahman, A. Seismic performance of self-centering structural walls incorporating energy dissipaters. *J. Struct. Eng.* **2007**, *133*, 1560–1570. [CrossRef]
13. Lu, X.L.; Wu, H. Study on seismic performance of prestressed precast concrete walls through cyclic lateral loading test. *Mag. Concr. Res.* **2017**, *69*, 1–14. [CrossRef]
14. Cui, H.; Wu, G.; Zhang, J.; Xu, J. Experimental study on damage-controllable rocking walls with resilient corners. *Mag. Concr. Res.* **2019**, *71*, 1113–1129. [CrossRef]
15. Tagliafierro, B.; Montuori, R.; Castellano, M.G. Shake table testing and numerical modelling of a steel pallet racking structure with a seismic isolation system. *Thin-Walled Struct.* **2021**, *164*, 107924. [CrossRef]
16. Jiang, L.; Li, X.M.; Zhang, F.W.; Dong, J.Z.; Jiang, L.X.; Xu, Q.F. Experimental investigation on seismic performance of frame-rocking wall structures using disc springs. *J. Build. Struct.* **2019**, *40*, 61–70. (In Chinese)
17. Dong, J.Z.; Li, X.M.; Zhang, F.W.; Jiang, L.X.; Jiang, L.; Xu, Q.F. Experimental study on seismic performance of frame-controlled rocking wall structures using SMA devices. *China Civil Eng. J.* **2019**, *52*, 41–51. (In Chinese)
18. Wu, S.J.; Pan, P.; Zhang, X. Characteristics of frame rocking wall structure and its application in a seismic retrofit. *Eng. Mech.* **2016**, *33*, 54–60+67. (In Chinese) [CrossRef]
19. *GB50010-2010(2015)*; Code for Design of Concrete Structures. China Architecture and Building Press: Beijing, China, 2015.
20. Ma, F.; Deng, M.; Ma, Y.; Lü, H.; Yang, Y.; Sun, H. Experimental study on interior precast concrete beam–column connections with lap-spliced steel bars in field-cast RPC. *Eng. Struct.* **2021**, *228*, 111481. [CrossRef]
21. Clough, R.W.; Penzien, J. *Dynamics of Structures*; McGraw-Hill Education: New York, NY, USA, 1993.

buildings

MDPI

Review

Experimental, Theoretical and Numerical Research Progress on Dynamic Behaviors of RC Structural Members

Rouhan Li [1], Mao Gao [1], Hongnan Li [2], Chao Li [2,*] and Debin Wang [3]

[1] Department of Civil Engineering, College of Transportation Engineering, Dalian Maritime University, Dalian 116026, China; lirouhan@dlmu.edu.cn (R.L.); gm1222@dlmu.edu.cn (M.G.)
[2] Faculty of Infrastructure Engineering, Dalian University of Technology, Dalian 116024, China; hnli@dlut.edu.cn
[3] School of Civil Engineering, Dalian Jiaotong University, Dalian 116028, China; wdb1215@163.com
* Correspondence: chaoli@dlut.edu.cn

Abstract: In this paper, research on dynamic behaviors of RC structural members was reviewed using experimental, theoretical and numerical perspectives. First, in a basic overview, measurement methods, main conclusions and current limitations of available dynamic loading tests were presented. Then, theoretical studies on the dynamic constitutive models of RC materials, the dynamic increase factor (DIF) model for concrete and reinforced steel and proposed modified models of dynamic behavior parameters at the structural member level were summarized. Finally, the available modeling approach and method for incorporating dynamic effects in numerical simulations of RC structures were reviewed. Moreover, the work involved a brief introduction to a dynamic hysteretic model established using experimental data, which was designed to provide an alternative approach to the commonly-used DIF method for considering these dynamic effects. This paper, therefore, aimed to provide a valuable reference for experimental studies and numerical simulations on the dynamic behaviors of RC structures—while also putting forward issues that need to be addressed by future work.

Keywords: reinforced concrete members; dynamic effect; experimental test; dynamic modified model; numerical modelling

Citation: Li, R.; Gao, M.; Li, H.; Li, C.; Wang, D. Experimental, Theoretical and Numerical Research Progress on Dynamic Behaviors of RC Structural Members. *Buildings* **2023**, *13*, 1359. https://doi.org/10.3390/buildings13051359

Academic Editor: Hugo Rodrigues

Received: 10 April 2023
Revised: 30 April 2023
Accepted: 15 May 2023
Published: 22 May 2023

1. Introduction

Reinforced concrete (RC) is one of the most widely-used building construction materials in civil engineering. In addition to static loads, RC structures may be subjected to different types of dynamic loads during their service life, such as explosion, impact, earthquake and wind load, etc. In past decades, a large number of RC structures have been damaged or even collapsed due to seismic hazards. Damage phenomena relating to different RC structures, e.g., public, residential and industrial buildings and bridges subjected to earthquake loads, are shown in Figure 1. According to statistical data, more than 10,000 people are impacted by earthquakes annually. This impacts are accompanied by economic losses totaling billions of U.S. dollars [1]. In order to reduce these human and economic losses, civil engineers and researchers have made great efforts to continually enhance the seismic performances of RC structures and to accurately evaluate their mechanical behaviors under earthquake load during the structural design, operation and maintenance stages.

It has been widely accepted that reinforcement and concrete exhibit different mechanical properties under static and dynamic loads, namely the strain rate sensitivity of the materials [2]. Consistent research findings concluded that, as the loading rate increased, the tensile strength and compressive strength of concrete—as well as the yield strength and ultimate strength of reinforcement were magnified. The elastic modulus and the strain, corresponding to the compressive strength of concrete, were also affected by the loading rate [3]. As for structural members consisting of RC materials, such as columns [4–7], beams [8–15], shear walls [16–19] and joints [20–24], changes in mechanical behaviors and failure patterns were observed for specimens under various loading rates, namely the dynamic effect at the member level. Different damage phenomena of RC structural members have been observed in response to seismic hazards, as shown in Figure 2. To acquire a better understanding of the dynamic behaviors of RC structural members under earthquake excitation, their performance must be comprehensively investigated, along with considerations of the seismically-induced loading rate. In the past half century, a number of dynamic loading tests have been carried out to deepen researchers' understanding of the mechanical behaviors of RC structural members that have been subjected to dynamic (i.e., blast, impact and seismic) loading rates. In addition, many researchers have performed numerical studies on the dynamic behaviors of RC structural members, specifically considering dynamic effects [25–28]. The advantages of numerical simulations, in relation to experimental tests, lie in the comparatively lesser manpower and material resources required for their execution. Additionally, they can more feasibly be applied to a wider range of structural parameters and loading rates. However, current seismic codes and most structural seismic analyses do not specifically consider the strain rate sensitivity of RC materials. The application of the dynamic increase factor (DIF) to modify the mechanical properties of RC materials, suggested by the CEB-FIP Model Code [29] and some other scholars [30–32], provided a common approach to consideration of the dynamic effects. Nevertheless, it should be noted that researchers have yet to achieve a deep understanding of the mechanisms by which the macro-mechanical behaviors of RC structural members under dynamic loads might be explained. To close the gap in this research area, several attempts have been made by scholars to establish dynamic modified models to consider dynamic effects at the member level, based on either experimental or numerical results [4,33,34].

Figure 1. Damage phenomena of RC structures under real earthquake load. (**a**) Damage to residential building; (**b**) Damage to public building; (**c**) Damage to industrial building; (**d**) Damage to bridge.

Figure 2. Damage patterns in RC structural members after seismic hazards. (**a**) Damage to joint elements; (**b**) Damage to column elements; (**c**) Damage to wall elements; (**d**) Damage to beam elements.

In this paper, the existing research works were systematically summarized from experimental, theoretical and numerical perspectives. Additionally, prospective directions for future efforts were also outlined. This review paper aimed to provide a significant reference for seismic design and analysis works, thereby improving the seismic performances of RC structures.

2. Experimental Studies on Dynamic Behaviors of RC Structural Members

2.1. Overview of Dynamic Loading Tests on Structural Members

In the civil engineering field, several methods for testing the dynamic behaviors of RC structural members have been adopted by scholars, including the pseudo-static test, the pseudo-dynamic test, the shaking table test and the earthquake observation test [3]. Among these test methods, the pseudo-static test method is the most commonly used. By employing monotonic or cyclic loading schemes, the dynamic behaviors of RC structural members in the elastic stage, the plastic stage and the final failure stage can be obtained. However, one shortcoming of this method is that it cannot reasonably reflect the influences of strain rates or loading rates on the mechanical behaviors of RC structural members. Of the dynamic test methods, the shaking table test method provides the most accurate and reliable results; as such, it is often used to evaluate the dynamic responses and failure mechanisms of structural members and systems under earthquake excitations. However, it requires significant time and financial resources. In the earthquake observation test, seismic instruments need to be installed onsite, i.e., on a building, in order to measure the building's structural dynamic response under real earthquake conditions. In contrast, the pseudo-dynamic test method is often used to obtain information on the seismic actions of structures using the controlling approach through computational analysis [35].

At present, most available tests on RC structural members are carried out using static loading rates. In recent years, with advances in experimental techniques and improvements in our base of empirical knowledge, a number of dynamic loading tests have been performed on different RC structural members. Figure 3 shows the strain rate range for RC structures under different loads. The most significant difference between the dynamic loading tests (10^{-4}/s $< \dot{\varepsilon} < 10^{1}$/s) and the pseudo-static loading tests (10^{-6}/s $< \dot{\varepsilon} < 10^{-5}$/s) was the magnitude of strain, or loading rate, exerted on the specimens.

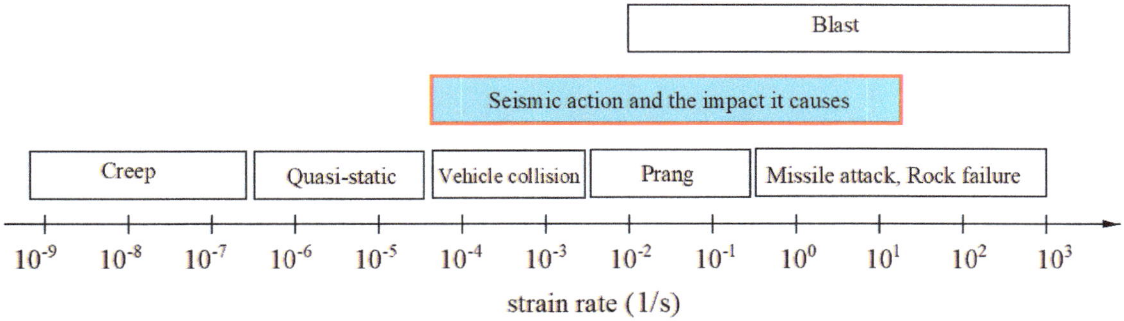

Figure 3. Strain rate range for RC structure under different dynamic loadings [36].

As an earthquake is a kind of dynamic load, dynamic loading tests provide results that more closely match the real-world seismic behaviors of RC structural members. Recent research on dynamic behaviors of RC structural members under impact and blast loading rates has been comprehensively reviewed [37–40]. However, research reviews of RC structural members upon subjection to seismic loading rates are, by comparison, lacking. Therefore, this paper mainly focused on dynamic loading tests of RC structural members carried out under earthquake-induced loading rates. Basic information on available dynamic loading tests (i.e., member type, specimen number, loading rate and scheme) is summarized in Table 1.

Table 1. Summary of dynamic loading tests performed on RC structural members subjected to earthquake loading rates.

No.	Reference	Type	Number	Loading Rate	Loading Scheme
1	Bertero et al. [41]	Beam	6	0.1, 10/s	Mono, cycl
2	Kulkarni and Shah [42]	Beam	14	0.0071–380 mm/s	Mono
3	White et al. [43]	Beam	4	0.0167–36 mm/s	Mono, Cycl
4	Zhang et al. [44]	Beam	36	1.05×10^{-5}, 1.25×10^{-3}/s	Mono
5	Marder et al. [13]	Beam	17	100 Hz	Mono, Cycl
6	Yan [45]	Beam	/	1×10^{-5}–1×10^{-3}/s	Cycl
7	Xiao et al. [46]	Beam	5	0.1–10 mm/s	Mono
8	Li and Li [11]	Beam	16	0.05–30 mm/s	Mono, Cycl
9	Zhou et al. [47]	Beam	7	0.06 mm–66 mm/s	Mono
10	Otani et al. [48]	Beam	8	0.1, 100 mm/s	Cycl
11	Guo [49]	Beam	12	0, 2, 6 m/s	Mono
12	Wu et al. [50]	Beam	3	87.89–135.8 Hz	Mono
13	Song et al. [5]	Beam	5	3.5–6 m/s	Mono
14	Adhikary et al. [8,10,15]	Beam	24	4×10^{-4}–2 m/s	Mono
15	Adhikary et al. [51]	Beam	30	0–5.6 m/s	Mono
16	Zeng [52]	Beam	6	10^{-2}/s–8.85 m/s	Mono
17	Feng et al. [53]	Beam	10	3–7.7 m/s	Mono
18	Mutsuyoushi and Machida [54]	Beam	14	0.1, 10, 100 cm/s	Mono, Cycl
19	Fukuda et al. [55]	Beam	48	4×10^{-4}–2 m/s	Mono
20	Yuan and Yi [56]	Beam	18	3.5×10^{-4}–1 m/s	Mono
21	Ye et al. [57]	Beam	14	0.8 m/s	Mono
22	Fujikake [14]	Beam	6	5×10^{-4} m/s, 2 m/s	Mono

Table 1. *Cont.*

No.	Reference	Type	Number	Loading Rate	Loading Scheme
23	Xiang et al. [58]	Column	7	/	Mono
24	Gutierrez et al. [59]	Column	3	0.02–1 Hz	Cycl
25	Bousias et al. [60]	Column	12	/	Cycl, Biax
26	Li et al. [61]	Column	30	0.000011–0.0167/s	Mono
27	Witarto et al. [6]	Column	4	0.05–5 Hz	Cycl
28	Perry et al. [62]	Column	4	0.7×10^{-4}–0.7×10^{-3}/s	Mono, Cycl
29	Yan [45]	Column	/	10^{-5}–10^{-2}/s	Mono
30	Zou et al. [63]	Column	/	10^{-5}–10^{-2}/s	Mono
31	Wang et al. [64]	Column	30	0.1–50 mm/s	Mono, Cycl, Biax
32	Jiang [65]	Column	12	0.1–20 mm/s	Mono, Cycl, Biax
33	Ghannoum et al. [35]	Column	10	0.25–1061 mm/s	Cycl
34	Liu et al. [66]	Column	10	0, 4.85, 6.86 m/s	Mono
35	Liu et al. [67]	Column	13	/	Mono
36	Lee et al. [68]	Column	6	/	Cycl
37	Wei et al. [69]	Column	6	4.95–5.42 m/s	Mono
38	Fan et al. [70]	Column	8	6.86, 5.42 m/s	Mono
39	Orozco and Ashford [71]	Column	3	0.22–1 m/s	Cycl
40	Shah et al. [72]	Joint	3	2.5×10^{-3}–1.0 Hz	Cycl
41	Chung and Shah [20]	Joint	12	0.0025–2.0 Hz	Cycl
42	Gibson et al. [73]	Joint	4	0–405 mm/s	Cycl
43	Pan [23]	Joint	10	0.1–10 mm/s	Cycl
44	Fan et al. [74]	Joint	3	0.4–40 mm/s	Cycl
45	Wang et al. [75]	Joint	8	0.4–40 mm/s	Cycl
46	Zhang [17]	Shear wall	7	10^{-5}–10^{-3}/s	Cycl
47	Xu et al. [16]	Shear wall	2	1–10 mm/s	Cycl
48	Chiu et al. [76]	Infill wall	6	0–0.4 g	Cycl
49	Yilmaz et al. [77]	Slab	9	4.43, 4.95, 5.42 m/s	Mono

Note: No information is provided in the original literature, which is represented by "/" in the table; 'Mono' and 'Cycl' denote the monotonic and the cyclic loading schemes, respectively; 'Biax' denotes the specimen is loaded in two horizontal directions, as opposed to the default situation in the table, i.e., the specimen is loaded in a single horizontal direction.

From the summarized results, it was noted that there were fewer dynamic loading tests, as compared with static loading tests, and that the investigations primarily focused on beam and column members [5,8,35,54,59,72]. The majority of tests were conducted using electro-hydraulic servo loading systems. A few were carried out using drop-hammer impact testing machines. As earthquake loads are multidimensional in nature, it is reasonable to experimentally study the seismic performances of RC members and structures in space [9]. Wang et al. [78] studied the multidimensional dynamic behaviors of RC columns using two horizontal, and one vertical, electro-hydraulic servo actuators. Due to the difficulty of multi-axis loading testing and the higher requirements for testing equipment, the available literature and experimental data of multi-axis dynamic loading tests were inadequate [60,65].

2.2. Measurement Methods for Dynamic Loading Test

In dynamic loading tests, the observed quantities upon which researchers have focused include bearing capacity, displacement, strain, failure mode and cracks that can be directly measured or observed, as well as stiffness, ductility, damage and energy dissipation capacities—which need to be acquired indirectly. In the following sections, measurements of the test data were summarized in detail. Figure 4 shows the primary measured quantities of RC structural members in the dynamic loading tests.

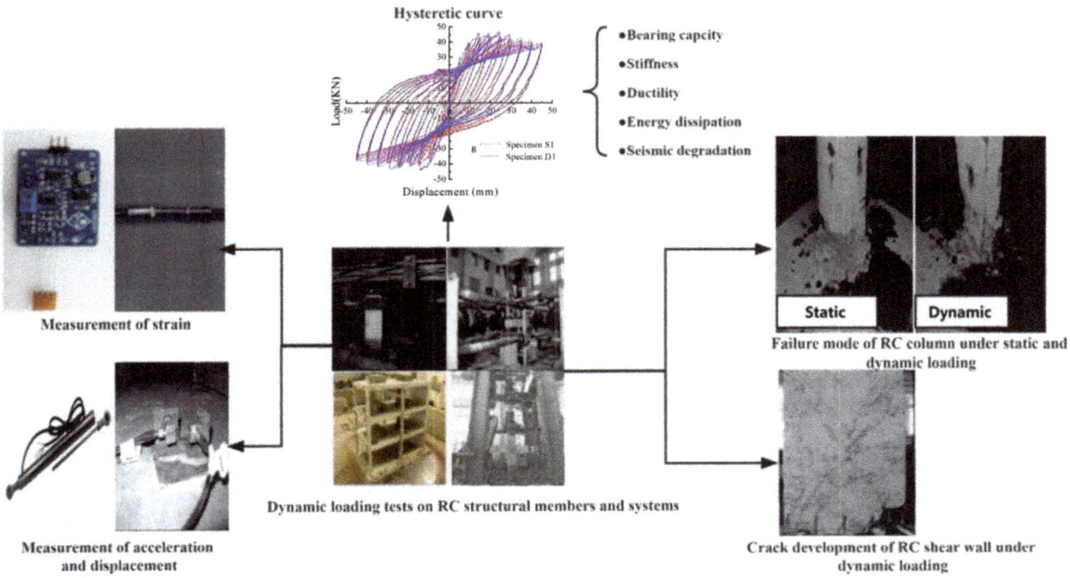

Figure 4. Measured quantities of RC structural members in dynamic loading tests.

To measure the force–displacement relationships between RC structural members, mechanical sensors, inside or outside of the loading device, have commonly been used to collect test data. For example, Wang et al. [78] used the force sensor and displacement sensor of a servo hydraulic actuator to measure the horizontal top displacement and bottom reaction force of the column specimens. Gutierrez et al. [59] used a mechanical sensor, installed in series with the piston rod of a servo device, to measure force, and used an LVDT sensor to determine displacement. In the dynamic loading test performed by Shunsuke et al. [48], a laser displacement sensor was used to measure lateral displacement and a strain gauge was used to measure deformation at a plastic hinge region.

In order to measure material strain on RC members, strain gauges are generally pasted either on the surfaces of structural members or on reinforcement inside them (Kenneth et al. [7], Wang [64], Long [79], Adhikary et al. [8]). Zhang [80] used a fiber Bragg grating strain sensor to measure concrete strain and further derive the real-time strain rate during the whole loading process. An acceleration sensor was employed to measure the horizontal and vertical acceleration of floors. Perry et al. [62] installed LVDT sensors between the two frames of a servo hydraulic testing machine to measure the longitudinal strain of columns members.

A few novel methods have been used by researchers to measure the displacement of RC structural members. For example, Liu [81] used planar trusses of LVDT sensors, arranged outside of the column specimen, to measure displacement. By using geometrical transformation, the flexural, shear and bond-slip deformation components of column specimens were indirectly determined. However, to the best of the authors' knowledge, few works have been reported on the changes in deformation components of RC structural members under dynamic loading rates. Zhang [17] arranged force and displacement meters at the four corners, as well as the bottom, of shear wall specimens in order to measure the displacement and shear deformation of the specimens under dynamic loading rates.

In general, failure modes and crack patterns in RC structural members can be directly observed with naked eyes [6,41,64,71]. However, in some dynamic loading tests, in which crack development was not feasibly or easily measurable, high performance measuring equipment has been employed as alternatives. For example, Adhikary et al. [51] used digital photography and high-speed cameras to capture the crack development and fracture process of RC beams during a drop hammer impact test. A similar approach was adopted by Ye et al. [57] who investigated the failure pattern and crack development of RC column members during impact loading.

Aside from the above physical qualities, which can be measured directly, damage and energy dissipation capacities are generally obtained using indirect methods. In most of the available dynamic tests, the hysteretic curve of force–displacement can be acquired by measuring the bearing capacity and displacement of structural members during the process of cyclic loading. The degradation of bearing capacity and stiffness, as well as the seismic damage and energy dissipation capacity, can be further derived by analyzing the test data of hysteretic loops [4,20,48,72]. By using a self-developed carbon nanofiber aggregate (CNFA) as an internal sensor—which was able to accurately capture the transient changes of structural force and stiffness with almost no time delay—Witarto et al. [6] detected seismic damage in RC column specimens under various loading rates.

2.3. Summary of Experimental Findings

As both concrete and reinforcing steel are rate-sensitive materials, their tensile and compressive mechanical properties are closely relevant to the loading rate. Consequently, the mechanical behaviors of RC structural members under different loading rates differ, which has been demonstrated by many experiments. Bertero et al. [41] experimentally studied the mechanical behaviors of RC simply-supported beams under high loading rates. They found that, with increased loading rates, the yielding bearing capacities of the members increased, whereas the ultimate bearing capacity did not change significantly. Additionally, the strain rate had a minor influence on the energy dissipation capacity, while members at higher loading rates were more likely to enter into brittle shear failure mode. Mutsuyoushi and Machida [54] found that, with increased loading rates, the failure of RC members tended to change, from flexural failure to shear failure. Kulkarni and Shah [42] carried out dynamic tests on RC simply-supported beams at different loading rates. As the loading rates increased, the failure modes of some specimens changed from shear to bending failure, contrary to the conclusions obtained by most researchers. Shah et al. [72] conducted cyclic loading tests of beam-column joints under different strain rates. It was observed that, with increased loading rates, the number of cracks lessened, while the damage intensified and the plastic deformation increased.

Available dynamic loading tests have shown that the mechanical properties of components under different loading rates are closely related to structural parameters. Chung and Shah [20] carried out experimental studies on cantilever beam members at different loading rates, considering the effects of shear span ratio and longitudinal reinforcement ratio. They determined that bearing capacity increased, while cracks and ductility decreased for specimens at higher loading rates. Additionally, the strain rate effect was more significant for specimens with lower reinforcement ratios. Li et al. [61] studied the mechanical behaviors of RC column members with different longitudinal reinforcement ratios, transverse stirrups and cross section shapes under uniaxial dynamic loading. It was observed that, with increased loading rates, the dynamic effects grew less obvious for specimens with higher strength concrete, while the influence of the cross-sectional shape was minor. Zhang et al. [44] conducted an experimental study on the fracture behavior of RC beams under different strain rates, considering the effects of the size of the specimens. The experimental results showed that the strain rate sensitivity values of specimens increased with increased specimen size. Fukuda et al. [55] conducted dynamic tests on 48 RC beams with varied shear span ratio and reinforcement ratio under different loading rates. It was found that the influence of the loading rates on the ultimate bearing capacity of specimens was more significant for shear failure specimens than for flexural failure specimens. Adhikary et al. [8,10,15] carried out tests on a large number of RC beams at different loading rates, concluding that the dynamic effects grew more pronounced along with decreasing longitudinal reinforcement ratios or increasing shear span ratios. A large number of dynamic tests were carried out on reinforced concrete beams and columns by various authors [4,11,78,82,83]. They concluded that increased material strength and stirrup ratios would lead to decreased dynamic effects. Moreover, the strain rate sensitivity of the monotonic loading member was more significant than that of the cyclic loading member, and the areas of concrete crushing and falling off, as well as reinforcement buckling, were more localized.

2.4. Discussion on Dynamic Loading Tests

According to the available dynamic loading tests, the following consistent conclusions can be obtained: (1) with increases of the loading rate, the bearing capacity, stiffness and energy dissipation capacity of members are enhanced, while ductility may be reduced and the degradation of stiffness and bearing capacity may be aggravated. However, existing research works suffered from the following shortcomings: (1) in most of the dynamic loading tests, specimens were tested under nonaxial loading conditions. In order to more accurately reveal the dynamic behaviors of RC structural members, further experimental studies would be required, under multidimensional loading conditions; (2) Currently, the primary physical quantities measured in dynamic loading tests are stress, strain, displacement and force. There has not been sufficient experimental study of the influence of dynamic effects on the deformation and failure mechanisms of structural members; (3) As dynamic loading tests are inadequate, compared to traditional static loading tests, in-depth research will be needed to elucidate the influence of dynamic effects on the seismic behaviors of structural members with various structural parameters.

3. Theoretical Studies on Dynamic Behaviors of RC Structural Members

3.1. Dynamic Modified Model at Material Level

The influence of loading rates on the mechanical properties of concrete [84–98] and reinforcing steel [30,99–103] has been investigated by a large number of experimental studies. The rate sensitivity of concrete materials are influenced by many factors, including: (1) internal causes, such as dispersion in material properties, humidity [104–107] and the temperature [108,109]; (2) exterior causes, such as test loading method [110,111], equipment instability and measurement error, etc. After collecting test data on concrete under a wide range of loading rates (Figure 5), Pajak [110] found that the ratio of dynamic compressive strength to the corresponding static strength reached 3.5, whereas the dynamic tensile strength to the corresponding static strength reached 13. Moreover, it was pointed out by Bischoff [112] that the ratio of strain of the dynamic compressive strength to the corresponding static strain was in the range of 70~140%.

As a multiphase composite material, the constitutive relationship of concrete is highly complex. Based on different theoretical backgrounds, i.e., the visco-elastic theory, visco-plastic theory, damage mechanics theory and fracture mechanics theory, a variety of dynamic concrete constitutive models have been established [113–119]. To reflect the influences of loading rates on the mechanical properties of concrete (e.g., the enhancement of compressive and tensile strength [2], the more brittle behavior for the descending branch of stress–strain curve [110]), the dynamic increase factor (DIF), which has been defined as the ratio of the mechanical behavior parameters of concrete under dynamic loading to the corresponding values under static loading, has become the most widely used. Notably, a few researchers removed the lateral inertia force and the end friction force of concrete specimens when obtaining DIF models [120,121]. Table 2 summarizes the commonly used models of dynamic increase factor (DIF) for concrete.

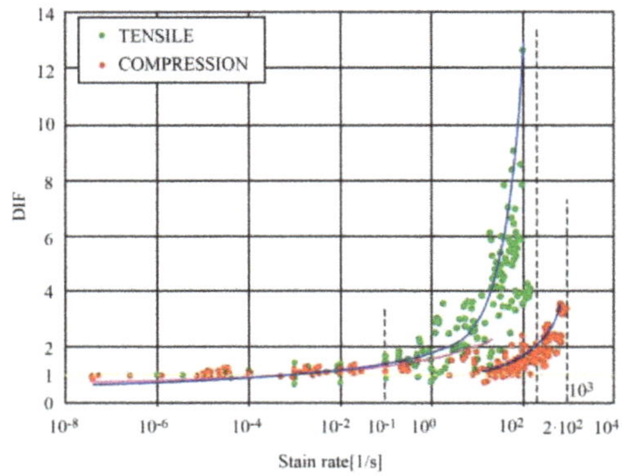

Figure 5. Statistical diagram of dynamic increase factor (DIF) for concrete tensile and compressive strength with variation in strain rate [110].

Table 2. Commonly used models of dynamic increase factor (DIF) for concrete.

Model	Range of Dynamic Strain Rate	Quasi-Static Strain Rate	Type of Formula	Modified Parameters
CEB model [29]	3.0×10^{-5}/s ~ 300/s	3.0×10^{-5}/s (compression) 3.0×10^{-6}/s (tension)	Exponential	$f_{cd} E_{cd}\ \varepsilon_{cfd}\ f_{td}\ E_{td}$
Malvar model [122]	10^{-6}/s ~ 160/s	1.0×10^{-6}/s	Exponential	f_{td}
Tedesco and Ross model [123]	10^{-7}/s ~ 10^{2}/s	10^{-7}/s	Linear logarithmic	$f_{cd} f_{td}$
Yan model [89]	10^{-5}/s ~ 10^{-2}/s	10^{-5}/s	Linear logarithmic	$f_{cd} E_{cd}\ f_{td}\ E_{td}$.
Xiao and Zhang model [124]	10^{-5}/s ~ 10^{-1}/s	10^{-5}/s	Linear logarithmic	$f_{cd} \varepsilon_{cfd}$
Li model [31]	10^{-5}/s ~ 10^{-2}/s	10^{-5}/s	Linear logarithmic	f_{cd}

Note: The values of quasi-static strain rate $\dot{\varepsilon}_0$ for compressive and tensile parameters are the same if not otherwise specified. The modified parameters f_{cd} and f_{td} denote the dynamic compressive and tensile strength of concrete; E_{cd} and E_{td} denote the elastic modulus of concrete under dynamic compressive and tensile loading conditions; ε_{cfd} denotes the dynamic strain, corresponding to the ultimate compressive strength of concrete.

To reflect the enhancement of yielding and ultimate strength under dynamic loading rates, researchers have established various dynamic constitutive models for reinforcing steel. For example, Johnson and Cook [125] developed the dynamic constitutive model of reinforcement, considering the combined influences of strain rate c effect and temperature. Morquio et al. [126] developed the predicted model for mechanical properties of reinforcement, considering strain rate sensitivity and size. Based on the thermo-visco-plastic theory, a dynamic constitutive model of reinforcement, applicable for a wide range of loading rates, was proposed by Rodríguez–Martínez [127]. Compared with the above models, the DIF models based on dynamic loading experimental results have been more widely employed. Table 3 summarizes commonly-used models of dynamic increase factor (DIF) for reinforcing steel.

As shown in Tables 2 and 3, the DIF models considered a variety of material strength degradation properties and a wide range of loading rates. Mechanical behavior parameters for dynamic modification include compressive strength (f_{cd}), tensile strength (f_{td}), elastic modulus of concrete (E_{cd} and E_{td}), and the yielding strength (f_{yd}) and ultimate strength (f_{ud}) of reinforcing steel. Generally, the exponential or linear logarithmic expressions are used for calibrating the DIF formulas. By modifying the quasi-static behavior parameters of material using the DIF models, the dynamic behavior parameters of material can be obtained. They can then be used to establish dynamic constitutive models. More importantly, dynamic modified models at the material level can be utilized to determine the influences of dynamic effects on RC structural members.

Table 3. Commonly-used models of dynamic increase factor (DIF) for reinforcing steel.

Model	Range of Dynamic Strain Rate	Quasi-Static Strain Rate	Type of Formula	Modified Parameters
CEB model [29]	5.0×10^{-5}/s ~ 10/s	5.0×10^{-5}/s	Linear logarithmic	$f_{yd} \cdot f_{ud} f_{nd}$
Malvar model [122]	10^{-4}/s ~ 10/s	3.0×10^{-4}/s	Exponential	$f_{yd} f_{ud}$
Lin Feng model [30]	< 2/s	3.0×10^{-4}/s	Linear logarithmic	$f_{yd} f_{ud}$
Li and Li model [103]	2.5 × 10^{-4}/s ~ 0.1/s	2.5×10^{-4}/s 10^{-5}/s	Linear logarithmic	$f_{yd} f_{ud} \varepsilon_{hd}$

Note: The modified parameters f_{yd}, f_{ud} and f_{nd} denote the dynamic yielding, ultimate and breaking strength of reinforcing steel; ε_{hd} denotes the dynamic strain at initial point of strain hardening stage.

3.2. Dynamic Modified Model at Member Level

Consensus was reached, among scholars, regarding the influence of loading rates on the mechanical behaviors of RC structural members. However, few works have focused on the mechanisms of the dynamic effects exhibited in experimental tests. These could be explained from different perspectives: (1) strain rate-sensitivity of materials, i.e., the physical mechanism of rate-sensitive concrete is attributed to the viscosity effect of the cement matrix [110]. (2) Inertial effects of member (Figure 6)—based on kinetic theory, the structural inertial force is magnified with the increasing loading rate and the constraints on the interior material are also intensified, resulting in the enhancement of macro bearing capacity and stiffness of structural members [128]. (3) Evolution of micro-cracks: due to limitations on time and space at higher loading rates, the probability of transfers of internal force in structural members and occurrences of bond-slip between concrete and reinforcement through stronger regions is increased [110].

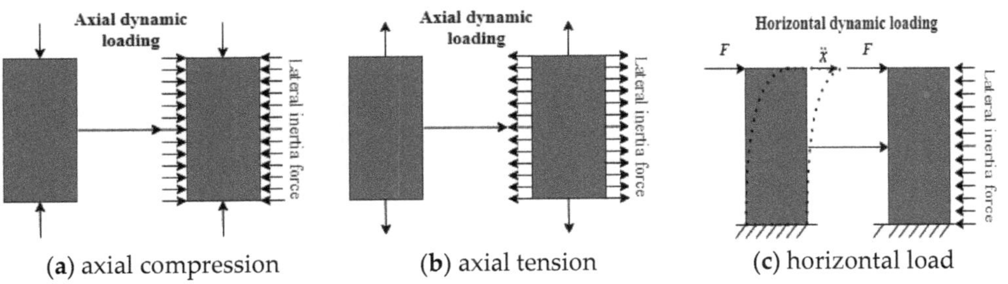

Figure 6. Schematic diagram of inertial effect for RC structural members under dynamic loading.

Due to the non-negligible dynamic effects on the mechanical behaviors of RC structural members, RC structural members exhibit different mechanical properties under static and dynamic loading (i.e., maximum bearing capacity, stiffness, ductility factor and hysteretic behavior). As scholars have done more comparative experiments on bearing capacities under various loading rates, the work counted dynamic increase factor (DIF) for bearing capacities of RC structural members with variations in strain rates, as shown in Figure 7. Data were taken from [5,8,10,11,14,15,17,19,23,45,47,49,51,52,55,56,64,65,67,74,129].

From the summarized results, it can be observed that, with increased loading rates, the dynamic increase factors (DIF) for the bearing capacities of RC structural members were enhanced. Additionally, as the orders of magnitude for strain rate increased, the increases for DIF for bearing capacities grew more and more significantly. However, different scholars studied different types of RC structural members (i.e., beams, columns and shear walls) with different design parameters. As such, there were certain disparities in the summary results. In addition, as bearing capacity impacts the macro-mechanical behavior of RC structural members, it was difficult to determine the influences of dynamic effects on the mechanical behaviors of RC structural members at the member level.

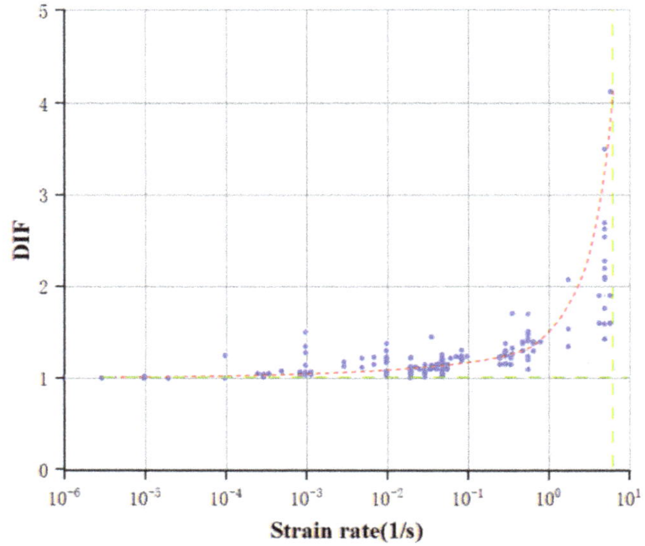

Figure 7. Statistical diagram of dynamic increase factor (DIF) for bearing capacity of RC structural members with variations in strain rates.

Scholars have made attempts to establish dynamic modified models to aid in considering dynamic effects at the member level. Zhan et al. [130] developed the dynamic modified model to predict the maximum and residual deflection of RC beam members based on a significant quantity of experimental data. Adhikary et al. [8] developed a dynamic modified model to evaluate the ultimate bearing capacity for RC beam members based on a large quantity of numerical simulation results utilizing the LS-DYNA software. They also studied the influences of longitudinal reinforcement ratios and transverse stirrup ratios on the dynamic modified factors of RC beam members. By using the dynamic modified material constitutive model, Wang et al. [64] established a finite element model of RC column members using the OpenSees software. Accordingly, the expressions of DIF (i.e., the ratios of dynamic mechanical behavior parameters to the corresponding static parameters at the member level) for ultimate bearing capacity of columns, considering axial load ratios, concrete strength and longitudinal reinforcement ratios, were obtained. Fan et al. [22] derived the calculation equations for shear strength of concrete and developed a modified model for predicting the dynamic shear bearing capacity of RC joints through multiple linear regression analyses of test data, considering the influences of dynamic effects and axial forces. Based on the dynamic loading test database of RC column members and the Bayesian update theory, Li et al. [33] proposed a probabilistic model of DMC (dynamic modified coefficient) to evaluate the yielding and ultimate bearing capacity, effective stiffness and displacement ductility ratios for RC column members under dynamic loading. The proposed modified models were able to accurately and reliably predict the mechanical behaviors of column members under seismic loading rates. Table 4 lists some of the representative dynamic modified models for RC structural members developed using finite element (FE) simulation or experimental results.

Table 4. Dynamic modified models for mechanical behavior parameters of RC structural members.

Reference	Equations of Dynamic Modified Model	Model Type
Adhikary et al. [8]	Maximum resistance of RC regular beams (1) With transverse reinforcements $DIF = [1.89 - 0.067\rho_g - 0.42\rho_v - 0.14(a/d)]e^{[-0.35-0.052\rho_g+0.179\rho_v+0.18(a/d)]\delta}$ (2) Without transverse reinforcements $DIF = [0.004\rho_g + 0.136(a/d) - 0.34]\log_e \delta + [0.009_{\rho_g} + 0.41(a/d) + 0.157]$	FE simulation results-based (Deterministic)
Adhikary et al. [15]	Maximum resistance of RC deep beams (1) With transverse reinforcements $DIF = [1.25 - 0.04\rho_g - 0.13\rho_v + 0.05(\frac{a}{d})]e^{[0.22-0.03\rho_g-0.17\rho_v+0.03(a/d)]\delta}$ (2) without transverse reinforcements $DIF = [0.45 + 0.09 + 0.48(\frac{a}{d})]e^{[0.30-0.05\rho_g-0.05(a/d)]\delta}$	FE simulation results-based (Deterministic)
Wang [64]	Ultimate bearing capacity of RC columns (1) Different axial load ratio $DIF = 1.0 + c_n \lg\frac{\dot{\varepsilon}_d}{\dot{\varepsilon}_s} \quad c_n = 0.1426n^2 - 0.0614n + 0.0337$ (2) Different concrete strength conditions $DIF = 1.0 + c_f \lg\frac{\dot{\varepsilon}_d}{\dot{\varepsilon}_s} \quad c_f = 1 \times 10^{-4} f_c^2 - 0.068 f_c + 0.153$ (3) Different longitudinal reinforcement ratios $DIF = 1.0 + c_\rho \lg\frac{\dot{\varepsilon}_d}{\dot{\varepsilon}_s} \quad c_\rho = 0.0129\rho^2 - 0.0643\rho + 0.1182$	FE simulation results-based (Deterministic)
Li et al. [33]	Mechanical behavior parameters of RC columns (including yielding and ultimate bearing capacity, effective stiffness and ductility coefficient) $DMC_j(\mathbf{x},\mathbf{\Theta}) = \sum_{i=1}^{6}\theta_i h_i(\mathbf{x}) + \sigma\varepsilon = \theta_1 f_y/f'_c + \theta_2 n_0 + \theta_3\lambda + \theta_4\rho_l + \theta_5\rho_s + \theta_6\lg(\dot{\varepsilon}_d/\dot{\varepsilon}_0) + \sigma\varepsilon$	Experimental date-based (Probabilistic)
Fan [74]	Shear bearing capacity of RC joints $DIF = 0.99679 + 0.1536n + 0.02326\lg\frac{\dot{\varepsilon}}{\dot{\varepsilon}_0}$	Experimental date-based (Deterministic)
Yan [45]	Elasticity modulus of RC beams (1) With transverse reinforcements $\frac{E_d}{E_s} = 1.3247(\dot{\varepsilon})^{0.027}$ (2) Without transverse reinforcements $\frac{E_d}{E_s} = 1.2486(\dot{\varepsilon})^{0.0213}$	Experimental date-based (Deterministic)
Song [5]	Dynamic increase factor in flexural strength of RC column $DIF_m \approx DIF_s \times \frac{1-\frac{1}{2}\frac{\sigma_y}{f_c}\frac{DIF_s}{DIF_c}\rho_s+\frac{1}{2}\frac{\sigma'_y}{f_c}\rho'_s-\eta}{1-\frac{1}{2}\frac{\sigma_y}{f_c}\rho_s+\frac{1}{2}\frac{\sigma'_y}{f_c}\rho'_s-\eta}$	FE simulation results-based (Deterministic)
Rouchette et al. [34]	Simplified formula for mid-span deflection of RC beams under impact loading $Di = Ds \times (1 + \frac{1.77E+18}{c^2}V^2)$	FE simulation results-based (Deterministic)

Note: the meaning of symbols in the each dynamic modified model can be referred from the relevant references.

3.3. Discussion on Dynamic Modified Models

To accurately evaluate the dynamic behaviors of RC structural members, quite a large number of research works have focused on the establishment of dynamic constitutive models and DIF models of concrete and reinforcing steel materials, as well as the development of dynamic modified models at the member level. Strictly speaking, many of these studies were carried out using methods that were partially theoretical and partially empirical. As such, they cannot be separated from experimental tests. Drawbacks of the available research works included: (1) The most commonly used method to determine dynamic effects on RC structural members is to modify static constitutive model parameters using DIF models at the material level. However, whether dynamic modification at material levels can effectively reflect dynamic effects on the mechanical behaviors of structural members has not been adequately verified. (2) The usage of dynamic modified models proposed at the member level provides a direct and efficient approach, reflecting the influences of dynamic effects on the mechanical behaviors of RC structural members. Due to inadequate test data, the suitability and accuracy of the models need to be improved. (3) The mechanisms underlying the dynamic effects on the mechanical behaviors of structural members remains an unsolved problem; it must be thoroughly investigated.

4. Numerical Studies on Dynamic Behaviors of RC Structural Members

4.1. Overview of Numerical Studies Considering Dynamic Effect

To date, the dynamic behaviors of RC structural members and structures have been numerically investigated, considering dynamic effects, by many researchers. The merits of numerical simulations, with respect to experimental tests, are primarily that they require fewer human and material resources, that they are repeatable, and that they can be applied to a broader range of structural parameters and loading rates. The computational accuracy and reliability of numerical results are directly dependent on the methods used to simulate structural dynamic behaviors.

Two currently-available methods through which to consider dynamic effects in numerical simulation of RC structures, i.e., the dynamic constitutive model (DCM) method and the dynamic increase factor (DIF) method [28]. The DCM method requires tedious and time-consuming computation; thus, it is used less frequently in engineering practice and research. The DIF method has been more frequently adopted by researchers. Quite a few studies have used this method to investigate the influences of dynamic effects on the seismic behaviors of RC members and structures [131–139]. The disadvantages of the DIF method include [28]: (1) it cannot fully reflect the adverse impacts of dynamic effects on the structural displacement ductility and performance degradation; (2) the influence of dynamic effects on the shear and bond-slip behaviors of RC structural members has generally been neglected.

Moreover, due to randomness in structural members (e.g., geometric sizes, material properties and reinforcement conditions) and external dynamic loads, a few attempts have been made to consider dynamic effects in a probabilistic manner [140]. Simplified or alternative methods for considering strain rates in materials have been proposed by researchers [25,141,142]. Through numerical simulations, the effectiveness and reliability of the proposed numerical models and methods have been validated with test data, and the influences of dynamic effects on the seismic behaviors of RC members and structures have been more comprehensively investigated.

4.2. Numerical Model for Simulating Structural Dynamic Behaviors

4.2.1. Finite Element Model Considering Dynamic Effect

To aid in developing reasonable FE models for RC structural members, different materials, element types and modeling techniques have been adopted by researchers, based on available FE software or self-compiled programs. Table 5 summarizes basic information on FE models (i.e., member type, element type, parameter and numerical effectiveness) of RC structural members subjected to dynamic loading rates.

Table 5. Summary of FE models on RC structural members subjected to varying loading rates.

Reference	Type	Elements	Parameter	Effectiveness
Wang [64]	Column	Solid element and truss element	Strain rates	Correlation between strain and strength under unidirectional dynamic loading test.
Wang [26]	Column	Three-dimensional fiber beam and birth–death element	Loading scheme Strain rate	User material subroutine for RC structural members considering the strain rate effect of materials.
Liu and Li [27]	Column	Three-dimensional fiber beam and birth–death element	Strain rates Damage	The dynamic behaviors of RC beams and column members.
Adhikary et al. [10]	Beam	Solid and beam element	Strain rates Inertia Longitudinal reinforcing ratio Stirrup ratio Shear span ratio Dynamic shear resistance	The dynamic shear resistance of RC deep beams was found to increase as the loading rates were increased.

Table 5. *Cont.*

Reference	Type	Elements	Parameter	Effectiveness
Zhao et al. [139]	Beam	Solid and Hughes–Liu beam elements	Strain rates Beam span Shear Impact mass Reinforcement ratio Sectional dimension	The resistance characteristics of localized shear failure of RC beam members subjected to varying loading rates.
Wang [64]	Column	Fiber beam-column element with plastic hinges	Strain rates Shear Bond-slip Axial compression ratio Longitudinal reinforcement ratio Shear span ratio Concrete strength	Reflected the bearing capacity and stiffness degradation of structural members under different loading rates.
Shi et al. [143]	Column	One-dimensional slide line model	Strain rates Shear Slip Damage	The blast-induced dynamic responses of RC column members considering the bond shear modulus, maximum elastic slip strain and damage curve exponential coefficient.
Rouchette et al. [34]	Beam	3-D spar element, solid element, bond-link element	Strain rates Corroded steel bar Flexural Bond-slip Impact mass Beam geometry Concrete strength Reinforcement ratio The solicitation force	Simulated the flexural behavior of reinforced concrete beams considering the bond between concrete and steel bar under impact loading. The accuracy of the FE numerical model could be improved, as compared with the no-bond-slip model.
Valipour et al. [131]	Beam	Fiber element	Strain rates Shear Impact mass	Dynamic analysis of reinforced concrete beams subjected to high strain rate loads considering the possible failure of shear.
Guner and Vecchio [144]	Shear wall	Secant-stiffness-based finite-element algorithm	Strain rates Shear	A simplified method for the dynamic analyses of shear-critical RC frame members under impact and seismic load. The influences of dynamic effects and the shear effect were incorporated based on the DIF models and the rotating smeared crack approach.
Jia et al. [137]	Beam	2DOF model	Strain rates Flexural Shear Impact mass Reinforcement ratio Concrete strength	Predicted the possible failure modes (i.e., the punching shear, shear, flexure, flexure-shear and instability) of RC structural members subjected to low-velocity impact load.
Adhikary et al. [15]	Beam	Hughes–Liu beam element and solid element	Strain rates Shear Bond-slip Impact mass	The relationship between failure mode and impact mass of RC beam members under impact load.
Li et al. [145]	Beam	Hughes–Liu beam element with 2 × 2 Gauss quadrature	Strain rates Impact energy Inclination angle of drop weight Concrete strength	Investigated the dynamic behavior of beams subjected to impact loading rates. The influences of dynamic effects and excessive distortion due to large deformations under impact loads were incorporated, based on the DIF models and a method to automatically remove the distorted elements, based on predefined criteria.

Table 5. *Cont.*

Reference	Type	Elements	Parameter	Effectiveness
Yang [138]	Shear wall	Solid and truss element	Strain rates Shear span ratio Reinforcement ratio Failure mode	Mechanical property and failure mode subjected to dynamic loading rates.
Song and Zhang [18]	Shear wall	Solid and truss element	Strain rates Shear span ratio Axial compression ratio	The response of RC shear wall with different shear span ratios and axial compression ratios under quasi-static load and dynamic load with high strain rate.

A summary is shown in Table 5, above, demonstrating finite element models of RC structural members subjected to varying loading rates. Most scholars drew unanimous conclusions with their experiments. Similar to the experiment, beam and column members are mostly investigated. Otherwise, RC structural members exhibited different failure modes, cracking patterns, and damages upon being subjected to static and dynamic loading rates. Many scholars have paid attention to these behaviors. Studies, like those above, have also demonstrated the efficacy of numerical analyses.

Based on the ABAQUS software, the detached model of RC column members was established by one of the authors [64], using the solid element and the truss element, respectively, to simulate concrete and reinforcing steel. The dynamic effects were included through modification of static material parameters in the damage plastic model of concrete and the ideal elastoplastic model of reinforcement with the corresponding DIF models [29,30]. For simplicity, the measured strain of longitudinal reinforcement at the bottom was used to derive the strain rate of the whole column member. Wang et al. [26] developed the user material subroutine for concrete and reinforcement, considering the strain rate effects of the materials. It was suitable for use with the three-dimensional fiber beam element on the ABAQUS software and could be further applied to nonlinear dynamic analyses and progressive collapse assessments of RC and steel structures. On the basis of this research work, the model was refined by Liu et al. [27], who incorporated the strength and stiffness deterioration levels induced by accumulated damage to the material. These have been shown to provide better simulation results for dynamic behaviors of RC beams and column members. The effectiveness of the subroutine and the proposed beam-column element (Figure 8a) were also verified by Zhang et al. [136], who numerically simulated the dynamic responses of a shaking table test frame structure. The DIFs of micro-concrete and iron wire, developed on the test data (Figure 8b), were used to represent the material dynamic properties in the beam-column models. Based on the ABAQUS software, the responses of RC shear walls with different shear span ratios and axial compression ratios, under quasi-static load and dynamic load with high strain rate, were studied. The failure modes and bearing capacities of shear walls under various shear span ratios, axial compression ratios and strain rates were also compared (Figure 9) by Zhang [17].

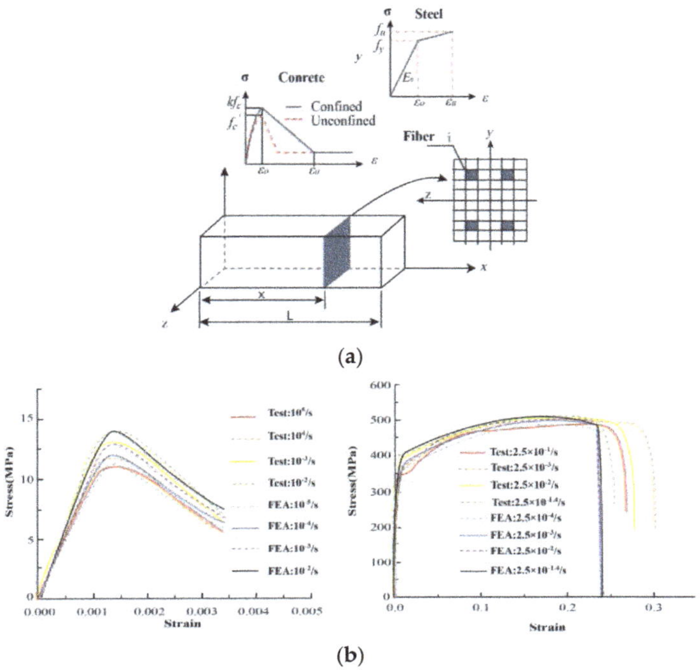

Figure 8. Schematic plot of dynamic fiber model for RC beam-column members employing the user material subroutine. (**a**) The proposed fiber beam-column element; (**b**) Stress–strain curves of micro-concrete and iron wire at different strain rates [136].

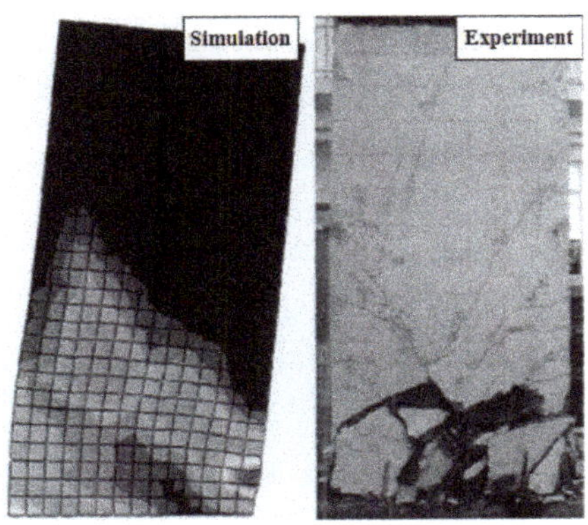

Figure 9. Damage to RC shear wall obtained from test and simulations [17].

Using LS-DYNA software, Adhikary et al. [8,15] established a three-dimensional numerical model for simulating the dynamic behaviors of RC beam members subjected to varying loading rates. The solid element and the beam element were adopted, respectively, for concrete and reinforcement. The material models in the software were used with the further incorporation of strain rate effects. Through numerical modeling, the load versus mid-span deflection and the cracking patterns of RC beam members were captured (Figure 10). Due to the assumption of complete compatibility of strains between concrete and steel, the bond-slip was not considered in this study. A similar method for development of dynamic numerical modes was proposed by Zhao et al. [139]. Moreover, a simplified three-degree-of-freedom (TDOF) model was proposed to facilitate investigation of the dynamic shear behavior of RC beam members subjected to impact loading.

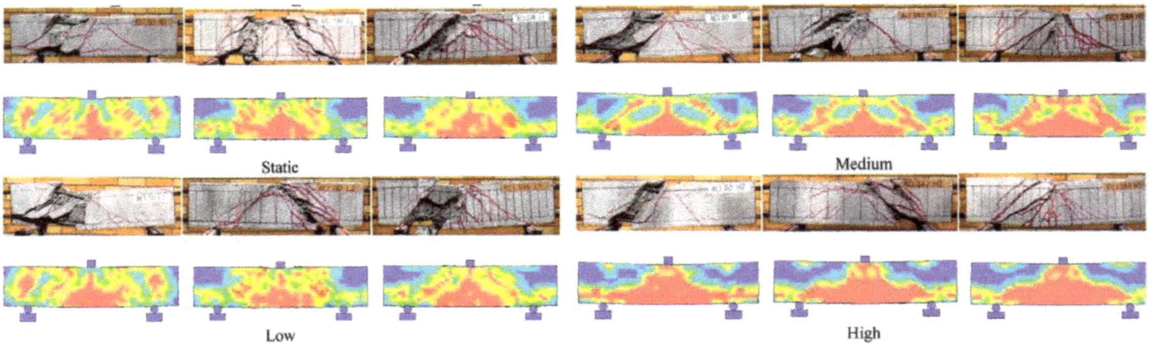

Figure 10. Comparisons between numerical and test results on cracking patterns for RC beam members under different loading rates [8].

It was noted that, for most of the numerical models considering dynamic effects, perfect bonds between the concrete and reinforcement materials were generally assumed. A few studies were conducted based on the establishment of dynamic numerical models for RC structural members. Using OpenSees software, a serial element model was developed by the authors of [64], using a fiber beam-column element with plastic hinges. The dynamic numerical model incorporated the shear and bond-slip springs. It was able to accurately reflect the bearing capacity and stiffness degradation values of structural members under different loading rates. Using LS-DYNA software, a one-dimensional slide line contact model was proposed by Shi et al. [143] that focused on modeling the blast-induced dynamic responses of RC column members, considering the bond-slip effect (Figure 11). A 3-D mesoscale numerical model was established by Jin et al. to investigate the impact resistances of RC beams under different combinations of mass and velocity [146]. The effects of the combination of impact mass and velocity on the failure modes of RC beams were simulated and compared with experimental results (Figure 12). Based on the available material models in the LS-DYNA software, Rouchette [34] further incorporated the strain rate effect and used two orthogonal springs to simulate the bond-slip between concrete and steel. It was found that the accuracy of the FE numerical model, as compared with the no-bond-slip model, left room for improvement.

In addition to the above mentioned research works, several scholars focused on modeling the shear failure of RC structural members under dynamic loading rates. Valipour et al. [131] used a fiber element to establish a numerical model and investigate the dynamic responses of RC beams and columns. DIF models were adopted to consider dynamic effects at the fiber level, and the shear cap was introduced at the section level to consider possible shear failure (Figure 13). Guner and Vecchio [144] developed a simplified method for dynamic analyses of shear-critical RC frame members under impact and seismic loads. In this study, the influences of dynamic effects and the shear effect were incorporated based on the DIF models and the rotating smeared crack approach, respectively. Recently, after introducing the combined dynamic flexural and shear resistance function, an improved two-degree-of-freedom (2DOF) model (Figure 14) was proposed by Jia et al. [137] that aimed to predict possible failure modes (i.e., punching shear, shear, flexure, flexure-shear and instability) of RC structural members subjected to low-velocity impact loads.

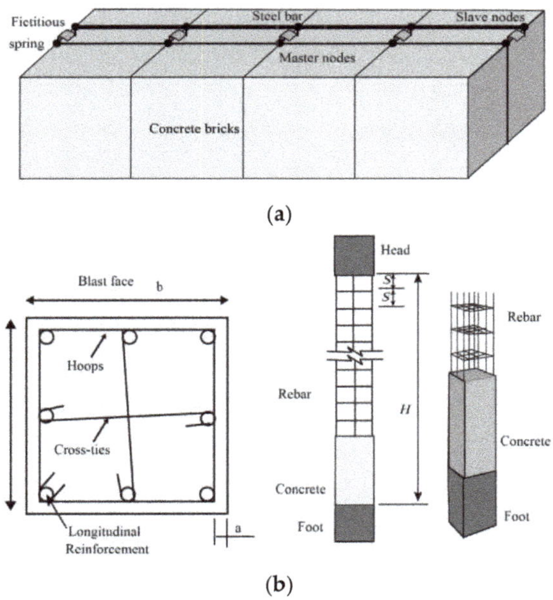

Figure 11. Schematic plot of numerical model for RC column members considering the dynamic effect and bond-slip between concrete and steel. (**a**) Sketch of fictitious spring between master and slave nodes in one-dimensional slide line model. (**b**) Detached numerical model for RC columns [143].

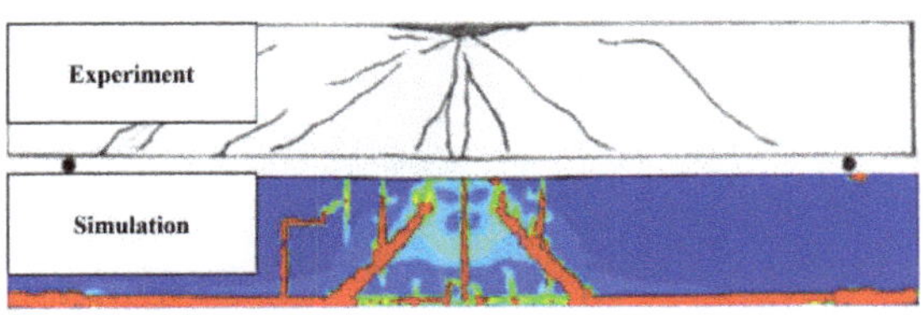

Figure 12. Failure patterns of RC beams obtained from tests and simulations [146].

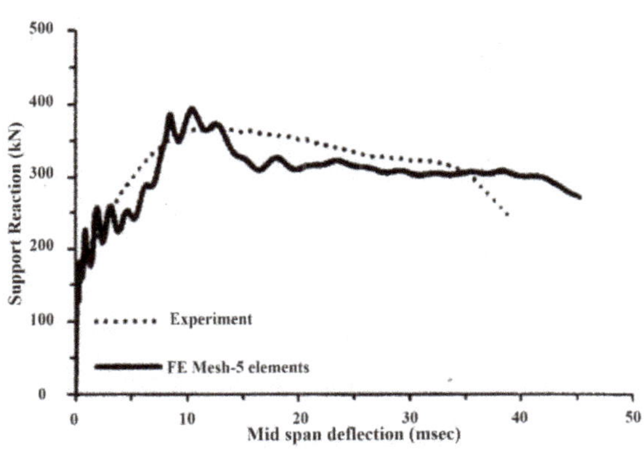

Figure 13. Comparison of experimental results and numerical simulations of mid-span deflection versus support reaction for RC beam [131].

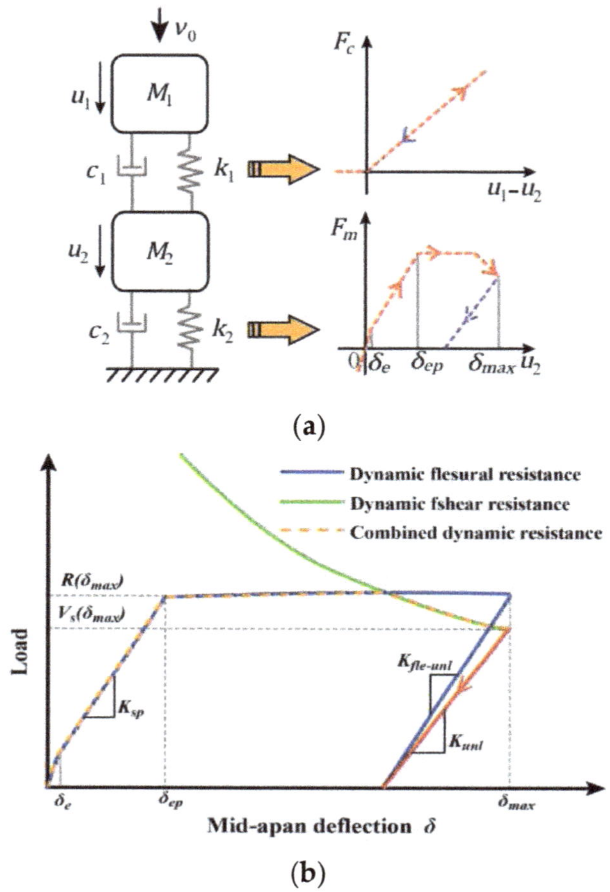

Figure 14. Schematic plot of 2DOF model for RC structural members considering the dynamic effect and different failure modes. (**a**) 2DOF numerical model; (**b**) Combined dynamic flexural and shear resistance function [137].

4.2.2. Hysteretic Model Considering Dynamic Effect

The hysteretic model was obtained by describing the load-deformation curve with skeleton and loading and unloading rules. Classical models include Clough [147], Takeda model [148], Ozcebe model [149], Park model [150], Bouc-Wen model [151] and others. With the continuous in-depth research on restoring force characteristics of structural members, researchers have determined that the degradation of bearing capacities and stiffness caused by material damage under cyclic dynamic loads significantly affects the structural seismic performance [33]. Many hysteretic models have been proposed which considered different degradation effect factors, including strength degradation, stiffness degradation, pinching effect and negative stiffness segment. A summary is shown in Table 6.

Table 6. Available hysteretic model considering different degradation effect factors.

Degradation Effect Factors		Relevant Studies
Single factor	• Stiffness degradation	Clough [147], Takeda [148], Wen [151], Takayanagi and Schnobrich [152], Saatcioglu et al. [153], Xu [154], Qu and Ye [155],
	• Pinching effect	Ambrisi and Filippou [156]
Multiple factor	• Stiffness degradation • Strength degradation	Gu et al. [157], Zheng et al. [158], Zheng et al. [159], Erberik [160], Wang et al. [161]
	• Stiffness degradation • Strength degradation • Pinching effect	Park and Ang [150], Ozcebeand Saatcioglu [149], Dowell et al. [162], Mostaghel and Byrd [163], Yan et al. [164], Wang et al. [165], Yu et al. [166], Sezen and Chowdhury [167], Leborgne and Ghannoum [168], Chao and Loh [169], Guo and Yang [170], Yu et al. [171], Cai et al. [172], Zhao and Dun [173], Huang et al. [174]
	• Stiffness degradation • Strength degradation • Negative stiffen • Pinching effect	Song and Pincheira [175], Ibarra et al. [176], Guo and Long [177], Li [33]

In addition to the above models of macroscopic force-displacement, hysteretic models of RC structural members reflecting different deformation mechanisms have been proposed by researchers [178,179]. It should be mentioned that the available hysteretic models were basically developed without consideration of dynamic effects. Under dynamic loading, hysteretic behaviors of RC structural members differ greatly from those under static loading. To consider the influences of dynamic effects, an effective approach might be to establish dynamic hysteretic models based on available dynamic loading test data. Li et al. [28] developed a damage index-based dynamic hysteretic model for RC column members (Figure 15), taking into account dynamic effects as well as different degradation modes (i.e., strength degradation, stiffness degradation, pinching effect and negative stiffness segment). Combined with the usage of concentrated plastic hinge elements, the numerical model could be applied to structural dynamic analyses considering the real-time dynamic effects and seismic degradation of members.

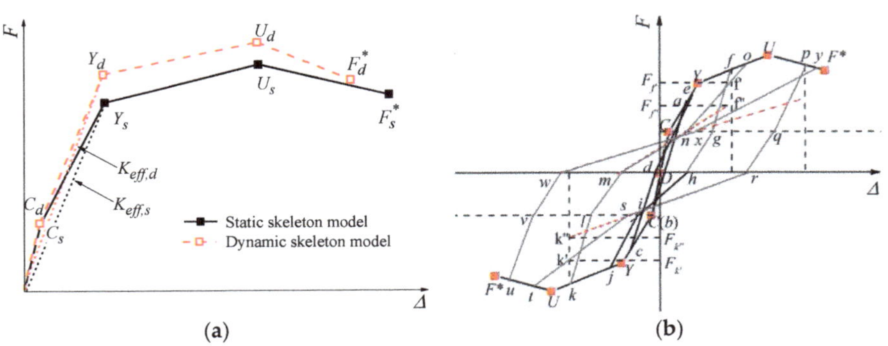

Figure 15. Illustration of hysteretic model for RC structural members considering dynamic effects and different degradation modes. (**a**) Static and dynamic skeleton models; (**b**) Hysteretic rules [28].

4.3. Discussion on Numerical Simulation Works

According to numerical simulation results involvingRC structural members under dynamic loading [64,129,132,134,136,180], it was concluded that, with increased loading rates, bearing capacities and stiffness values of structural members were enhanced, whereas deformation ductility were likely to be decreased. These numerical findings were in good agreement with available dynamic loading test observations. It has also been shown, in multiple studies, that cracking patterns, damage and failure modes can be accurately reflected through numerical analysis [64,134,137,144]. Moreover, parametric studies have been carried out to investigate the influences of structural parameters on the dynamic behaviors of RC structural members [16,64,129,133].

As for numerical analyses at overall structural levels, it was demonstrated by available studies that the measured dynamic responses of RC structures in experiments could be more accurately predicted if dynamic effects were included for consideration [136,181]. More importantly, dynamic effects could exert significant influence on seismic responses, collapse assessments and fragility analyses of RC structures [1,25,135,182–185].

In terms of numerical simulation, some shortcomings remain: (1) most of the FE numerical models have failed to effectively consider the shear and bond-slip behaviors between concrete and reinforcement of RC structural members. Due to the lack of relevant models for RC members under dynamic loading rates, there is a need to develop numerical models of RC structural members that effectively considering dynamic effects on shear and bond-slip behaviors. (2) The development of hysteretic models is largely dependent on limited dynamic loading test data and mathematical simplification. Thus, it will be necessary to improve model applicability and computational efficiency. (3) Employing refined numerical models and methods that consider the dynamic effects, further works must be undertaken to reveal the seismic damage evolution and failure mechanisms of RC structural members and structures.

5. Concluding Remarks

As RC buildings have been widely constructed and used in civil engineering, enhancing their seismic performances and improving the accuracy of seismic evaluations would play a very important role in reducing the huge human and economic losses induced by earthquakes. The relatively large strain rates found in RC materials may be observed in structural members under seismic load and compared with those observed under static load. Meanwhile, the strain rate-sensitivity of materials could result in changes in the dynamic behaviors of RC structural members—changes that must not be neglected. However, most current seismic designs and dynamic analyses of RC structures have been based on a large number of quasi-static experimental results without considering dynamic effects. Moreover, there is still a lack of consensus regarding whether the dynamic effects of RC members need to be considered for more reliable structural design and analysis. To date, a large number of experimental and numerical studies have focused on the dynamic behaviors of RC structural members under impact and blast load. During the past few decades, many research works have focused on the dynamic behaviors of RC structural members subjected to seismic-induced loading rates. In this paper, research progress on this topic was comprehensively reviewed from experimental, theoretical and numerical perspectives. The main conclusions have been summarized as follows:

(1) According to the statistical results of available experiments on RC structural members under dynamic loading rates and seismic load, many tests have been performed on RC beams and column members under uniaxial loading schemes and static and dynamic loading rates. As compared with high loading rate tests, the experiments under median loading rates have been inadequate.

(2) In several experimental studies, structural parameters were designed to be different in order to facilitate investigation of their influences on the dynamic behaviors of RC structural members. Most dynamic loading tests measured bearing capacity, displacement, strain, crack development and failure patterns. In addition, seismic damage and energy dissipation were indirectly acquired in a number of experiments.

(3) Based on the results of available dynamic loading tests, the following conclusion was reached: with increased loading rates, the bearing capacity, stiffness and energy dissipation capacity of members were enhanced, while ductility might be reduced, and the degradation of stiffness and bearing capacity aggravated. As for failure mode, research findings have not led to consistency or consensus.

(4) To reflect the influences of loading rates on the mechanical properties of RC materials, the DIF models established on the dynamic loading tests have been the most widely

used. By summarizing the DIF models for concrete and reinforcing steel, it was determined that the mechanical behavior parameters for general dynamic modification included compressive strength, tensile strength, elastic modulus of concrete and the yielding strength and ultimate strength of reinforcing steel.

(5) The mechanism of dynamic effects on RC structural members under seismic load could be explained by the strain rate-sensitivity of materials, the inertial effects of members and evolutions of micro-cracks. However, few research works have focused on this issue. Dynamic modified models for mechanical behavior parameters of RC structural members have been developed using finite element (FE) simulation or experimental results. These models considered the influences of loading rates and different structural parameters, and could be directly applied to estimate the dynamic behaviors of RC structural members.

(6) Base on available FE software and self-compiled programs, various numerical methods have been undertaken by researchers to establish FE models to simulate the dynamic behaviors of RC structural members under different loading rates. Moreover, the dynamic hysteretic model established on the dynamic loading test data provided an effective approach to reasonably consider the influences of dynamic effects.

(7) Through comparison with the test data, it was noted that more accurate results could be obtained using numerical models and methods that considered dynamic effects. In a few studies, cracking patterns, damage and failure modes of RC structural members were accurately captured through numerical simulations. Moreover, numerical studies could be applied to a broader range of structural parameters and loading rates, facilitating parametric analyses of the dynamic behaviors of RC structural members.

Given the research gaps in the available literature, the following could be suggested directions for future research:

(1) For dynamic loading tests, more research on RC structural members subjected to multidimensional dynamic loads should be carried out. Moreover, more tests should focus on the influence of dynamic effects on the deformation and damage mechanisms of structural members. Furthermore, in-depth studies are required to elucidate the influence of dynamic effects on structural members with different parameters and failure modes.

(2) Among dynamic modified models, DIF models are the most commonly used to consider the impact of dynamic effects on RC structural members. Due to randomness in structural members and external dynamic loads, the capability of dynamic modification, at the material level, to reliably reflect dynamic effects at the member level should be verified. In addition, the suitability and accuracy of the models proposed at the member level need to be improved based on supplementary data test data and advanced theoretical methodologies.

(3) For numerical simulation analysis, researchers should refine the available FE numerical models of RC structural members by incorporating shear and bond-slip behaviors with their consideration of dynamic effects. Moreover, more effort should be applied to improving model applicability and computational efficiency. Furthermore, the seismic damage evolution and failure mechanisms of RC structural members and structures must be deeply investigated, utilizing refined models and methods for numerical simulation.

Author Contributions: Conceptualization, H.L. and C.L.; methodology, R.L. and D.W.; writing—original draft preparation, M.G.; writing—review and editing, M.G. and R.L.; visualization, R.L. and M.G.; supervision, H.L. and D.W. All authors have read and agreed to the published version of the manuscript.

Funding: National Natural Science Foundation of China: 52108438, 52078106.

Data Availability Statement: Data sharing not applicable. No new data were created or analyzed in this study. Data sharing is not applicable to this article.

Conflicts of Interest: The authors declare no conflict of interest.

References

1. Cao, X.-Y.; Feng, D.-C.; Li, Y. Assessment of various seismic fragility analysis approaches for structures excited by non-stationary stochastic ground motions. *Mech. Syst. Signal Process.* **2023**, *186*, 109838. [CrossRef]
2. Fu, H.C.; Erki, M.A.; Seckin, M. Review of Effects of Loading Rate on Concrete in Compression. *J. Struct. Eng.* **1991**, *117*, 3645–3659. [CrossRef]
3. Li, H.; Huo, L. *Multiaxial Seismic Dynamic Effect of RC Structure*; Science Press: Beijing, China, 2021.
4. Wang, D.B.; Fan, G.X.; Zhang, H. Dynamic behaviors of reinforced concrete columns under multi-dimensional dynamic loadings. *J. Vib. Shock.* **2016**, *35*, 35–41.
5. Song, Y.; Wang, J.; Han, Q. Dynamic performance of flexure-failure-type rectangular RC columns under low-velocity lateral impact. *Int. J. Impact Eng.* **2023**, *175*, 104541. [CrossRef]
6. Witarto, W.; Lu, L.; Roberts, R.H.; Mo, Y.L.; Lu, X. Shear-critical reinforced concrete columns under various loading rates. *Front. Struct. Civ. Eng.* **2014**, *8*, 362–372. [CrossRef]
7. Reinschmidt, K.F.; Hansen, R.J.; Yang, C.Y. Dynamic tests of reinforced concrete columns. *J. Proc.* **1964**, *61*, 317–334.
8. Adhikary, S.D.; Li, B.; Fujikake, K. Dynamic behavior of reinforced concrete beams under varying rates of concentrated loading. *Int. J. Impact Eng.* **2012**, *47*, 24–38. [CrossRef]
9. Hongnan, L.; Li, M. Dynamic test and numerical simulation of reinforced concrete beams. *J. Vib. Shock.* **2015**, *34*, 110–115.
10. Adhikary, S.D.; Li, B.; Fujikake, K. Effects of high loading rate on reinforced concrete beams. *Aci Struct. J.* **2014**, *111*, 651–660. [CrossRef]
11. Li, M.; Li, H.N. Effects of Loading Rate on Reinforced Concrete Beams. In Proceedings of the 15th World Conference on Earthquake Engineering, Lisboa, Portugal, 24–28 September 2011.
12. Xiao, S.; Zhang, H. Effects of Loading Rate on the Shear Behaviors of RC Beams. *J. Water Resour. Archit. Eng.* **2018**, *16*, 7–13+19.
13. Marder, K.J.; Motter, C.J.; Elwood, K.J.; Clifton, G.C. Effects of variation in loading protocol on the strength and deformation capacity of ductile reinforced concrete beams. *Earthq. Eng. Struct. Dyn.* **2018**, *47*, 2195–2213. [CrossRef]
14. Fujikake, K.; Li, B.; Soeun, S. Impact response of reinforced concrete beam and its analytical evaluation. *J. Struct. Eng.* **2009**, *135*, 938–950. [CrossRef]
15. Adhikary, S.D.; Li, B.; Fujikake, K. Strength and behavior in shear of reinforced concrete deep beams under dynamic loading conditions. *Nucl. Eng. Des.* **2013**, *259*, 14–28. [CrossRef]
16. Xu, B.; Chen, J.M.; Xu, N. Test on strain rate effects and its simulation with dynamic damaded plasticity model for RC shear walls. *Eng. Mech.* **2012**, *29*, 39–287.
17. Zhang, X. Effects of Loading Rates on Seismic Behavior of Reinforced Concrete High Shear Walls. Master's Thesis, Dalian University of Technology, Dalian, China, 2012.
18. Yupu, S.; Xiaoli, Z. Influence of Strain Rate on Earthquake Resistance Effect of Reinforced Concrete Shear Wall. *Ind. Constr.* **2012**, *42*, 7–11.
19. Xu, N. Study on the Performance Test and Numerical Simulation of Reinforced Concrete Shear Wall under Rapid Loading. Ph.D. Thesis, Hunan University, Changsha, China, 2012.
20. Chung, L.; Shah, S.P. Effect of loading rate on anchorage bond and beam-column joints. *Struct. J.* **1989**, *86*, 132–142.
21. Wang, L.; Fan, G.; Qin, Q.; Song, Y. Experimental study on seismic behavior of reinforced concrete frame joints with consideration of strain rate effect. *J. Build. Struct.* **2014**, *35*, 38.
22. Guoxi, F.; Yupu, S.; Licheng, W. Study on the Seismic Performance of Reinforced Concrete Beam-Coulumn Joints under Different Strain Rates. *Ind. Constr.* **2014**, *44*, 1.
23. Pan, H. Effects of Strain Rate Sensitivity on Dynamic Properties of Reinforced Concrete Materials and Beam-Column Joints. Master's Thesis, Dalian University of Technology, Dalian, China, 2012.
24. Qin, Q. Study on Seismic Behavior of Reinforced Concrete Frame Joints with the Consideration of Strain Rate Effect. Master's Thesis, Dalian University of Technology, Dalian, China, 2013.
25. Asprone, D.; Frascadore, R.; Ludovico, M.D.; Prota, A.; Manfredi, G. Influence of strain rate on the seismic response of RC structures. *Eng. Struct.* **2012**, *35*, 29–36. [CrossRef]
26. Wang, W. Study on the Seismic Analysis Method of Structure Considering Strain Rate Effect. Ph.D. Thesis, Dalian University of Technology, Dalian, China, 2013.
27. Liu, H.; Li, H. Seismic-Collapse Analysis of Reinforced Concrete Frame Structure Considering Cumulative Damage and Strain Rate Effect. *Eng. Mech.* **2018**, *35*, 87. [CrossRef]
28. Li, H.N.; Li, R.H.; Li, C.; Wang, D.B. Development of hysteretic model with dynamic effect and deterioration for seismic-performance analysis of reinforced concrete structures. *J. Struct. Eng.* **2020**, *146*, 04020215. [CrossRef]
29. CEB. *CEB-FIP Model Code 1990, First Draft*; CEB Bull. Committee Euro-International du Beton: Lausanne, Switzerland, 1990; p. 195.
30. Lin, F.; Gu, X.; Kuang, X.; Yin, X. Constitutive Models for Reinforcing Steel Barsunder High Strain Rates. *J. Build. Mater.* **2008**, *11*, 14–20.

31. Li, M. Effects of Rate-Dependent Properties of Material on Dynamic Properties of Reinforced Concrete Structural. Ph.D. Thesis, Dalian University of Technology, Dalian, China, 2011.
32. Ross, C.A.; Tedesco, J.W.; Kuennen, S.T. Effects of Strain-Rate on Concrete Strength. *Aci. Mater. J.* **1995**, *92*, 37–47.
33. Li, R. Study on Seismic Performance of Reinforced Concrete Structure Considering Dynamic Effects. Ph.D. Thesis, Dalian University of Technology, Dalian, China, 2020.
34. Rouchette, A.; Zhang, W.P.; Chen, H. Simulation of Flexural Behavior of Reinforced Concrete Beams under Impact Loading. *Appl. Mech. Mater.* **2013**, *351–352*, 1018–1023. [CrossRef]
35. Ghannoum, W.; Saouma, V.; Haussmann, G.; Polkinghorne, K.; Eck, M.; Kang, D.H. Experimental investigations of loading rate effects in reinforced concrete columns. *J. Struct. Eng.* **2012**, *138*, 1032–1041. [CrossRef]
36. Thomas, R.J.; Sorensen, A.D. Review of strain rate effects for UHPC in tension. *Constr. Build. Mater.* **2017**, *153*, 846–856. [CrossRef]
37. Gholipour, G.; Zhang, C.; Mousavi, A.A. State-of-the-Art Review on Responses of RC Structures Subjected to Lateral Impact Loads. *Arch. Comput. Methods Eng.* **2020**, *28*, 2477–2507.
38. Hao, H.; Hao, Y.; Li, J.; Chen, W. Review of the current practices in blast-resistant analysis and design of concrete structures. *Adv. Struct. Eng.* **2016**, *19*, 1193–1223. [CrossRef]
39. Ullah, A.; Ahmad, F.; Jang, H.W.; Kim, S.W.; Hong, J.W. Review of analytical and empirical estimations for incident blast pressure. *KSCE J. Civ. Eng.* **2017**, *21*, 2211–2225. [CrossRef]
40. Das, N.; Nanthagopalan, P. State-of-the-art review on ultra high performance concrete—Ballistic and blast perspective. *Cem. Concr. Compos.* **2022**, *127*, 104383. [CrossRef]
41. Bertero, V.; Rea, D.; Mahin, S.; Atalay, M. Rate of loading effects on uncracked and repaired reinforced concrete members. In Proceedings of the Fifth World Conference on Earthquake Engineering, Rome, Italy, 25–29 June 1973; pp. 1461–1470.
42. Kulkarni, S.M.; Shah, S.P. Response of reinforced concrete beams at high strain rates. *ACI Struct. J.* **1998**, *95*, 705–715.
43. White, T.W.; Soudki, K.A.; Erki, M.A. Response of RC Beams Strengthened with CFRP Laminates and Subjected to a High Rate of Loading. *J. Compos. Constr.* **2001**, *5*, 153–162. [CrossRef]
44. Zhang, X.X.; Ruiz, G.; Yu, R.C. Experimental study of combined size and strain rate effects on the fracture of reinforced concrete. *J. Mater. Civ. Eng.* **2008**, *20*, 544–551. [CrossRef]
45. GuiLan, Y. The Research on Strain Rate Effect of Dynamic Behaviors of Reinforced Concrete Remembers. Master's Thesis, Harbin Institute of Technology, Harbin, China, 2010.
46. Shiyun, X.; Wenbo, C.; Haohao, P. Experimental study on mechanical behavior of reinforced concrete beams at different loading rates. *J. Build. Struct.* **2012**, *33*, 142–146.
47. Zhou, H.; Xing, X.; Xie, Y.; Yuan, H. Experimental Study on Dynamic Loading Size Effect of RC Beam-Column. *J. Basic Sci. Eng.* **2020**, *28*, 1187–1196.
48. Otani, S.; Kaneko, T.; Shiohara, H. *Strain Rate Effect on Performance of Reinforced Concrete Members*; Kajima Technical Research Institute, Kajima Corporation: Tokyo, Japan, 2003.
49. Guo, J.; Cai, J.; Chen, Q.; Liu, X.; Wang, Y.; Zuo, Z. Dynamic behaviour and energy dissipation of reinforced recycled aggregate concrete beams under impact. *Constr. Build. Mater.* **2019**, *214*, 143–157. [CrossRef]
50. Wu, X.-Q.; Zhong, B.; Lv, Y.; Li, Z.-X.; Chouw, N. Experimental Study on Dynamic Amplification Factor of Simple-Supported Reinforced Concrete Beams Under Impact Loading Generated by an Impulse Hammer. *Int. J. Struct. Stab. Dyn.* **2020**, *21*, 2150036. [CrossRef]
51. Adhikary, S.D.; Li, B.; Fujikake, K. Residual resistance of impact damaged reinforced concrete beams. *Mag. Concr. Res.* **2015**, *67*, 364–378. [CrossRef]
52. Zeng, X. Experimental and Numerical Study of Behaviors of RC Beams and Columns under Impact Loadings and Rapid Loadings. Ph.D. Thesis, Hunan University, Changsha, China, 2014.
53. Feng, Z.; Wang, X.; Zhang, S.; Chu, Y. Experimental investigation on cantilever square CFST columns under lateral continuous impact loads. *J. Constr. Steel Res.* **2022**, *196*, 107416. [CrossRef]
54. Mutsuyoshi, H.; Machida, A. Dynamic properties of reinforced concrete piers. In Proceedings of the 8th World Conference on Earthquake Engineering, San Francisco, CA, USA, 21–28 July 1984.
55. Fukuda, T.; Sanuki, S.; Miyakawa, M.; Fujikake, K. Influence of loading rate on shear failure resistance of RC beams. *Appl. Mech. Mater.* **2011**, *82*, 229–234. [CrossRef]
56. Yuan, J.; Weijian, Y.I. Tests for effects of loading rate on shear behaviors of RC beams. *J. Vib. Shock.* **2019**, *38*, 119–127.
57. Ye, J.-B.; Cai, J.; Chen, Q.-J.; Liu, X.; Tang, X.-L.; Zuo, Z.-L. Experimental investigation of slender RC columns under horizontal static and impact loads. *Structures* **2020**, *24*, 499–513. [CrossRef]
58. Xiang, S.; Zeng, L.; Liu, Y.; Mo, J.; Ma, L.; Zhang, J.; Chen, J. Experimental study on the dynamic behavior of T-shaped steel reinforced concrete columns under impact loading. *Eng. Struct.* **2020**, *208*, 110307. [CrossRef]
59. Gutierrez, E.; Magonette, G.; Verzeletti, G. Experimental studies of loading rate effects on reinforced concrete columns. *J. Eng. Mech.* **1993**, *119*, 887–904. [CrossRef]
60. Bousias, S.N.; Verzeletti, G.; Fardis, M.N.; Gutierrez, E. Load-Path Effects in Column Biaxial Bending with Axial Force. *J. Eng. Mech.* **1995**, *121*, 596–605. [CrossRef]
61. Li, B.; Park, R.; Tanaka, H. Constitutive behaviour of high strength concrete under dynamic loading. *ACI Struct. J.* **2000**, *97*, 619–629.

62. Perry, S.H.; Alshaikh, A.H.; Cheong, H.K. Influence of strain rate on laterally confined concrete columns subjected to cyclic loading. *J. Mater. Res.* **1986**, *1*, 382–389. [CrossRef]
63. Zou, D.J.; Liu, T.J.; Teng, J.; Yan, G.L. Strain rate effect on uniaxial compressive behaviour of concrete columns. *J. Vib. Shock.* **2012**, *31*, 145–150.
64. Wang, D.B. Dynamic Effects on Seismic Behavior of Reinforced Concrete Column. Ph.D. Thesis, Dalian University of Technology, Dalian, China, 2013.
65. Jiang, Z. Dynamic Loading Experiment and Numerical Simulation Study on Reinforced Concrete Columns. Master's Thesis, Hunan University, Changsha, China, 2012. (In Chinese).
66. Liu, B.; Fan, W.; Guo, W.; Chen, B.S.; Liu, R. Experimental investigation and improved FE modeling of axially-loaded circular RC columns under lateral impact loading. *Eng. Struct.* **2017**, *152*, 619–642. [CrossRef]
67. Liu, L.; Zong, Z.; Ma, Z.J.; Qian, H.; Gan, L. Experimental Study on Behavior and Failure Mode of PSRC Bridge Pier under Close-In Blast Loading. *J. Bridge Eng.* **2021**, *26*, 04020124. [CrossRef]
68. Lee, T.-H.; Choi, S.-J.; Yang, D.-H.; Kim, J.-H.J. Experimental Seismic Structural Performance Evaluations of RC Columns Strengthened by Stiff-Type Polyurea. *Int. J. Concr. Struct. Mater.* **2022**, *16*, 65. [CrossRef]
69. Wei, J.; Li, J.; Wu, C. An experimental and numerical study of reinforced conventional concrete and ultra-high performance concrete columns under lateral impact loads. *Eng. Struct.* **2019**, *201*, 109822. [CrossRef]
70. Fan, W.; Shen, D.; Yang, T.; Shao, X. Experimental and numerical study on low-velocity lateral impact behaviors of RC, UHPFRC and UHPFRC-strengthened columns. *Eng. Struct.* **2019**, *191*, 509–525. [CrossRef]
71. Orozco, G.; Ashford, S. *Effects of Large Velocity Pulses on Reinforced Concrete Bridge Columns*; PEER Report 2002/23; Pacific Earthquake Engineering Research Center, University of California: Berkeley, CA, USA, 2002.
72. Shah, S.P.; Wang, M.L.; Lan, C. Model concrete beam-column joints subjected to cyclic loading at two rates. *Mater. Struct.* **1987**, *20*, 85–95. [CrossRef]
73. Gibson, N. *Performance of Beam to Column Bridge Joints Subjected to a Large Velocity Pulse*; PEER Report 2002/24; Pacific Earthquake Engineering Research Center, University of California: Berkeley, CA, USA, 2002.
74. Fan, G.X. Study on the Dynamic Mechanical Properties of Reinforced Concrete Beam-Column Joints. Ph.D. Thesis, Dalian University of Technology, Dalian, China, 2015.
75. Wang, L.; Qin, Q.; Fan, G.; Song, Y. Finite element analysis on mechanical behavior of reinforced concrete beam-column joint considering strain rate effect. *J. Build. Struct.* **2014**, *35*, 131.
76. Chiu, C.K.; Sung, H.F.; Chiou, T.C. Experimental quantification of seismic damage for RC infill walls using static and dynamic loading test. *J. Build. Eng.* **2022**, *50*, 104177. [CrossRef]
77. Yılmaz, T.; Kıraç, N.; Anil, Ö.; Erdem, R.T.; Kaçaran, G. Experimental Investigation of Impact Behaviour of RC Slab with Different Reinforcement Ratios. *KSCE J. Civ. Eng.* **2019**, *24*, 241–254. [CrossRef]
78. Wang, D.B.; Li, H.N.; Li, G. Experimental study on dynamic mechanical properties of reinforced concrete column. *J. Reinf. Plast. Compos.* **2013**, *32*, 1793–1806. [CrossRef]
79. Long, Y. Study on Fast Loading Test and Dynamic Hysteresis Law of Reinforced Concrete Column. Master's Thesis, Hunan University, Changsha, China, 2010.
80. Zhang, H. Strain Rate Effect of Materials on Seismic Response of Reinforced Concrete Frame-Wall Structure. Master's Thesis, Dalian University of Technology, Dalian, China, 2012.
81. Liu, L. Study on Seismic Behavior of RC Columns in Flexural-Shear Failure. Ph.D. Thesis, Hunan University, Changsha, China, 2018.
82. Wang, D.B.; Li, H.N.; Li, G. Experimental tests on reinforced concrete columns under multi-dimensional dynamic loadings. *Constr. Build. Mater.* **2013**, *47*, 1167–1181. [CrossRef]
83. Wang, D.; Li, H. Effects of strain rate on dynamic behavior of reinforced concrete column. *J. Earthq. Eng. Eng. Vib.* **2011**, *31*, 67–72.
84. Zielinski, A.J.; Reinhardt, H.W.; Körmeling, H.A. Experiments on concrete under uniaxial impact tensile loading. *Matériaux Constr.* **1981**, *14*, 103–112. [CrossRef]
85. Yon, J.H.; Hawkins, N.M.; Kobayashi, A.S. Strain-rate sensitivity of concrete mechanical properties. *ACI Mater. J.* **1992**, *89*, 146–153.
86. Cadoni, E.; Labibes, K.; Berra, M. High-strain-rate tensile behaviour of concrete. *Mag. Concr. Res.* **2000**, *52*, 365–370. [CrossRef]
87. Lin, G.; Yan, D.; Yuan, Y. Response of concrete to dynamic elevated-amplitude cyclic tension. *ACI Mater. J.* **2007**, *104*, 561.
88. Chen, X.; Bu, J.; Xu, L. Effect of strain rate on post-peak cyclic behavior of concrete in direct tension. *Constr. Build. Mater.* **2016**, *124*, 746–754. [CrossRef]
89. Yan, D.; Lin, G.; Wang, Z.; Zhang, Y. A Study on Direct Tensile Properties of Concrete at Different Strain Rates. *China Civ. Eng. J.* **2005**, *38*, 97–103.
90. Xiao, S.; Tian, Z. Experimental study on the uniaxial dynamic tensile damage of concrete. *China Civ. Eng. J.* **2008**, *41*, 14–20.
91. Chen, Y.; Wang, Y.; Chen, X. Experimental research on effects of strain rate on tensile stress-strain curve of concrete in axial tension condition. *Build. Struct.* **2015**, *45*, 4.
92. Yan, D.M.; Lin, G. Experimental Study on the Uniaxial Dynamic Compression Properties of Concrete. *Water Sci. Eng. Technol.* **2005**, *6*, 8–10.
93. Goldsmith, W. Dynamic behavior of concrete. *Exp. Mech.* **1966**, *6*, 65–79. [CrossRef]

94. Watson, A.J.; Hughes, B.P. Compressive strength and ultimate strain of concrete under impact loading. *Mag. Concr. Res.* **1978**, *30*, 189–199.
95. Grote, D.L.; Park, S.W.; Zhou, M. Dynamic behavior of concrete at high strain rates and pressures: I. experimental characterization. *Int. J. Impact Eng.* **2001**, *25*, 869–886. [CrossRef]
96. Sun, J.S.; Ma, L.J.; Dou, Y.M.; Zhou, J. Effect of strain rate on the compressive mechanical properties of concrete. *Adv. Mater. Res.* **2012**, *450–451*, 244–247. [CrossRef]
97. Guo, Y.B.; Gao, G.F.; Jing, L.; Shim, V.P.W. Response of high-strength concrete to dynamic compressive loading. *Int. J. Impact Eng.* **2017**, *108*, 114–135. [CrossRef]
98. Dong, Y.; Xie, H.; Zhao, P. Experimental study and constitutive model of the whole process of concrete compression under different strain rates. *J. Hydraul. Eng.* **1997**, *7*, 72–77.
99. Chang, K.C.; Lee, G.C. Strain rate effect on structural steel under cyclic loading. *J. Eng. Mech.* **1987**, *113*, 1292–1301. [CrossRef]
100. Restrepo-Posada, J.I.; Dodd, L.L.; Park, R.; Cooke, N. Variables affecting cyclic behavior of reinforcing steel. *J. Struct. Eng.* **1994**, *120*, 3178–3196. [CrossRef]
101. Soroushian, P.; Choi, K.B. Steel mechanical properties at different strain rates. *J. Struct. Eng.* **1987**, *113*, 663–672. [CrossRef]
102. Cadoni, E.; Fenu, L.; Forni, D. Strain rate behaviour in tension of austenitic stainless steel used for reinforcing bars. *Constr. Build. Mater.* **2012**, *35*, 399–407. [CrossRef]
103. Li, M.; Li, H. Dynamic test and constitutive model for reinforcing steel. *China Civ. Eng. J.* **2010**, *43*, 70–75.
104. Reinhardt, H.W.; Rossi, P.; Mier, J.G.M.V. Joint investigation of concrete at high rates of loading. *Mater. Struct.* **1990**, *23*, 213–216. [CrossRef]
105. Ross, C.A.; Jerome, D.M.; Tedesco, J.W.; Hughes, M.L. Moisture and strain rate effects on concrete strength. *ACI Mater. J.* **1996**, *93*, 293–300.
106. Cadoni, E.; Labibes, K.; Albertini, C.; Berra, M.; Giangrasso, M. Strain-rate effect on the tensile behaviour of concrete at different relative humidity levels. *Mater. Struct.* **2001**, *34*, 21. [CrossRef]
107. Ranjith, P.G.; Jasinge, D.; Song, J.Y.; Choi, S.K. A study of the effect of displacement rate and moisture content on the mechanical properties of concrete: Use of acoustic emission. *Mech. Mater.* **2008**, *40*, 453–469. [CrossRef]
108. Filiatrault, A.; Holleran, M. Stress-strain behavior of reinforcing steel and concrete under seismic strain rates and low temperatures. *Mater. Struct.* **2001**, *34*, 235–239. [CrossRef]
109. Yan, D.M.; Xu, W. Strain-rate sensitivity of concrete: Influence of temperature. *Adv. Mater. Res.* **2011**, *243–249*, 453–456. [CrossRef]
110. Pajak, M. The influence of the strain rate on the strength of concrete taking into account the experimental techniques. *Laval Théologique Philos.* **2011**, *39*, 423–428.
111. Chen, X.; Wu, S.; Zhou, J.; Chen, Y.; Qin, A. Effect of testing method and strain rate on stress-strain behavior of concrete. *J. Mater. Civ. Eng.* **2013**, *25*, 1752–1761. [CrossRef]
112. Bischoff, P.H.; Perry, S.H. Compressive behaviour of concrete at high strain rates. *Mater. Struct.* **1991**, *24*, 425–450. [CrossRef]
113. Barpi, F. Impact behaviour of concrete: A computational approach. *Eng. Fract. Mech.* **2004**, *71*, 2197–2213. [CrossRef]
114. Georgin, J.F.; Reynouard, J.M. Modeling of structures subjected to impact: Concrete behavior under high strain rate. *Cem. Concr. Compos.* **2003**, *25*, 131–143. [CrossRef]
115. Lorefice, R.; Etse, G.; Carol, I. Viscoplastic approach for rate-dependent failure analysis of concrete joints and interfaces. *Int. J. Solids Struct.* **2008**, *45*, 2686–2705. [CrossRef]
116. Pedersen, R.R.; Simone, A.; Sluys, L.J. An analysis of dynamic fracture in concrete with a continuum visco-elastic visco-plastic damage model. *Eng. Fract. Mech.* **2008**, *75*, 3782–3805. [CrossRef]
117. Eibl, J.; Schmidt-Hurtienne, B. Strain-rate-sensitive constitutive law for concrete. *J. Eng. Mech.* **1999**, *125*, 1411–1420. [CrossRef]
118. Ragueneau, F.; Gatuingt, F. Inelastic behavior modelling of concrete in low and high strain rate dynamics. *Comput. Struct.* **2003**, *81*, 1287–1299. [CrossRef]
119. Ren, X.D.; Li, J. Stochastic damage constitutive model for concrete considering strain rate effect. *Key Eng. Mater.* **2009**, *400–402*, 251–256. [CrossRef]
120. Hao, Y.; Hao, H. Influence of the concrete DIF model on the numerical predictions of RC wall responses to blast loadings. *Eng. Struct.* **2014**, *73*, 24–38. [CrossRef]
121. Lee, S.; Kim, K.M.; Cho, J.Y. Investigation into pure rate effect on dynamic increase factor for concrete compressive strength. *Proc. Eng.* **2017**, *210*, 11–17. [CrossRef]
122. Malvar, L.J. Review of Static and Dynamic Properties of Steel Reinforcing Bars. *ACI Mater. J.* **1998**, *95*, 609–616.
123. Tedesco, J.W.; Ross, C.A. Strain-rate-dependent constitutive equations for concrete. *J. Press. Vessel. Technol.* **1998**, *120*, 398–405. [CrossRef]
124. Xiao, S.; Zhang, J. Experimental study on dynamic compression of concrete under load history. *J. Dalian Univ. Technol.* **2011**, *51*, 6.
125. Johnson, G.R.; Cook, W.H. A Constitutive Model and Data for Metals Subjected to Large Strains, High Strain Rates and High Temperatures. In Proceedings of the 7th International Symposium on Ballistics, Hague, The Netherlands, 19–21 April 1983; pp. 541–547.
126. Morquio, A.; Riera, J.D. Size and strain rate effects in steel structures. *Eng. Struct.* **2004**, *26*, 669–679. [CrossRef]
127. Rodríguez-Martínez, J.A.; Rusinek, A.; Klepaczko, J.R. Constitutive relation for steels approximating quasi-static and intermediate strain rates at large deformations. *Mech. Res. Commun.* **2009**, *36*, 419–427. [CrossRef]

128. Cotsovos, D.M.; Stathopoulos, N.D.; Zeris, C.A. Behavior of RC beams subjected to high rates of concentrated loading. *J. Struct. Eng.* **2008**, *134*, 1839–1851. [CrossRef]
129. Xiao, S.; Xu, D. Influence of strain rate effect on reinforced concrete column. *J. Disaster Prev. Mitig. Eng.* **2009**, *29*, 668–675.
130. Zhan, T.; Wang, Z.; Ning, J. Failure behaviors of reinforced concrete beams subjected to high impact loading. *Eng. Fail. Anal.* **2015**, *56*, 233–243. [CrossRef]
131. Valipour, H.R.; Luan, H.; Foster, S.J. Analysis of RC beams subjected to shock loading using a modified fibre element formulation. *Comput. Concr.* **2009**, *6*, 377–390. [CrossRef]
132. Xu, B.; Liang, T.; Long, Y. Strain rate effects on dynamic behavior of reinforcement concrete columns with various reinforcement ratio. *J. Civ. Archit. Environ. Eng.* **2012**, *47*, 24–38.
133. Sharma, A.; Ožbolt, J. Influence of high loading rates on behavior of reinforced concrete beams with different aspect ratios—A numerical study. *Eng. Struct.* **2014**, *79*, 297–308. [CrossRef]
134. Xiao, S.Y. Numerical study of dynamic behaviour of RC beams under cyclic loading with different loading rates. *Mag. Concr. Res.* **2015**, *67*, 325–334. [CrossRef]
135. Li, R.H.; Li, H.N.; Li, C. Seismic performance assessment of RC frame structures subjected to far-field and near-field ground motions considering strain rate effect. *Int. J. Struct. Stab. Dyn.* **2018**, *18*, 1850127. [CrossRef]
136. Zhang, H.; Li, H.N.; Li, C.; Cao, G.W. Experimental and numerical investigations on seismic responses of reinforced concrete structures considering strain rate effect. *Constr. Build. Mater.* **2018**, *173*, 672–686. [CrossRef]
137. Jia, P.C.; Wu, H.; Fang, Q. An improved 2DOF model for dynamic behaviors of RC members under lateral low-velocity impact. *Int. J. Impact Eng.* **2023**, *173*, 104460. [CrossRef]
138. Yang, Z. Numerical Simulation Study on the Dynamic Mechanical Behaviors of Reinforced Concrete Shear Walls. Master's Thesis, Dalian University of Technology, Dalian, China, 2019.
139. Zhao, D.B.; Yi, W.J.; Kunnath, S.K. Numerical simulation and shear resistance of reinforced concrete beams under impact. *Eng. Struct.* **2018**, *166*, 387–401. [CrossRef]
140. Li, R.H.; Li, C.; Li, H.N.; Yang, G.; Zhang, P. Improved estimation on seismic behavior of RC column members: A probabilistic method considering dynamic effect and structural parameter uncertainties. *Struct. Saf.* **2023**, *101*, 102308. [CrossRef]
141. Stochino, F.; Carta, G. SDOF models for reinforced concrete beams under impulsive loads accounting for strain rate effects. *Nucl. Eng. Des.* **2014**, *276*, 74–86. [CrossRef]
142. Krauthammer, T.; Shanaa, H.M.; Assadi, A. Response of structural concrete elements to severe impulsive loads. *Comput. Struct.* **1994**, *53*, 119–130. [CrossRef]
143. Shi, Y.; Li, Z.X.; Hao, H. Bond slip modelling and its effect on numerical analysis of blast-induced responses of RC columns. *Struct. Eng. Mech.* **2009**, *32*, 251–267. [CrossRef]
144. Guner, S.; Vecchio, F.J. Analysis of shear-critical reinforced concrete plane frame elements under cyclic loading. *J. Struct. Eng.* **2011**, *137*, 834–843. [CrossRef]
145. Li, H.; Chen, W.; Hao, H. Dynamic response of precast concrete beam with wet connection subjected to impact loads. *Eng. Struct.* **2019**, *191*, 247–263. [CrossRef]
146. Jin, L.; Lan, Y.; Zhang, R.; Du, X. Impact resistance of RC beams under different combinations of mass and velocity: Mesoscale numerical analysis. *Arch. Civ. Mech. Eng.* **2020**, *20*, 119. [CrossRef]
147. Clough, R.W. Effect of Stiffness Degradation on Earthquake Ductility Requirements. In *Proceedings of Japan Earthquake Engineering Symposium*; Macuze Company: Macuse, Mozambique, 1966.
148. Takeda, T. Reinforced Concrete Response to Simulated Earthquakes. *J. Struct. Div. Proc. Am. Soc. Civ. Eng.* **1970**, *96*, 2557–2573. [CrossRef]
149. Ozcebe, G.; Saatcioglu, M. Hysteretic shear model for reinforced concrete members. *J. Struct. Eng.* **1989**, *115*, 132–148. [CrossRef]
150. Park, Y.J.; Ang, A.H.S. Mechanistic seismic damage model for reinforced concrete. *J. Struct. Eng.* **1985**, *111*, 722–739. [CrossRef]
151. Wen, Y.K. Method for Random Vibration of Hysteretic Systems. *J. Eng. Mech. Div.* **1976**, *102*, 249–263. [CrossRef]
152. Takayanagi, T.; Schnobrich, W.C. Non-linear analysis of coupled wall systems. *Earthq. Eng. Struct. Dyn.* **2010**, *7*, 1–22. [CrossRef]
153. Saatcioglu, M.; Alsiwat, J.M.; Ozcebe, G. Hysteretic Behavior of Anchorage Slip in R/C Members. *J. Struct. Eng.* **1992**, *118*, 2439–2458. [CrossRef]
154. Xu, S.Y. Modeling Axial-Shear-Flexure Interaction of RC Columns for Seismic Response Assessment of Bridges. Ph.D. Thesis, University of California, Los Angeles, CA, USA, 2010.
155. Qu, Z.; Ye, L. Strength Deterioration Model Based on Effective Hysteretic Energy Dissipation for RC members under Cyclic Loading. *Eng. Mech.* **2011**, *28*, 45–051.
156. D'Ambrisi, A.; Filippou, F.C. Modeling of cyclic shear behavior in RC members. *J. Struct. Eng.* **1999**, *125*, 1143–1150. [CrossRef]
157. Gu, X.; Huang, Q.; Wu, Z. Analysis of load-displacement relationship for RC columns under reversed load considering accumulative damage. *Earthq. Eng. Eng. Vib.* **2006**, *24*, 7.
158. Zheng, H.; Zheng, S.; He, J.; Zhang, Y.; Shang, Z. Research on restoring force model of reinforced concrete rectangular beams considering reinforcement corrosion. *J. Cent. South Univ. Sci. Technol.* **2020**, *51*, 10.
159. Zheng, S.; Rong, X.; Zhang, Y.; Dong, L. Restoring force model of freeze-thaw injury reinforced concrete shear walls. *Huazhong Univ. Sci. Technol. Nat. Sci. Ed.* **2019**, *47*, 6.

160. Erberik, M.A. Importance of Degrading Behavior for Seismic Performance Evaluation of Simple Structural Systems. *J. Earthq. Eng.* **2011**, *15*, 32–49. [CrossRef]
161. Wang, B.; Zheng, S.; Guo, X.; Li, L. Study on restoring force model of SRHSHPC frame columns considering damage effects. *J. Build. Struct.* **2012**, *33*, 69–76.
162. Dowell, R.K.; Seible, F.; Wilson, E.L. Pivot Hysteresis Model for Reinforced Concrete Members. *Aci Struct. J.* **1998**, *95*, 607–617.
163. Mostaghel, N.; Byrd, R.A. Analytical Description of Multidegree Bilinear Hysteretic System. *J. Eng. Mech.* **2000**, *126*, 588–598. [CrossRef]
164. Yan, C.; Yang, D.; Ma, Z.J.; Jia, J. Hysteretic model of SRUHSC column and SRC beam joints considering damage effects. *Mater. Struct.* **2017**, *50*, 88. [CrossRef]
165. Wang, P.H.; Ou, Y.C.; Chang, K.C. A new smooth hysteretic model for ductile flexural-dominated reinforced concrete bridge columns. *Earthq. Eng. Struct. Dyn.* **2017**, *46*, 2237–2259. [CrossRef]
166. Yu, B.; Ning, C.L.; Li, B. Hysteretic Model for Shear-Critical Reinforced Concrete Columns. *J. Struct. Eng.* **2016**, *142*, 04016056. [CrossRef]
167. Sezen, H.; Chowdhury, T. Hysteretic Model for Reinforced Concrete Columns Including the Effect of Shear and Axial Load Failure. *J. Struct. Eng. Asce* **2009**, *135*, 139–146. [CrossRef]
168. Leborgne, M.R.; Ghannoum, W.M. Calibrated analytical element for lateral-strength degradation of reinforced concrete columns. *Eng. Struct.* **2014**, *81*, 35–48. [CrossRef]
169. Chao, S.H.; Loh, C.H. A biaxial hysteretic model for a structural system incorporating strength deterioration and pinching phenomena. *Int. J. Non Linear Mech.* **2009**, *44*, 745–756. [CrossRef]
170. Guo, Z.; Yang, Y. State-of-the-art of restoring force models for RC structures. *World Earthq. Eng.* **2004**, *20*, 47–51.
171. Yu, J.; Li, Y.; Xia, H. Hysteretic shear model for RC columns with construction joint. *J. Build. Struct.* **2011**, *32*, 84–91.
172. Cai, M.; Gu, X.; Hua, J.; Lin, F. Seismic response analysis of reinforced concrete columns considering shear effects. *J. Build. Struct.* **2011**, *32*, 97–108.
173. Zhao, J.; Dun, H. A restoring force model for steel fiber reinforced concrete shear walls. *Eng. Struct.* **2014**, *75*, 469–476. [CrossRef]
174. Huang, C.; Li, Y.; Gu, Q.; Liu, J. Machine learning–based hysteretic lateral force-displacement models of reinforced concrete columns. *J. Struct. Eng.* **2022**, *148*, 04021291. [CrossRef]
175. Song, J.; Pincheira, J. Spectral Displacement Demands of Stiffness and Strength-Degrading Systems. *Earthq. Spectra* **2000**, *16*, 817–851. [CrossRef]
176. Ibarra, L.F.; Medina, R.A.; Krawinkler, H. Hysteretic models that incorporate strength and stiffness deterioration. *Earthq. Eng. Struct. Dyn.* **2010**, *34*, 1489–1511. [CrossRef]
177. Guo, Y.; Long, M. Parameter Identification and Application of Reinforced Concrete Column Based on Modified Ibarra-Medina-Krawinkler Hysteretic Model. *J. Hunan Univ. Nat. Sci.* **2021**, *48*, 126–134.
178. Galal, E.M.; Ghobarah, A. Flexural and shear hysteretic behaviour of reinforced concrete columns with variable axial load. *Eng. Struct.* **2003**, *25*, 1353–1367. [CrossRef]
179. Zhang, Y.; Han, S. Hysteretic Model for Flexure-shear Critical Reinforced Concrete Columns. *J. Earthq. Eng.* **2018**, *22*, 1639–1661. [CrossRef]
180. Zhang, H.; Li, H.N.; Wang, Z. Study on Strain Rate Effect in High-Rise Reinforced Concrete Shear Wall Structure. *Adv. Mater. Res.* **2011**, *243–249*, 5854–5857. [CrossRef]
181. Wang, C.Q.; Xiao, J.Z.; Sun, Z.P. Seismic analysis on recycled aggregate concrete frame considering strain rate effect. *Int. J. Concr. Struct. Mater.* **2016**, *10*, 307–323. [CrossRef]
182. Pankaj, P.; Lin, E. Material modelling in the seismic response analysis for the design of RC framed structures. *Eng. Struct.* **2005**, *27*, 1014–1023. [CrossRef]
183. Iribarren, B.S.; Berke, P.; Bouillard, P.; Vantomme, J.; Massart, T.J. Investigation of the influence of design and material parameters in the progressive collapse analysis of RC structures. *Eng. Struct.* **2011**, *33*, 2805–2820. [CrossRef]
184. Zhang, H.; Li, H.N. Dynamic analysis of reinforced concrete structure with strain rate effect. *Mater. Res. Innov.* **2011**, *15*, s213–s216. [CrossRef]
185. Cao, X.-Y.; Feng, D.-C.; Beer, M. Consistent seismic hazard and fragility analysis considering combined capacity-demand uncertainties via probability density evolution method. *Struct. Saf.* **2023**, *103*, 102330. [CrossRef]

MDPI
St. Alban-Anlage 66
4052 Basel
Switzerland
www.mdpi.com

Buildings Editorial Office
E-mail: buildings@mdpi.com
www.mdpi.com/journal/buildings

www.ingramcontent.com/pod-product-compliance
Lightning Source LLC
Chambersburg PA
CBHW061513210926
51850CB00008B/1898
* 9 7 8 3 7 2 5 8 1 3 8 5 8 *